PASSING THE Georgia EOC Test in Physical Science

Adheres to Georgia GPS Standards

Liz Thompson

American Book Company
PO Box 2638
Woodstock, GA 30188-1383
Toll Free: 1 (888) 264-5877 Phone: (770) 928-2834
Fax: (770) 928-7483 Toll Free Fax 1 (866) 827-3240
Website: www.americanbookcompany.com

ACKNOWLEDGEMENTS

The authors would like to gratefully acknowledge the formatting and technical contributions of Marsha Torrens.

We also want to thank Mary Stoddard for her expertise in developing the graphics for this book.

A special thanks to Becky Wright for her editing assistance.

Printed in the United States of America

01/06 08/06

Table of Contents

Chapter 6 Nuclear Processes 121

Chapter 7 Chemical Equations and Reactions 135

Chapter 8 Matter and Energy 155

Chapter 9 Solutions 179

Chapter 10 Forces and Motion 195

Chapter 11 Energy, Work, and Power 211

Chapter 12 Waves 225

Chapter 13 Electricity and Magnetism 245

Post Test 1 265

Post Test 2 285

Index 313

Preface

Passing the Georgia End of Course Exam in Physical Science will help students who are learning or reviewing material for the Georgia test that is now required for each gateway or benchmark course. **The materials in this book are based on the Georgia Performance Standards as published by the Georgia Department of Education.**

This book contains several sections. These sections are as follows: 1) General information about the book; 2) A Diagnostic Test and Evaluation Chart; 3) Chapters that teach the concepts and skills that improve readiness for the end of course exam in Physical Science; 4) Two Practice Tests. Answers to the tests and exercises are in a separate manual. The answer manual also contains a Chart of Standards for teachers to make a more precise diagnosis of student needs and assignments.

We welcome comments and suggestions about the book. Please contact us a

American Book Company
PO Box 2638
Woodstock, GA 30188-1383

Toll Free: 1 (888) 264-5877
Phone: (770) 928-2834
Fax: (770) 928-7483
web site: www.americanbookcompany.com

About the Author

Liz A. Thompson holds a B.S. in Chemistry and an M.S. in Analytical Chemistry, both from the Georgia Institute of Technology. Research conducted as both an undergraduate and graduate student focused on the creation and fabrication of sensors based on conducting polymers and biomolecules. Post graduate experience includes work in radioanalytical chemistry. Her publications include several articles in respected scientific journals, as well as authorship of two chapters in the textbook *Radioanalytical Chemistry* (in press). At every educational level, Mrs. Thompson has enjoyed teaching, tutoring and mentoring students in the study of science.

PREPARING FOR THE GEORGIA EOCT TESTS

Introduction

If you are a student in a Georgia school district, the End-of-Course Test (EOCT) program requires you to take a test at the end of each gateway or benchmark course. Gateway courses currently include Algebra 1, Geometry, United States History, Economics, Biology, Physical Science, Ninth Grade Literature and Composition, and American Literature and Composition. The EOCT will count for 15% of the student's grade in each gateway course. The Physical Science EOCT was first administered in the 2003–2004 school year.

This book will help students prepare for the Georgia Physical Science EOCT. The following section will provide general information about the physical science test.

How long do I have to take the exam?

The test is given in two 45 – 60 minute sessions.

What materials will I be allowed to use during the exam?

You may not use a calculator for the physical science test. You will be provided with an equation reference sheet and a periodic table of the elements.

How is the exam organized?

There are 45 multiple choice questions in each section of the test, for a total of 90 questions.

The questions for the test will be linked to the Georgia Performance Standards (GPS) for Physical Science released by the Department of Education. The GPS are divided into two broad categories; these are Characteristics of Science (SCsh) and the Physical Science Content (SPS) Standards. SCSh standards, of which there are 9, address the learning, reasoning and application skills necessary to study science. SPS standards, of which there are 10, constitute the factual basis of physical science.

Each question on the Diagnostic Test, Practice Test 1 and Practice Test 2 in this book is correlated to one of these standards; the specific standard is noted to the right of each question on each of the three tests.

Georgia Physical Science Diagnostic Test

Georgia Physical Science Reference Sheet

Motion and Force

$v = \frac{d}{t}$	v = speed or velocity $\left(\frac{m}{s}\right)$	d = distance (meter, m)	t = time (second, s)
$a = \frac{\Delta v}{\Delta t} = \frac{v_f - v_i}{t_f - t_i}$	a = acceleration	v = velocity t = time	f = final i = initial
weight = m*g*	m = mass (kilogram, kg)	*g* = gravitational acceleration $(9.8 \frac{m}{s^2})$	
F = ma	F = force (newton, N)	m = mass (kilogram, kg)	a = acceleration $\left(\frac{m}{s^2}\right)$

Electricity

V = IR	V = potential (volt, V)	I = current (ampere, A)	R = resistance (ohm, Ω)

Gas Laws

$P_1V_1 = P_2V_2$	P = pressure (atm)	V = volume (liter, L)
$\frac{V_1}{T_1} = \frac{V_2}{T_2}$	V = volume (liter, L)	T = temperature (kelvin, K)

Energy, Work, Power, and Efficiency

$KE = \frac{1}{2}mv^2$	KE = kinetic energy (joule, J)	m = mass (kg)	v = velocity $\left(\frac{m}{s}\right)$
PE = m*g*h	PE = potential energy (J)	m = mass *g* = gravitational acceleration	h = height (m)
W = Fd	W = work (joule, J)	F = force (newton, N)	d = distance (meter, m)
$MA = \frac{F_r}{F_e}$	MA = mechanical advantage	F_r = resistance force	F_e = effort force
$P = \frac{W}{t}$	P = power (watt, W)	W = work (joule, J)	t = time (second, s)

$$\text{Efficiency} = \frac{\text{Work output}}{\text{Work input}} \times 100\%$$

Waves

$v = f\lambda$	v = velocity $\left(\frac{m}{s}\right)$	f = frequency (hertz, Hz)	λ = wavelength (meter, m)
$f = \frac{1}{T}$	f = frequency (hertz, Hz)	T = period (second, s)	

NOTE: You may make copies of pages 1 and 2 to use as handouts.

THE PERIODIC TABLE OF THE ELEMENTS

Key: Atomic Number → 36; Symbol → Kr; Name → Krypton; Atomic Mass → 83.80

Noble Gases: 18 VIIIA

Period	1 IA	2 IIA	3 IIIB	4 IVB	5 VB	6 VIB	7 VIIB	8 VIIIB	9 VIIIB	10 VIIIB	11 IB	12 IIB	13 IIIA	14 IVA	15 VA	16 VIA	17 VIIA	18 VIIIA
1	1 H Hydrogen 1.007g																	2 He Helium 4.0026
2	3 Li Lithium 6.941g	4 Be Beryllium 9.01218											5 B Boron 10.81	6 C Carbon 12.011	7 N Nitrogen 14.0067	8 O Oxygen 15.9994	9 F Fluorine 18.998403	10 Ne Neon 20.179
3	11 Na Sodium 22.9898	12 Mg Magnesium 24.305											13 Al Aluminum 26.98154	14 Si Silicon 28.0855	15 P Phosphorus 30.97376	16 S Sulfur 32.06	17 Cl Chlorine 35.453	18 Ar Argon 39.948
4	19 K Potassium 39.0983	20 Ca Calcium 40.08	21 Sc Scandium 44.9559	22 Ti Titanium 47.90	23 V Vanadium 50.9415	24 Cr Chromium 51.996	25 Mn Manganese 54.9381	26 Fe Iron 55.847	27 Co Cobalt 58.9332	28 Ni Nickel 58.69	29 Cu Copper 63.546	30 Zn Zinc 65.38	31 Ga Gallium 69.723	32 Ge Germanium 72.61	33 As Arsenic 74.9216	34 Se Selenium 78.96	35 Br Bromine 79.904	36 Kr Krypton 83.80
5	37 Rb Rubidium 85.4678	38 Sr Strontium 87.62	39 Y Yttrium 88.9059	40 Zr Zirconium 91.22	41 Nb Niobium 92.9064	42 Mo Molybdenum 95.94	43 Tc Technetium 97.91	44 Ru Ruthenium 101.07	45 Rh Rhodium 102.9055	46 Pd Palladium 106.4	47 Ag Silver 107.868	48 Cd Cadmium 112.41	49 In Indium 114.82	50 Sn Tin 118.71	51 Sb Antimony 121.75	52 Te Tellurium 127.60	53 I Iodine 126.9045	54 Xe Xenon 131.30
6	55 Cs Cesium 132.9054	56 Ba Barium 137.33		72 Hf Hafnium 178.49	73 Ta Tantalum 180.9479	74 W Tungsten 183.84	75 Re Rhenium 186.2	76 Os Osmium 190.2	77 Ir Iridium 192.22	78 Pt Platinum 195.09	79 Au Gold 196.9665	80 Hg Mercury 200.59	81 Tl Thallium 204.383	82 Pb Lead 207.2	83 Bi Bismuth 208.9808	84 Po Polonium 208.98244	85 At Astatine 209.98704	86 Rn Radon 222.02
7	87 Fr Francium 223.01976	88 Ra Radium 226.0254		104 Rf Rutherfordium 261.1	105 Db Dubnium 262.11	106 Sg Seaborgium 263.12	107 Bh Bohrium 262.12	108 Hs Hassium 264.13	109 Mt Meitnerium 266.14	110 Ds Darmstadtium 271	111 Rg Roentgenium 272	112	113	114	115	116	117	118

Series															
Lanthanide Series	57 La Lanthanum 138.9055	58 Ce Cerium 140.12	59 Pr Praseodymium 140.9077	60 Nd Neodymium 144.24	61 Pm Promethium 144.91279	62 Sm Samarium 150.4	63 Eu Europium 151.96	64 Gd Gadolinium 157.25	65 Tb Terbium 158.9254	66 Dy Dysprosium 162.50	67 Ho Holmium 164.9304	68 Er Erbium 167.26	69 Tm Thulium 168.9342	70 Yb Ytterbium 173.04	71 Lu Lutetium 174.967
Actinide Series	89 Ac Actinium 227.02779	90 Th Thorium 232.0381	91 Pa Protactinium 231.0359	92 U Uranium 238.029	93 Np Neptunium 234.0482	94 Pu Plutonium 244.06424	95 Am Americium 243.06139	96 Cm Curium 247.07035	97 Bk Berkelium 247.07030	98 Cf Californium 251.0796	99 Es Einsteinium 252.08	100 Fm Fermium 257.09515	101 Md Mendelevium 258.1	102 No Nobelium 259.100	103 Lr Lawrencium 262.11

SECTION 1

Karen conducted an experiment to investigate whether two new foods increase the rate at which tomato plants grow. She obtained three small tomato plants. She placed the plants so each received the same amount of sunlight and was at the same temperature. She gave each plant the same amount of water. Karen added plant food A to the first plant. She added the same amount of plant food B to the second plant. She added no plant food to the third plant. Karen measured the growth of each plant once per week for six weeks.

1. Select the correct description of the plant food in Karen's experiment. SCSh3 Ch 1

   A. independent variable
   B. dependent variable
   C. constant
   D. control

2. Stefan observed goldfish in a lighted fish tank. He noticed the fish were more active with the light on than with it off. Stefan decided to investigate whether colored light affected the activity of the fish. He planned an experiment in which he would first use a white bulb, then a red bulb, and finally, a green bulb to light the tank. He would observe the activity of the goldfish when each bulb was switched on. Select the best hypothesis for Stefan's experiment. SCSh3 Ch 1

   A. Goldfish cannot survive under colored lights.
   B. Goldfish prefer red light to green light or white light.
   C. Goldfish are more active under white light than colored light.
   D. Goldfish live longer under white light than under either red or green light.

3. Select the equipment that would be used to measure the mass of sample of sodium chloride. SCSh2 Ch 2

   A. beam balance
   B. graduated cylinder
   C. beaker
   D. meter stick

4. Sean was conducting a chemistry experiment that overheated and caught fire. Select the safety rule Sean most likely failed to follow.

SCSh2 Ch 2

   A. wear safety goggles
   B. follow the teacher's instructions
   C. use tongs to handle hot glassware
   D. keep the work space clean and uncluttered

5. Fig. 1 shows a graduated cylinder that contained 25 ml of water. The mass of the cylinder and water was 68.0 g. Fig. 2 shows the same cylinder after a small stone was lowered into the cylinder. The water level rose to the 30.0 ml mark, and the mass of the cylinder, water, and stone was 78.0 g. select the density of the small stone.

SPS2 Ch 7

   A. 2.00 g/ml   C. 0.50 ml/g
   B. 10.0 g   D. 5.00 ml

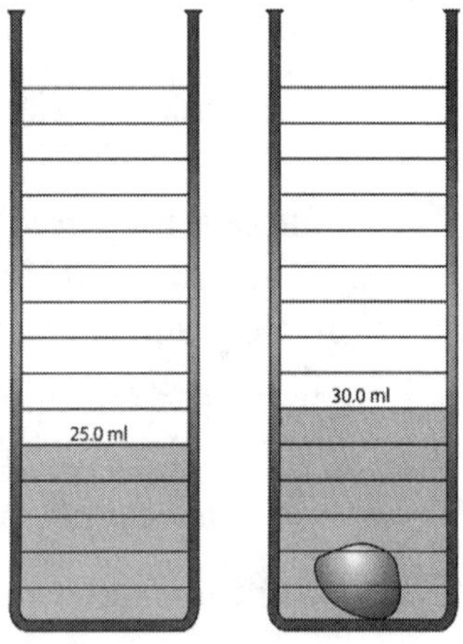

6. Select the perimeter of a triangle whose sides measure 2.2×10^2 cm, 4.8×10^2 cm, and 1.1×10^1 cm.

SCSh5 Ch 4

   A. 8.1×10^{10} cm   C. 8.1×10^2 cm
   B. 7.1×10^2 cm   D. 1.2×10^4 cm

Answer questions 7 – 9 based on the summary of the following experiment.

A group of students investigated how temperature affects the rate of chemical reactions. They used hydrogen peroxide, which breaks down into oxygen and water, for their experiment. The students measured how long it took to obtain 50 mL of oxygen gas from a given volume of hydrogen peroxide heated to different temperatures. Their data are shown in the table to the right.

Temperature (°C)	Time (minutes)
10	33
20	16
30	8
40	4
50	2

7. Based on the students' data, select the correct conclusion about the rate of chemical reactions.

SCSh3 Ch 1

   A. Reaction rates increase as time increases.
   B. Reaction rates decrease as time decreases.
   C. Reaction rates increase as temperature increases.
   D. Reaction rates decrease as temperature increases.

8. Other students argued the investigation was not repeatable and the data were insufficient to support the investigators' conclusions. Select the investigators' best response. SCSh1 Ch 3

   A. Tell the skeptical students they received an excellent grade for their investigation.

   B. Provide a full description of the experimental method to the skeptical students.

   C. Explain to the skeptical students the reason for conducting the experiment.

   D. Tell the skeptical students to mind their own business.

9. Select the best way for the students to reduce the experimental error in their investigation. SCSh5 Ch 4

   A. test more than one variable at a time

   B. perform repeated trials

   C. change their answers if they do not match their hypothesis

   D. perform the experiment only one time

10. A video-rental store offered four rental plans. In Plan A, customers pay a $10 annual fee plus a $3 per video rental fee. In Plan B, the annual fee is $25 and the rental fee is $2.00. Plan C customers pay a $100 annual fee but no rental fee. Under Plan D, there is no annual fee and the rental fee is $5. A graph of cost against number of videos rented for each plan is shown below. Select the least expensive annual plan for Mark, who plans to rent two videos per month. SCSh3 Ch 1

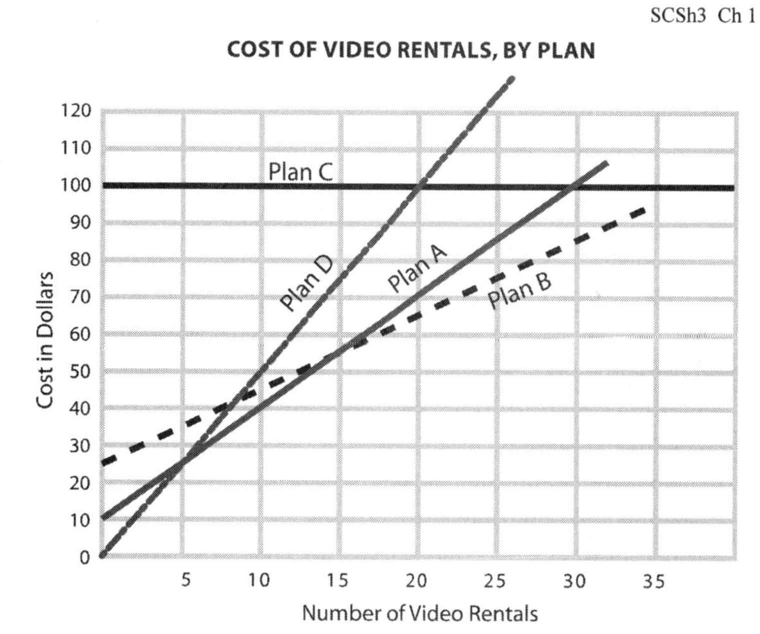

   A. Plan A

   B. Plan B

   C. Plan C

   D. Plan D

Use the information to the right to answer questions 11 and 12.

Andre conducted an investigation about a variable that could affect the growth rate of pea plants. He grew one pea plant in a room where hip-hop music was continually played. He grew a second plant in a room where country music was played. He grew a third pea plant in a room where classical music was played. He grew a fourth pea plant in a room with no music. Andre controlled all other variables that could affect the growth of the plants. The results are shown to the right.

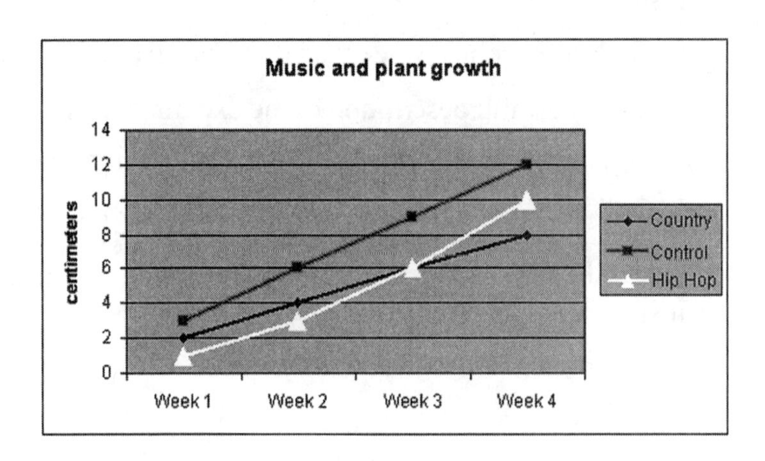

11. Select the variable Andre was testing to see if it affected the growth rate of pea plants — SCSh3 Ch 1

    A. type of music
    B. loudness of music
    C. amount of sunlight
    D. temperature of soil

12. Based on the results, in which week will the height of the plants in the room where hip-hop is played be the same as in the room where country is played? — SCSh3 Ch 4

    A. Week 1
    B. Week 2
    C. Week 3
    D. Week 4

13. The table shown presents data from an experiment that examined the effect of incubation temperatures on (a) the sex of the baby turtles that hatched, and (b) the number of eggs that hatched. — SCSh3 Ch 1

Group	Temperature °C	Number of Eggs	Number of Hatchlings		
			Male	Female	Total
1	26	25	19	2	21
2	28	25	14	9	23
3	30	25	4	16	20
4	32	25	2	22	24

Select the best conclusion about incubation temperatures above 29°C.

    A. more eggs hatch
    B. fewer eggs hatch
    C. more male than female turtles hatch
    D. fewer male than female turtles hatch

14. Visualize the thread of a screw being unwrapped, as in the figure to the right. Which statement correctly describes the simple machine represented in the screw? SCSh3 Ch 1

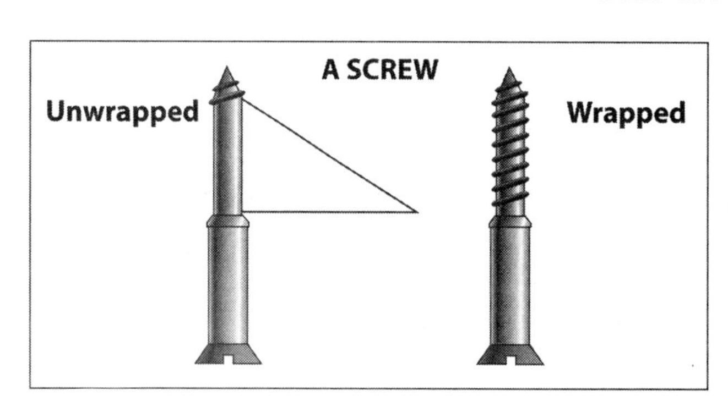

A. The thread of a screw is a wedge wrapped around a shaft.

B. The thread of a screw is a lever wrapped around a shaft.

C. The thread of a screw is an inclined plane wrapped around a shaft.

D. The thread of a screw is a pulley wrapped around a shaft.

15. Select the BEST reason for scientists to replace an existing theory with a new theory. SCSh7 Ch 3

A. Several senior scientists announced their support of the new thory.

B. A group of important religious leaders announced its support of the new theory.

C. Several senior politicians announced that the new theory would boost the economy.

D. A research team announced observations that were better explained by the new thory.

16. Select the product of radioactive decay that has the greatest ability to penetrate matter. SPS3 Ch 6

A. alpha particles
B. beta particles
C. gamma rays
D. neutrons

17. A skier traveling at 30.0 m/s falls and comes to rest 10.0 seconds later. What is her average acceleration? SPS8 Ch 10

A. 300 m/s^2 B. 3.00 m/s^2 C. -3.00 m/s^2 D. -300 m/s^2

18. Select the reaction force to the downward gravitational force exerted on a body by the earth. SPS8 Ch 10

A. an upward force exerted on the body by the earth

B. an upward force exerted on the earth by the body

C. a downward force exerted on the body by the earth

D. a downward force exerted on the earth by the body

19. The distance-time graph at right describes the rate of a person skiing down a mountain. Select the skier's average speed. SPS8 Ch 10

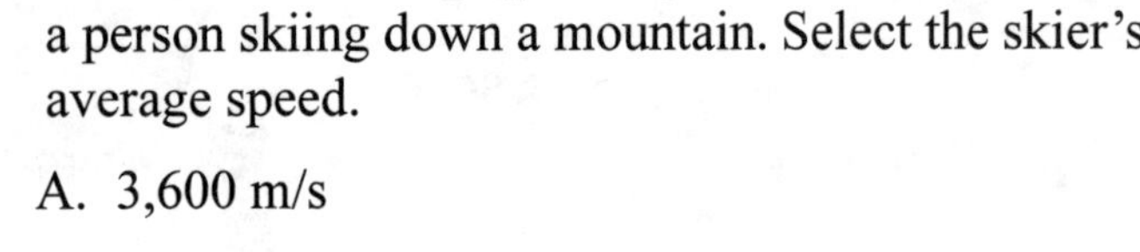

A. 3,600 m/s

B. 200 m/s

C. 18 m/s

D. 0.067 m/s

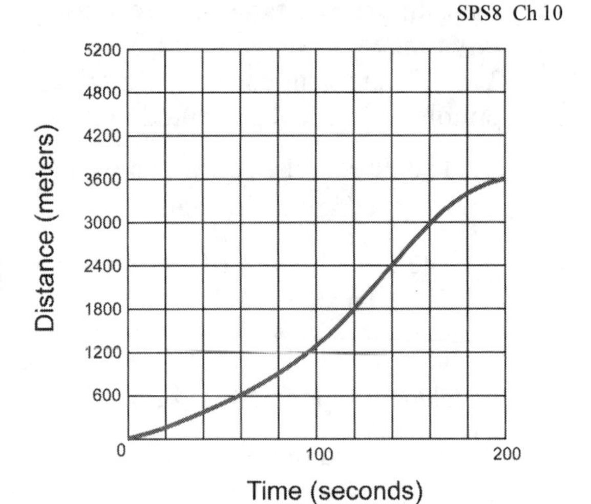

20. Select the situation that will result in the greatest gravitational force between two bodies. SPS8 Ch 10

A. large combined mass and small distance apart

B. large combined mass and great distance apart

C. small combined mass and great distance apart

D. small combined mass and small distance apart

21. The flow of electricity through a light bulb filament can be compared to the flow of water down a waterfall. In such a comparison, select the property of the waterfall that would be analogous to the voltage across the light bulb. SPS8 Ch 11

A. water depth

B. water temperature

C. rate of flow of water

D. height of the waterfall

22. The diagram below shows an electrical circuit made up of a cell and three loads. The voltage established by the cell and the resistance of each load is marked on the diagram. Select the current flowing through the 3-ohm load. SPS10 Ch 13

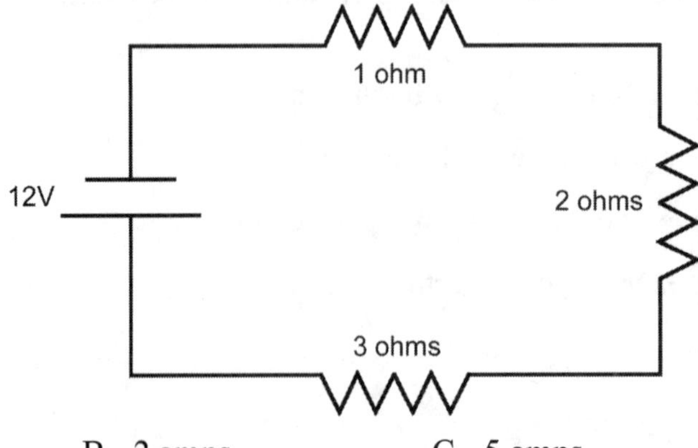

A. 4 amps

B. 2 amps

C. 5 amps

D. 0.25 amps

23. Select the voltage that will result in a current of 10 amps in a circuit with a resistance of 12 ohms. SPS10 Ch 13

    A. 1.2 volts B. 2 volts C. 22 volts D. 120 volts

24. Select the result of an electrical current flowing through a wire. SPS10 Ch 13

    A. emission of X-rays
    B. emission of alpha particles
    C. gravitational force between the wire and a nearby metal ball
    D. electromagnetic force between the wire and a nearby compass needle

25. Select the scenario in which an airplane does have kinetic energy but does *not* have gravitational potential energy. SPS8 Ch 11

    A. accelerating down the runway
    B. waiting in line on the taxiway
    C. parked at the loading gate
    D. in flight

26. Select the method of heat transfer that causes the iron handle of an iron skillet to get hot when the skillet is heated on a stove. SPS7 Ch 8

    A. convection B. radiation C. conduction D. friction

27. The diagram at the right shows a ball at the top of a ramp. Select the statement that correctly describes the changes to the ball's kinetic energy (KE) and gravitation potential energy (PE) as it rolls down the ramp. SPS8 Ch 11

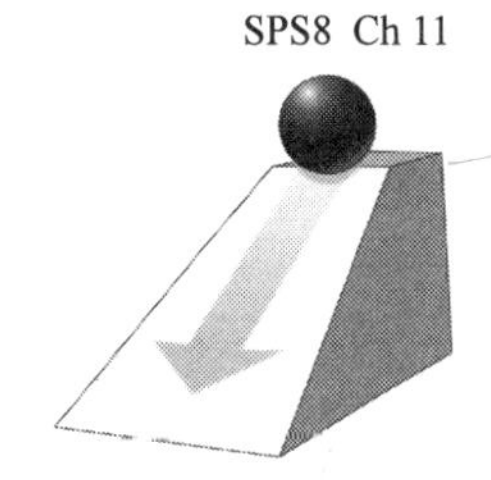

    A. KE and PE both decrease
    B. KE and PE both increase
    C. KE decreases and PE increases
    D. KE increases and PE decreases

28. The diagram of a wave indicates two important properties of a wave: its wavelength and amplitude. Wavelength is inversely related to frequency: a longer wavelength indicates a lower frequency. Amplitude is proportional to the intensity of the energy that the wave can impart to matter: a greater amplitude indicates a greater intensity. Consider the following situation: Wave A has a wavelength 5λ and an amplitude 2γ, while Wave B has a wavelength 3λ and an amplitude 4γ. Which statement best describes these two waves? SPS 9 Ch 12

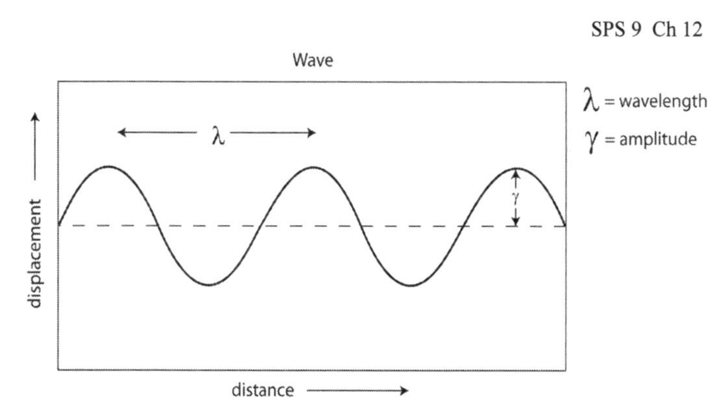

    A. Wave A has a lower frequency and greater intensity than Wave B.
    B. Wave A has a higher frequency and greater intensity than Wave B.
    C. Wave A has a higher frequency and lower intensity than Wave B.
    D. Wave A has a lower frequency and lower intensity than Wave B.

29. Select the behavior of electrons that is the source of the orange-red light given off by neon lights. SPS9 Ch 12

    A. Electrons drop from one energy level to one lower energy level.

    B. Electrons jump from one energy level to one higher energy level.

    C. Electrons drop from one energy level to multiple lower energy levels.

    D. Electrons jump from one energy level to multiple higher energy levels.

30. Select the correct statement about different isotopes of a given element. SPS3 Ch 6

    A. same number of protons and same number of neutrons

    B. same number of protons but different number of neutrons

    C. different number of protons but same number of neutrons

    D. different number of protons and different number of neutrons

31. Use the diagram below to tell where the most reactive metals are located. SPS4 Ch 5

PERIODIC TABLE OF THE ELEMENTS

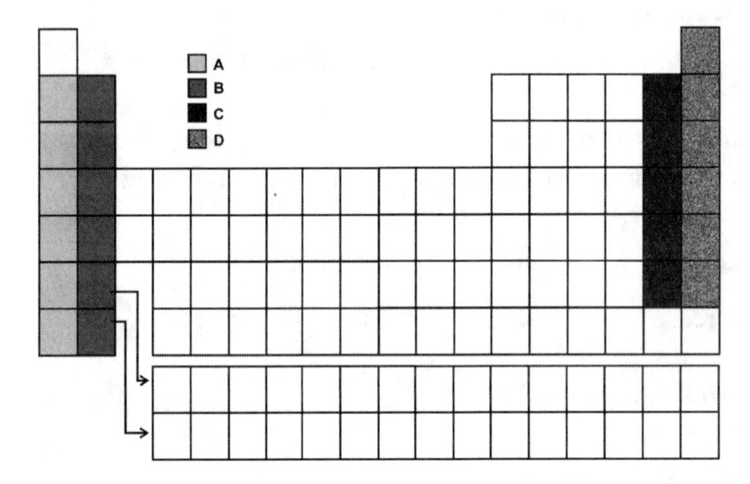

    A. light gray section A

    B. dark gray section B

    C. black section C

    D. textured section D

32. Using the diagram to the right, determine the number of neutrons that most copper atoms contain. SPS1 Ch 5

    A. 29   B. 35   C. 64   D. 93

29
Cu
Copper
63.546
2,8,18,1

33. Magnesium (Mg) looses two electrons to form magnesium ions. Chlorine (C1) gains one electron to form chloride ions. Select the correct formula for magnesium chloride. SPS2 Ch 7

    A. $MgC1$ B. Mg_2C1 C. $MgC1_2$ D. Mg_2C1_2

34. Which of the diagrams correctly represents an ionic bond between two elements?

SPS1 Ch 5

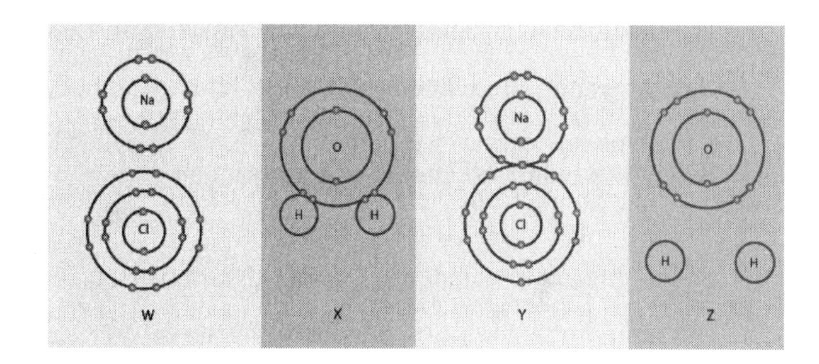

A. diagram W

B. diagram X

C. diagram Y

D. diagram Z

35. How many oxygen atoms are there in one formula unit of aluminum carbonate [$A1_2(CO_3)_3$]?

SPS1 Ch 5

A. 1 B. 3 C. 6 D. 9

36. Select the reason why water molecules are polar molecules.

SPS4 Ch 5

A. Hydrogen atoms are much smaller than oxygen atoms.

B. Water molecules have a bond angle greater than 100 degrees.

C. Water molecules form hydrogen bonds with other water molecules.

D. Oxygen nuclei attract electrons more strongly than hydrogen nuclei attract electrons.

37. Select the properties of a solution of salt in water that will allow additional salt to dissolve quickest.

SPS6 Ch 9

A. high temperature and high salt concentration

B. high temperature and low salt concentration

C. low temperature and low salt concentration

D. low temperature and high salt concentration

38. Select the phase or phases of matter that can be compressed.

SPS5 Ch 8

A. gas only

B. liquid only

C. liquid and gas

D. solid, liquid, and gas

39. Salt is an ionic compound, methanol is a polar covalent compound, and octane, the major component of gasoline, is a non-polar, covalent compound. Select the substance or substances that are soluble in water.

SPS6 Ch 9

A. salt only

B. octane only

C. salt and methanol

D. octane and methanol

40. Select the observation that does NOT support the conclusion that a chemical reaction took place. SPS2 Ch 7

A. Liquid water changed to water vapor.
B. A precipitate was formed.
C. An odor developed.
D. Temperature rose.

41. Select the balanced chemical equation. SPS2 Ch 7

A. $Zn + HC1 \rightarrow ZnC1_2 + H_2$
B. $C_2H_4 + 2O_2 \rightarrow 2CO_2 + 2H_2O$
C. $2Na + 2H_2O \rightarrow 2NaOH + H_2$
D. $2CaCO_3 + H_2SO_4 \rightarrow 2CaSO_4 + H_2O + CO_2$

42. As illustrated in the diagram below, Carl investigated the change of mass during a chemical reaction. He massed a baloon, two selzer tablets and a plastic bottle that contained 50 mL of water. He recorded a total mass of 200g. Carl put the seltzer tablets inside the balloon and pulled the balloon over the neck of the bottle. He shook the balloon so the seltzer tablets fell into the water. Carl observed the tablets fizzing and the balloon expanding. Carl again massed his apparatus after the fizzing and the expansion of the balloon stopped. SPS2 Ch 7

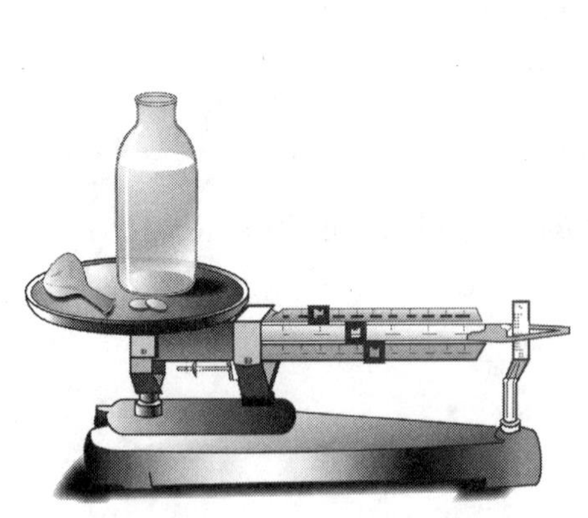

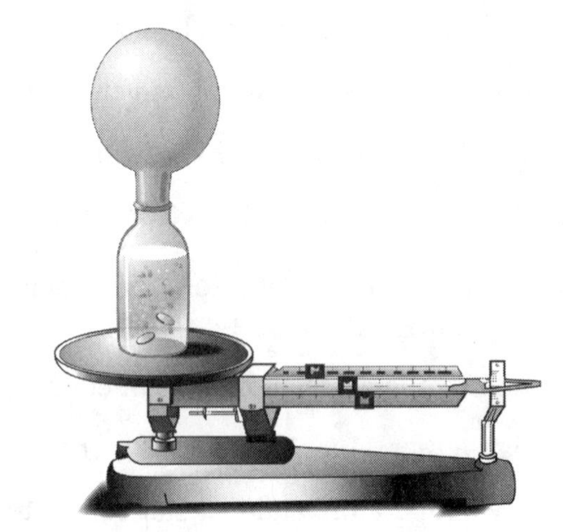

If the combined mass of the two seltzer tablets before they were dropped into the water was 5 g, select the mass of Carl's apparatus at the end of his investigation.

A. 195 g
B. between 195 g and 200 g
C. 200 g
D. more than 200 g

43. Select the type of ion produced when a strong acid is added to water. SPS6 Ch 9

A. hydrated
B. hydride
C. hydroxide
D. hydronium

44. Select the particle or particles that account for more than 99.99% of the mass of atoms other than hydrogen. SPS1 Ch 5

A. protons only
B. electrons only
C. protons and neutrons
D. electrons and neutrons

45. Select the repeating units from which polymers are made. SPS2 Ch 7

A. monomers B. hydrocarbons C. carbohydrates D. macromolecules

This is the halfway point in the test. With the approval of your instructor, you may take a 5-minute stretch break.

SECTION 2

46. The chemical equation for the rusting of iron is: $4Fe + 3O_2 \rightarrow 2Fe_2O_3$. Select the number of electrons gained or lost by every iron atom that is converted to iron (III) oxide in that reaction. SPS2 Ch 7

A. gain 3 B. gain 2 C. lose 2 D. lose 3

47. Select the change that will DECREASE the rate of a chemical reaction SPS7 Ch 8

A. cooling the reactants
B. shaking the reactants
C. adding a catalyst to the reactants
D. increasing the concentration of the reactants

48. The atomic number of beryllium is 4. Select the number of electrons a beryllium atom will gain or lose when it forms a beryllium ion. SPS4 Ch 5

A. gain 2 B. gain 4 C. lose 2 D. lose 4

49. X-rays and microwaves traveling in space have the same SPS9 Ch 12

A. wavelength. B. frequency. C. speed. D. energy.

50. The attractive force between protons and electrons within an atom is called the SPS10 Ch 13

A. gravitational force.
B. nuclear force.
C. magnetic force.
D. electric force.

51. As water changes phase from a liquid to a gas, what is expected to happen? SPS5 Ch 8

A. The distance between particles will increase.
B. The distance between particles will decrease.
C. The mass of the sample will increase.
D. The mass of the sample will decrease.

52. When a neutral metal sphere is charged by contact with a positively charged glass rod, the sphere SPS10 Ch 13

A. loses electrons.
B. loses protons.
C. gains electrons.
D. gains protons.

53. Select the correct classification of silicon, the material used in the “chips” that power electronic devices such as computers and cell phones. SPS10 Ch 13

A. insulator B. conductor C. semiconductor D. superconductor

54. Which of the following web addresses is *most likely* to contain reliable information on cleaning up a hazardous waste site? SCSh8 Ch 1

A. http://en.wikipedia.org/wiki/Cleanup
B. http://www.epa.gov/
C. http://www.lets-clean-up.com/
D. http://www.bobspage.com/

55. Choose the answer that correctly organizes four metric prefixes from highest to lowest. SCSh4 Ch 2

A. kilometer, hectometer, dekameter, decimeter
B. kilometer, decimeter, dekameter, centimeter
C. kilometer, hectometer, centimeter, decimeter
D. kilometer, decimeter, millimeter, centimeter

56. When Henry walked into school in the morning, the sky was dark and cloudy. If Henry states that "it will rain soon," he is making a SCSh3 Ch 1

A. hypothesis. B. prediction. C. inference. D. theory.

57. The diagram below shows a compass placed next to a powerful bar magnet. SPS10 Ch 13

Permanent Magnet

Identify the arrow that shows the direction of the compass needle.

A. B. C. D.

58. Laboratory instructions indicate that 3 grams of salt should be added to 20 mL of water. The mixture should be heated to boiling, and boiled until the volume is reduced by 10 mL. The beaker should then be slowly cooled. Which of the following best describes the effects of this activity on the salt solution? SPS6 Ch 9

A. The salt will evaporate, causing the concentration of salt in the water to decrease.
B. The water will evaporate, causing concentration of salt in the water to increase.
C. The salt will evaporate, causing the concentration of salt in the water to increase.
D. The water will evaporate, causing concentration of salt in the water to decrease.

59. Which group has eight valence electrons in elemental form? SPS4 Ch 5

A. noble gases
B. alkaline earth metals
C. halogens
D. alkali metals

60. Which of the following is an organic compound? SPS2 Ch 7

A. potassium permanganate, $KMnO_4$
B. sodium sulfate, Na_2SO_4
C. ethane, C_2H_6
D. ammonia, NH_3

61. What process does the following description refer to: "a wet sidewalk begins to steam in the sun after a spring shower." SPS5 Ch 8

A. melting
B. sublimation
C. evaporation
D. fumigation

62. Which of the following dissociates to produce hydrogen (H+) cations when dissolved in water? SPS6 Ch 9

A. bases
B. acids
C. salts
D. sugars

63. Which of the following elements has the smallest total number of electrons? SPS1 Ch 5

A. nitrogen (N)
B. sodium (Na)
C. helium (He)
D. lead (Pb)

64. Which of the following is an element? SPS4 Ch 5

A. water
B. ozone
C. oxygen
D. carbon dioxide

65. Based on the graph below, which of the following accurately describes the roller coaster ride? SPS8 Ch 10

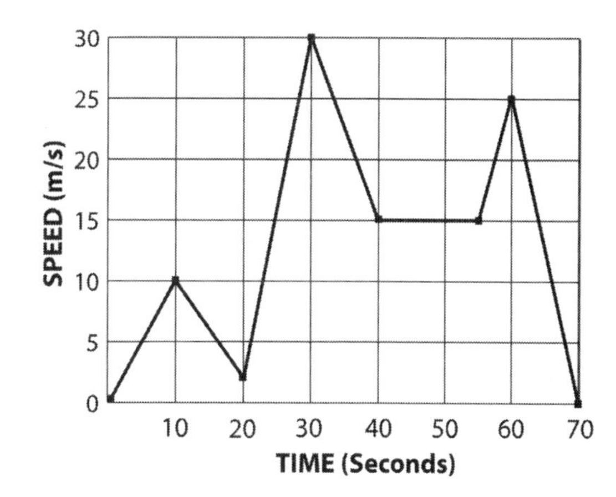

A. The roller coaster pulls away from the platform and accelerates for 10 seconds. Then it starts up a steep hill, going slower and slower, for 10 seconds. Over the next 10 seconds it goes down a hill very fast then slows down to half its high speed. It then maintains its speed for 25 seconds until it maneuvers to the top of a small hill. It then goes fast down that hill for 10 seconds and then decelerates, coming to a stop, over the last 5 seconds of the ride.

B. The roller coaster pulls away from the platform and accelerates for 10 seconds. Then it starts up a steep hill, going slower and slower, for 10 seconds. Over the next 20 seconds it climbs another hill very fast then slows down to half its high speed. It then maintains its speed for 15 seconds until it maneuvers to the top of a small hill. It then goes fast down that hill for 10 seconds and then decelerates coming to a stop over the last 5 seconds of the ride.

C. The roller coaster pulls away from the platform and accelerates for 10 seconds. Then it starts up a steep hill, going slower and slower, for 10 seconds. Over the next 10 seconds it accelerates down the hill very fast, then slows down to half its high speed. It then maintains its speed for 15 seconds until it accclerates down another hill. Over the last ten seconds of the ride, the roller coaster decelerates at a steady rate until it pulls into the platform.

D. The roller coaster pulls away from the platform and accelerates for 10 seconds. Then it starts up a steep hill, going slower and slower, for 10 seconds. Over the next 20 seconds it comes down the hill very fast, then slows down to half its high speed. It then maintains its speed for 15 seconds until it deccelerates going up a small hill. Over the last ten seconds of the ride, the roller coaster decelerates at a steady rate until it pulls into the platform.

66. Consider the following reaction: SPS2 Ch 7

$$N_2\,(g) + 3H_2\,(g) \rightarrow 2NH_3\,(g)$$

Which of the following statements correctly describes the reaction?

A. Two mols of N_2 react with three mols H_2 to form two mols of NH_3.

B. One mol of N_2 reacts with three mols H_2 to form two mols of NH_3.

C. One mol of N_2 reacts with six mols H_2 to form six mols of NH_3.

D. Two mols of N_2 react with six mols H_2 to form six mols of NH_3.

67. An example of using a lever is SPS8 Ch 11

A. using a big spoon to stir a pot of soup.

B. using a fork to scramble an egg.

C. using a ramp to push a barrel into a pickup truck .

D. using a screwdriver to pry the lid off a can of paint.

68. Seawater contains sodium chloride and other dissolved salts. Identify the term that correctly describes those dissolved salts. SPS6 Ch 9

A. mixtures B. elements C. solvents D. solutes

69. A solution is made by dissolving 10 grams of salt in 500 mL of water. Identify the volume of the resulting solution. SPS6 Ch 9

A. 500 ml

B. more than 500 mL but less than 510 mL

C. 510 mL

D. more than 510 mL

70. Kyle charged a glass rod negatively by rubbing it with a piece of cloth. Identify the statement that correctly describes how the rod became negatively charged. SPS10 Ch 13

A. Friction stripped electrons from the rod.

B. Protons were transferred from the rod to the cloth.

C. Friction produced heat that in turn produced the charge.

D. Electrons were stripped from the cloth and transferred to the rod.

71. Corundrum is a hard mineral of aluminum oxide, Al_2O_3. The pure oxide is colorless, but impurities in the crystal structure of the mineral can impart different colors to it. The blue sapphire is an example of gem-quality corundrum. The red ruby is also an example of gem-quality corundrum. Which of the following statements is true about the color of sapphires and rubies? SCSh3 Ch 1

A. Sapphires and rubies are made of different minerals.

B. Sapphires and rubies are made of the same mineral with different impurities.

C. Sapphires and rubies are made of different minerals with the same impurity.

D. Sapphires and rubies are actually colorless, but jewelers dye them.

72. What is the fastest way to separate 10 grams of sand from 50 mL of salt water? SPS5 Ch 8

A. Use electrolysis to break down the water into hydrogen and oxygen.

B. Evaporate the salt water off of the sand by heating the whole sample.

C. Pour off most of the water and let the rest of the sample sit in the fume hood to dry.

D. Pour off most of the water and then filter the remaining mixture using filter paper.

73. What number should precede O_2 in the chemical equation below in order for the equation to be balanced? SPS2 Ch 7

$$H_2O_2(l) \rightarrow H_2O(l) + \_\_\_ O_2(g)$$

A. 1/2 B. 2 C. 3 D. 4

74. Which of the following is the best heat conductor? SPS7 Ch 8

A. copper B. wood C. glass D. silicon

75. Identify the statement that describes the objects that will be attracted to a negatively charged object. SPS10 Ch 13

A. negatively charged objects only
B. positively charged objects only
C. both negatively charged and neutral objects
D. both positively charged and neutral objects

76. Identify the type of current that powers a lamp plugged into an electrical outlet in an American home. SPS10 Ch 13

A. static current
B. direct current
C. potential current
D. alternating current

77. ROYGBIV describes visible light portion of the electromagnetic spectrum, with Red light (R) at one end and Violet light (V) at the other. Which of the following statements is true? SPS9 Ch 12

A. Violet light has a higher frequency and shorter wavelength than red light.
B. Violet light has a lower frequency and shorter wavelength than red light.
C. Violet light has a higher frequency and longer wavelength than red light.
D. Violet light has a lower frequency and longer wavelength than red light.

78. Identify the graph that correctly shows the effect on the current (I) flowing through a static 5-ohm resistor when the voltage (V) across the resistor is gradually increased. SPS10 Ch 13

A.

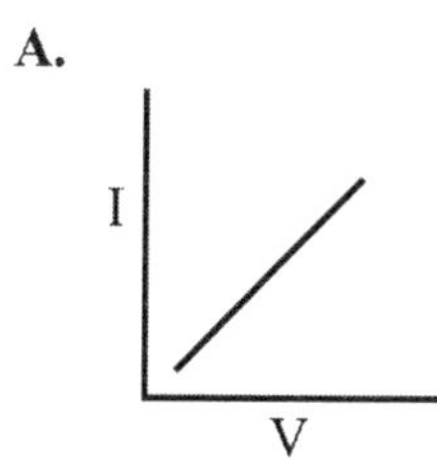

C.

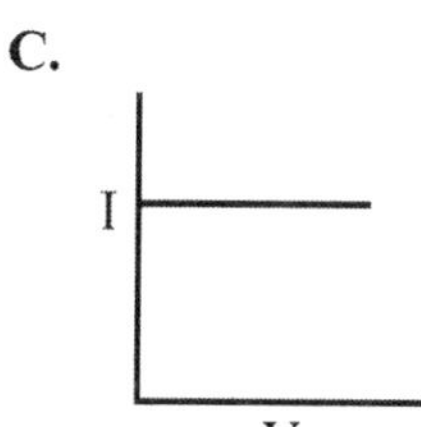

B.

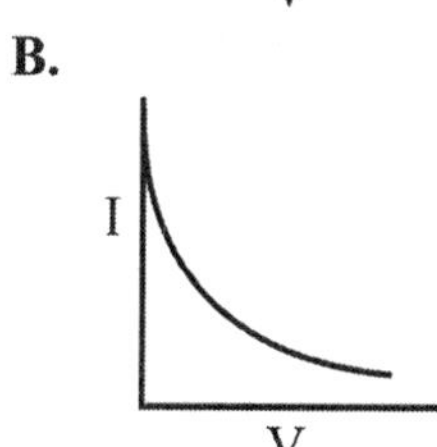

D.

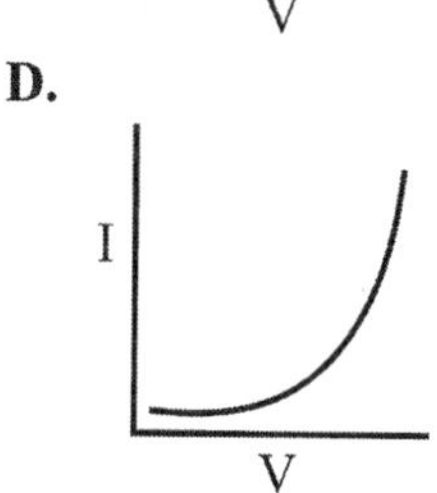

79. Identify the voltage needed to establish a current of 0.25 amps in a circuit with a total resistance of 10 ohms? SPS10 Ch 13

A. 10.25 volts B. 9.75 volts C. 2.5 volts D. 0.025 volts

80. The oil that is added to a car engine has several functions, one of which is lubrication. The purpose of lubrication is SPS8 Ch 10

A. to reduce friction between metal components of the engine.

B. to increase friction between metal components of the engine.

C. to eliminate friction of the engine.

D. to transform frictional force into chemical energy.

81. Lisa needs to move a load of bricks across the yard. The bricks weigh 650 N, which is too heavy for Lisa to carry. She decides to load the bricks into a wheelbarrow to take them across the yard. She can now lift the bricks with only 130 N of effort force. What is the mechanical advantage of the wheelbarrow? SPS8 Ch 11

A. 1/5 B. 5 C. 0.20 D. 5.00

82. The following diagram depicts heat movement by  SPS7 Ch 8

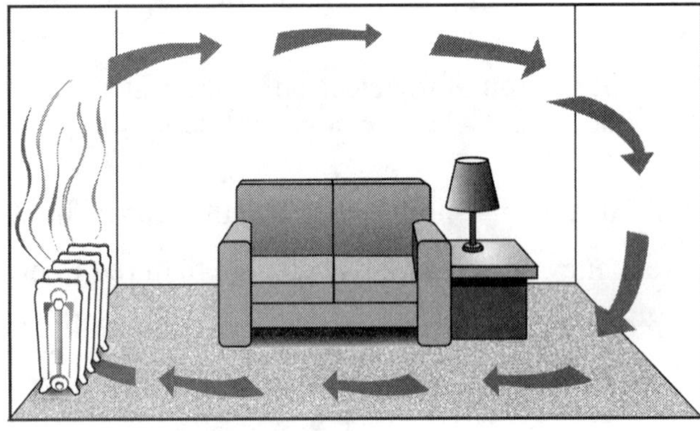

A. conduction B. convection C. radiation D. insulation

83. Fusion describes the process of SPS3 Ch 6

A. small nuclei joining together to produce heavier nuclei.

B. small nuclei joining together to produce lighter nuclei.

C. small nuclei joining together to form a macromolecule.

D. large nuclei breaking apart to form lighter nuclei.

84. The wavelength of blue light is around 475 nanometers. A blue sky appears blue because it SPS9 Ch 12

A. absorbs light with a wavelength of 475 nm.

B. reflects all visible light except light with a wavelength of 475 nm.

C. reflects only visible light with a wavelength of 475 nm.

D. reflects only ultraviolet light.

85. The Doppler effect results in the listener hearing SPS9 Ch 12

A. increased loudness. C. increased pitch.

B. destructive interference. D. better sound quality.

86. Which of the following will give the user a mechanical advantage? SPS8 Ch 11

A. an inclined plane C. a wheel and axle

B. an oven D. A and C only

87. Which wave interaction is characterized by a wave bending in response to a change in speed? SPS9 Ch 12

A. reflection B. refraction C. diffraction D. interference

88. Litmus paper is used to test for SPS6 Ch 9

A. metals B. acidity C. salt content D. oxygen levels

89. Which element has the highest electronegativity? SPS4 Ch 5

A. oxygen B. chlorine C. hydrogen D. fluorine

90. Which of the following is not a strong acid? SPS6 Ch 9

A. hydrochloric acid B. sulfuric acid C. acetic acid D. nitric acid

EVALUATION CHART

GEORGIA PHYSICAL SCIENCE DIAGNOSTIC TEST

Directions: On the following chart, circle the question numbers that you answered incorrectly, and evaluate the results. Then turn to the appropriate topics (listed by chapters), read the explanations, and complete the exercises. Review other chapters as needed. Finally, complete the Practice Tests to prepare for the Georgia EOCT in Physical Science.

Chapters	Question Numbers	Pages
Chapter 1: Scientific Method	1, 2, 7, 10, 11, 13, 56, 71	23 – 44
Chapter 2: Safety, Equipment, and Scientific Measurement	3, 4, 55	45 – 62
Chapter 3: Scientific Inquiry and Communication	8, 15, 54	63 – 76
Chapter 4: Math in Science	6, 9, 12	77 – 102
Chapter 5: Structure, Properties, and Bonding of Elements	31, 32, 34, 35, 36, 44, 48, 59, 63, 64, 89	103 – 120
Chapter 6: Nuclear Processes	16, 30, 83	121 – 134
Chapter 7: Chemical Equations and Reactions	5, 33, 40, 41, 42, 45, 46, 60, 66, 73	135 – 154
Chapter 8: Matter and Energy	26, 38, 47, 51, 61, 72, 74, 82	155 – 178
Chapter 9: Solutions	37, 39, 43, 58, 62, 68, 69, 88, 90	179 – 194
Chapter 10: Forces and Motion	14, 17, 18, 19, 20, 25, 65, 80	195 – 210
Chapter 11: Energy, Work, and Power	11, 21, 27, 67, 81, 86	211 – 224
Chapter 12: Waves	28, 29, 49, 77, 84, 85, 87	225 – 244
Chapter 13: Electricity and Magnetism	22, 23, 24, 50, 52, 53, 57, 70, 75, 76, 78, 79	245–264

Chapter 1
The Scientific Method

PHYSICAL SCIENCE STANDARDS COVERED IN THIS CHAPTER INCLUDE:

SCSh3	Students will identify and investigate problems scientifically.
SCSh4	Students will use tools and instruments for observing, measuring and manipulating scientific equipment and materials.
SCSh5	Students will demonstrate the computation and estimation skills necessary for analyzing and developing reasonable scientific explanations.
SCSh6	Students will communicate scientific investigations and information clearly.
SCSh8	Students will understand important features of the nature of scientific inquiry.

DESIGNING AND CONDUCTING AN EXPERIMENT

The Latin root for the word "science" is scientia, meaning knowledge. The discipline of science consists of many activities including the observation, identification, description, and explanation of occurrences in the world around us. Through the study of science, we ask questions, develop hypotheses (educated guesses), design and carry out experiments, record data and analyze results in order to gain a better understanding of the universe. The information collected during the course of these activities is added to the body of knowledge collected by scientists throughout recorded history and forms the basis for future investigations.

In order for these investigations to be fruitful, however, they must follow some analytical process. The scientific method is just such a process, defining a systematic method of inquiry about the natural universe based on the collection and analysis of data. Said another way, the scientific method is a way of asking questions that helps to ensure that the answers have real meaning. It consists of several steps listed in Table 1.1 and shown in Figure 1.1 on the following page. Each of these steps will be discussed in detail in this section.

Table 1.1 Steps in the Scientific Method

1. Make an observation:	Observe birds in flight.
2. Ask questions:	How do birds fly?
3. Form the hypothesis:	The wing designs of birds catch the air differently.
4. Set up an experiment:	Make and fly different designs of paper airplanes to test the hypothesis
5. Collect the data:	Take notes on flight patterns of paper airplanes. Create a data table on how each airplane flies.
6. Draw a conclusion:	The size and shape of the wing gives lift to the bird.
7. Make a prediction:	Different types of birds have different wing designs.

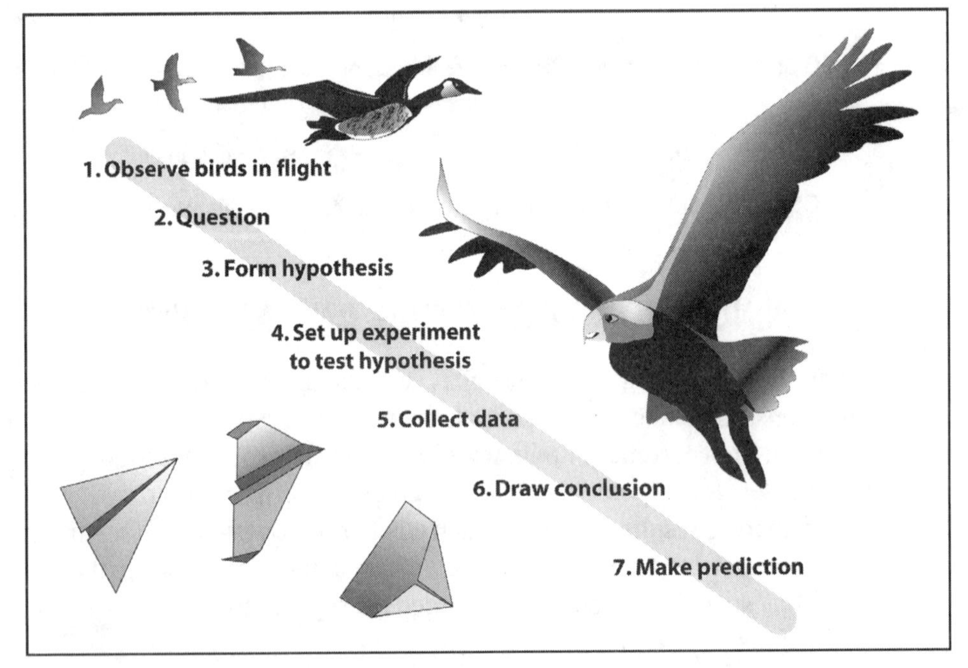

Figure 1.1 Steps in the Scientific Method

Making Observations and Defining the Problem

Scientists believe all natural phenomena in the universe have logical, verifiable explanations. A **natural phenomenon** is something occurring in nature that we experience through our senses. The scientific method defines the process of inquiry that we use to research and explain natural phenomena. The first step in the scientific method is to observe the things around us. *Observations* are made by using the five senses (sight, touch, smell, sound and taste) to obtain information. Examples of observations include

Figure 1.2 Different Bird Behaviors for Observation

watching a bird build a nest, listening to the bird's call and touching the nest's material. All of these observations help describe or explain some aspect of the bird's behavior. Making observations may lead to identifying problems as well. For example, you might observe that pigeons are making nests on the tops of city buildings. You might also observe the problem that pigeon droppings are damaging shingles and defacing city property, as well as possibly spreading disease. These observations lead to the next step in the process.

Asking Questions

Asking appropriate questions is the second step in solving a problem. By asking questions, we can search for logical explanations of what we observe and find ways to solve problems.

We observe birds in flight. There are large and small birds. Some flap their wings quickly while others seem to hardly move their wings at all. The question we might ask is, "How do birds fly?"

Questions can be asked about every facet of life, and the activities and organisms that it encompasses. Some questions may have already been answered. Learning how to search online and in the library for answers generated by other scientists will help you refine your question. As the rest of the steps in the scientific method are discussed, think about how you could apply them to a scientific question of interest to you. Some questions that may be of interest are "What is the effect of sunlight on green plants?" "Does water temperature affect the oxygen intake of goldfish?" "Will aspirin help cut roses last longer?"

Forming the Hypothesis

Everything we experience takes place in an ordered universe. As our knowledge of the universe grows, we recognize patterns. After observing these patterns and asking questions, we can form an opinion about how or why something happens. A **hypothesis** is a statement that gives the best possible response to the question and should be based on already known facts. It is an educated guess. For example, the question might be "What is the effect of sunlight on green plants?" The hypothesis might be "Green plants need sunlight to grow." The process of developing a hypothesis is important, because an observation cannot be tested. One must think of an explanation for that observation (the hypothesis) and then test that explanation.

Another form of the hypothesis is the more inclusive term, the scientific theory. The **theory** is an explanation of a broad range of related observations. Scientists test theories by creating testable predictions of what should be observed under certain conditions. Such predictions are another form of hypothesis. A hypothesis that is tested and shown to be accurate provides support for the theory. A hypothesis that is proven incorrect may indicate a flaw in the theory.

The hypothesis can be developed by using inductive reasoning. **Inductive reasoning** is a logical process in which the scientist draws from his knowledge and experience to make a general explanation. For example, large amounts of petrified wood (stoney, deadwood) are found in the desert area of a southwestern state. By using inductive reasoning, it is reasonable to assume that at one time large forests grew in that desert.

Setting Up the Experiment

Scientists conduct investigations using a range of methods. One such method is the **scientific experiment**. A scientific experiment is an investigation in which researchers change one **variable** at a time, to determine the effect of that change on another variable. The goal is to prove (or to disprove) the hypothesis.

For example, assume that the hypothesis is that more radish seeds will sprout in a lighted environment than in a dark environment. An experiment could be conducted to examine the effect of light intensity on the percentage of radish seeds that sprout in a given time. Consider the following experimental parameters.

Experiment #1: Researchers set out three sets of 100 radish seeds to sprout. They exposed the first set (Group A) to direct light. The second set (Group B) was shielded by a screen that blocked 50% of the light. The third set (Group C) was covered with a screen that blocked 100% of the light, leaving the seeds entirely in the dark. The temperature was kept constant at 23 degrees C for all of the plants. The moisture level was kept the same for each plant. After the pre-determined experimental time period was over, the researchers counted the number of seeds in each set that sprouted from each set, and then converted the number into a percentage: (# seeds sprouted) divided by (the original # of seeds) times 100.

Table 1.2 Conditions (and Variables) for Radish Seed Experiment

Group	Temperature	Light	Moist
A	23 ° C	full	yes
B	23 ° C	partial	yes
C	23 ° C	none	yes

There are three types of variables seen during the course of an experiment:

1. **Independent variable** – The factor that is changed or manipulated by the researchers, in order to determine the effect of the change. Light intensity was the independent variable in our example.
2. **Dependent variable** – The factor that the experimenter is measuring or counting. This variable changes in response to the independent variable. The percentage of seeds that sprouted was the dependent variable in our example.
3. **Control variable** – All other factors in the experiment. These are things that the investigator attempts to control, with the goal of keeping them the same for all samples. By controlling these variables, the investigator hopes to be able to isolate the source of the response during an experiment to the manipulation of the independent variable. Identifying all of the control variables that could possibly affect the experiment is one of the most difficult aspects of designing an experiment. Neglecting even one can invalidate your results. Temperature and moisture level were two control variables in our example. Can you think of any others?

The concept of the control variable — an element of the experiment that is outside of the field of inquiry, that is maintained at a constant state for the course of the experiment — can be expanded into a slightly different concept: the **control group**. The concept of the control group is illustrated by a continuation of our example.

Experiment #2: The outcome of the radish seed experiment is that 87% of Group A seeds sprouted, 38% of Group B seeds sprouted and 1% of Group C seeds sprouted. The investigator now wants to know what would make the seeds that have sprouted grow faster. The seedlings from Group A are all of similar sizes. He decides to plant each of the seedlings from that group in a separate pot. He divides the Group A seedlings into four new divisions; he will fertilize three of the divisions with three different fertilizers, and leave one division unfertilized. He will measure the height of the plants after one week.

In this case, the experimenter has established three variable groups (fertilized seedlings) and one control group (unfertilized seedling). Its rate of growth reveals the background of the experiment by demonstrating the change in the height of the plant (dependent variable) when the independent variable (type of fertilizer) is not allowed to operate. Using the control group, the investigator will now be able to determine two things: if the addition of fertilizer makes radish seedling grow faster than they would without fertilizer and which fertilizer makes radish seedlings grow fastest. Without the control group, the investigator could not be certain that fertilizer improved the growth rate of the radish seedlings.

Given these definitions, go back to Experiment #1. Can you identify the control group now?

Another kind of experiment is the **double blind experiment**. Double blind experiments are often done when the experimental subjects are humans, in order to eliminate bias. In this kind of experiment, neither the subjects in the experiment (the first "blind") nor the researchers measuring the dependent variable (the second "blind") know which subjects are in the experimental group and which are in the control group. This experimental methodology is often used during drug trials. Slightly less scientifically rigorous is the **single blind study**, where the subjects are unaware of what group they are in, while the researchers do know. This type of study allows more bias into the experiment; the researchers may consciously or unconsciously favor the group that is exhibiting the outcome that they want.

Yet another experimental methodology is the **scientific study**. This type of experiment is necessary when the scientist cannot, for practical, ethical or legal reasons, control any variables during the course of the investigation. As an example of practical limitations, consider researchers who are investigating the effects of ocean temperature on the migration of whales. The researchers cannot manipulate the ocean temperature. They can, however, measure ocean temperatures and record the movement of the whales over time, and correlate the two in order to understand the relationship.

Collecting Data

Data is gathered from the observations and measurements taken during a scientific experiment. **Qualitative data** is information that cannot be assigned a numerical value. It is often collected using the five senses. Examples of qualitative data can include shades of color, texture, taste, or smell. **Quantitative data** is anything that can be expressed as a number, or quantified. Quantitative data can include lengths, weights, masses, volumes, time, or anything else expressed as a number.

The observations from the experiment must be recorded. If the data collected are organized in a logical manner, they can be more easily analyzed to determine the results of the experiment. A table is a good way to organize data. A table presents information in an orderly fashion, usually in rows and columns. The independent variable and dependent variable(s) are listed at the top of each column. The measurements are then written under the appropriate column headings.

Latisha, a student, conducted an experiment to investigate the effect of dissolved salt on the freezing point of water. She dissolved five different amounts of salt in 50ml water in five 100 ml beakers. She placed each beaker in a freezer at –30°C. Latisha opened the freezer at 10 minute intervals. She removed a beaker when ice could be clearly seen forming on the surface of the water in that beaker. She measured the temperature of the water in that beaker. Latisha recorded that temperature as the freezing point.

Independent variable:	mass of dissolved salt
Dependent variable:	freezing point
Control group:	sample with 0g dissolved salt
Control variables:	volume of water, type of beaker, temperature of the freezer

Table 1.3 below shows how Latisha organized and presented the quantitative data she collected.

Control: 50 mL of tap water at room temperature.

Variable: 50 mL of tap water at room temperature with 1 g of table salt added.

Both were placed in a freezer. The time it took for each to freeze (as indicated by an asterisk *) was recorded.

Table 1.3 Mass of Dissolved Salt and Freezing Point

Mass Dissolved Salt (g)	Freezing Point (°C)
0	0.5
2.5	–6.5
5.0	–10.0
7.5	–14.0
10.0	–15.5
12.5	–17.5

Section Review 1: Designing and Conducting an Experiment

A. Define the following terms.

hypothesis	inductive reasoning	control group	dependent variable
double blind experiment	control variable	theory	qualitative data
single blind study	scientific study	independent variable	quantitative data

B. Choose the best answer.

1. The ability to draw from previous knowledge and experience to make an explanation is called
   A. a hypothesis.
   B. inductive reasoning.
   C. deductive reasoning.
   D. processing.

2. Something occurring in nature that is experienced through our senses is
   A. a scientific experiment.
   B. inductive reasoning.
   C. a hypothesis.
   D. a natural phenomenon.

3. An educated guess is a/an

   A. hypothesis. B. puzzle. C. answer. D. reason.

4. The group that experiences no change of the independent variable is the

   A. variable group.
   B. hypothetical group.
   C. control group.
   D. independent group.

5. The group designed with a variable to test or invalidate the hypothesis is called the

   A. control group.
   B. experimental group.
   C. questionable group.
   D. hypothetical group.

6. Both qualitative and quantitative data collected during an experiment can be organized in a

   A. group. B. variable. C. table. D. question.

A nutritional supplement manufacturer conducted an experiment to determine how creatine supplements affect muscle growth in body builders. The change in muscle mass over the course of the experiment was determined by using a special machine that distinguishes muscle mass from bone and fat mass by electrical conductance. One hundred body builders were divided into four equal groups, and their muscle mass was measured. All groups were given the same diet and the same strict workout schedule. Group 1 was given a placebo sugar pill for 4 weeks. Group 2 was given 5 grams of creatine twice a day for 4 weeks. Group 3 was given 5 grams of creatine four times a day for 4 weeks. Group 4 was given 5 grams of creatine four times a day for the first two weeks and then 5 grams two times a week for the second two weeks. None of the body builders knew which group they belonged to, but the researchers conducting the experiment did. At the end of 4 weeks, the muscle mass of each body builder was measured again and compared to the initial measurement.

7. Which group was used as the control group?

   A. Group 1 B. Group 2 C. Group 3 D. Group 4

8. What was the dependent variable?

   A. amount of creatine
   B. equivalent diets
   C. muscle mass
   D. electrical conductance

9. Which of the following is a control variable?

   A. amount of creatine
   B. equivalent diets
   C. equivalent workout routines
   D. B and C only

10. What kind of experimental methodology is used in this example?

    A. a single blind experiment
    B. a double blind experiment
    C. a scientific study
    D. a scientific investigation

C. Answer the following questions.

1. Judy went to a picnic at the lake. A jar of dill pickles fell off the table and broke on a rock. When she cleaned up the spill, she noticed small foaming bubbles on the rock. What scientific questions might she ask?
2. When Judy went to clean up the spilled pickles in question 1, she noticed a strong smell of vinegar. She learned in science class that vinegar is an acid. If the rock on which the pickles landed was a limestone rock, what hypothesis might she form?
3. Ryan noticed that his cola loses its carbonation as it warms. He knows that it is carbon dioxide that causes cola to fizz. Ryan decided to do a scientific experiment to research this phenomenon. Write a hypothesis for Ryan based on the knowledge and observation given.
4. The hypothesis of an experiment states: "Exposure to music changes the growth rate of plants." Three groups of plants are grown in the same conditions, and their growth rates are compared. Group 1 has plants that are exposed to heavy metal music for 3 hours per day. Group 2 is composed of plants that are exposed to love songs for 3 hours per day. Group 3 is not exposed to any music. Which of the groups discussed above is/are the experimental group(s) and why?
5. Why should an experiment test only one variable at a time? Can you think of another way to set up the experiment in question 4?
6. Allen wants to know the relationship between the temperature of a gas and the volume of a gas. He used a spherical balloon filled with air to conduct his experiment. Explain how he should obtain the data and what measurements he should record in his data table.

PRESENTING AND ANALYZING DATA

PRESENTING DATA

Once experimental data are recorded they must be reported in an organized manner. **Tables, graphs, charts, models**, and **diagrams** are often the most effective way of presenting the results of an experiment. Data presented in this way are easier to analyze and interpret. By carefully evaluating the data, you can interpret the true results of the experiment.Now we will look at several methods of data presentation.

According to a *National Geographic* article, two men manned a hot-air balloon and ascended to the edge of the atmosphere. They recorded the data in Table 1.4 during their ascent. Time and altitude are independent variables. Temperature and pressure are dependent variables.

Table 1.4 Temperature and Pressure as Altitude Increases

time	altitude (ft)	temperature (^{o}F)	pressure (psi)
7:08 am	0	74	14.7
7:34 am	26,000	-27	6.8
7:50 am	43,000	-73	2.4
8:10 am	53,000	-94	1.4
8:25 am	65,000	-80	0.74
9:05 am	95,000	-41	0.2
9:47 am	113,740	-29	0.09

Practice Exercise 1: Tables

Use Table 1.4 to answer the following questions.

1. How did the pressure change as the altitude increased?
2. At what altitude did they experience the coldest recorded temperature?
3. How long did it take them to go from an altitude of 0 feet to their maximum altitude of 113,740 feet?

Line graphs are good to use when you are trying to show a trend. A line graph is best used to show how one variable changes with respect to another. Data recorded in a table can often be graphed to show the relationship between the data in a way that is easier to analyze. When plotting line graphs, the independent variable is plotted on the *x*-axis (horizontal axis), and the dependent variable is plotted on the *y*-axis (vertical axis). The axes are always labeled, and the units used in measuring are shown in parentheses.

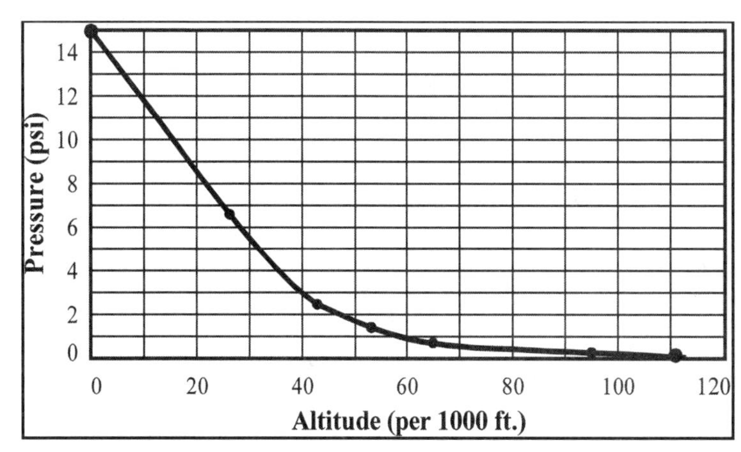

Figure 1.3 Pressure Versus Altitude

Look again at the data in Table 1.4 of the manned balloon flight. We can graph pressure versus altitude to better analyze the relationship between the two. We plot the points given in the data table and then connect a smooth line between the points

Practice Exercise 2: Line Graphs

Use Figure 1.3 to answer questions 1 – 3.

1. How much did the pressure drop between 40,000 feet and 60,000 feet?
2. Give the range of altitudes during which the pressure drops the fastest.
3. Give a reason why graphing this data is easier to analyze than comparing the same data in the Table 1.4.

Line graphs are also used to compare multiple groups of data. These are called **multiple line graphs**.

Solubility (as used in Figure 1.4) measures how many grams of a substance will dissolve in a given amount of water. Figure 1.4 shows how the solubilities of sodium chloride (table salt), alum, and baking soda change with temperature.

From Figure 1.4, you can easily observe that the solubility of each substance increases as temperature increases. In other words, more grams of the substance will dissolve in 100 grams of water at higher temperatures than at lower temperatures.

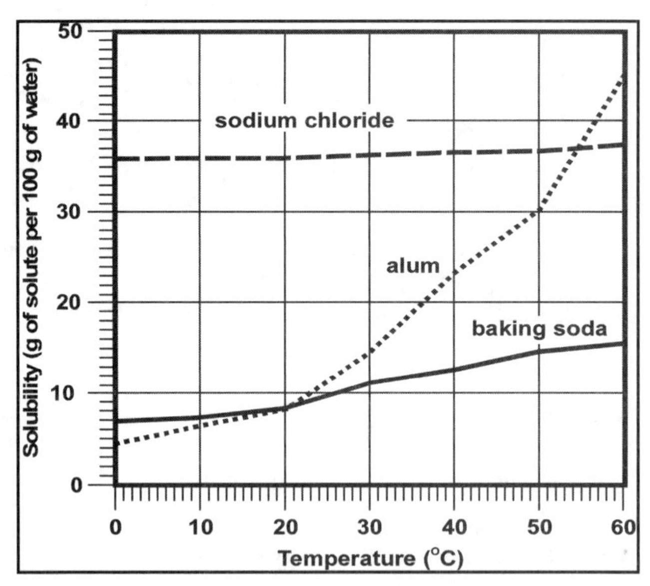

Figure 1.4 Solubility Versus Temperature

Practice Exercise 3: Multiple Line Graphs

Use Figure 1.4 to answer questions 1 – 5.

1. At 20°C, which two substances have the same solubility?
2. At about what temperature does alum have the same solubility as sodium chloride?
3. Which substance's solubility is least affected by temperature?
4. At 50°C, how many grams of alum will dissolve in 100 grams of water?
5. About how much does the solubility of sodium chloride increase from 0°C to 60°C?

Bar graphs are used to show easy-to-read, unconnected bars which represent a quantity of information. The quantities represented by the bars can then be compared and contrasted. As with line graphs, the independent variable is plotted on the *x*-axis, and the dependent variable is plotted on the *y*-axis. Figure 1.5 shows how fast sound waves travel through different mediums. Using a bar graph to present this information can help visualize the relationship between the data.

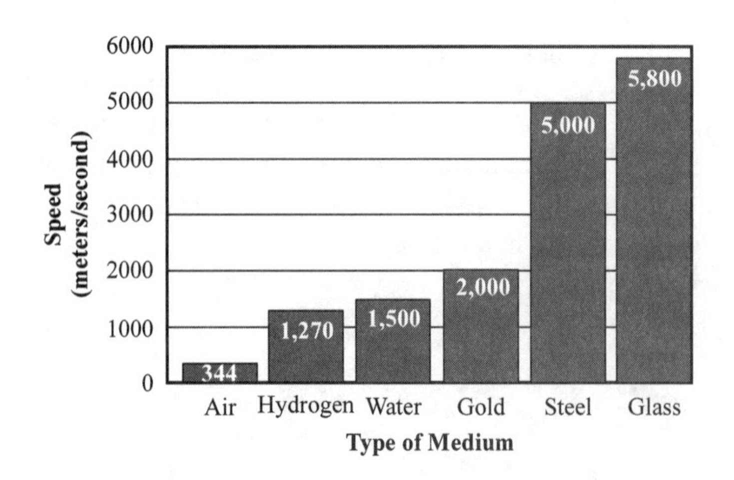

Figure 1.5 The Speed of Sound Waves Through a Medium

Practice Exercise 4: Bar Graphs

Use Figure 1.5 to answer questions 1 – 5.

1. What is the speed of sound through water?
2. How much faster does sound travel through steel than through gold?
3. In which medium does sound travel the slowest?
4. How does the method of presenting information in a bar graph differ from that in a line graph? Could you use a line graph to present the information shown in Figure 1.5?
5. A group of biology students roped off a 3 yd^2 area of forest floor and counted all of the animals they saw. Their results are as follows: 1 rabbit, 8 spiders, 96 ants, 67 termites, and 12 grasshoppers. Construct a bar graph using the students' results.

A **circle graph**, also known as a **pie chart**, is used to show parts of a whole. Many times circle graphs show percentages of a total.

Figure 1.6 shows the percentages of gases in the atmosphere.

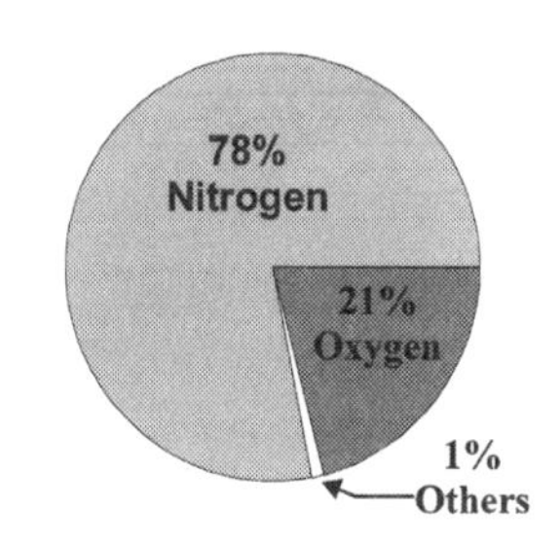

Figure 1.6

Practice Exercise 5: Circle Graphs

Use Figure 1.6 to answer the following questions.

1. What is the total percent of the circle graph?
2. Why would a circle graph not measure the growth of a plant?
3. A person breathing at rest inhales about 500 mL of atmospheric air per breath. How many mL of nitrogen are in that 500 mL?

Mathematical expressions are equations that are used to predict a phenomenon based on experimental data and/or observations. Figure 1.7 to the right shows the relationship between force (F), mass (m), and acceleration (a). Force is a push or pull, and acceleration is a measure of how fast an object speeds up or slows down. The equation allows the scientist to predict what is going to happen without having to measure it experimentally.

$$\mathbf{F} = \boldsymbol{m} \times \boldsymbol{a}$$

Figure 1.7 Mathematical Relationship between Force and Acceleration

Practice Exercise 6: Mathematical Expressions

Use Figure 1.7 to answer the following questions.

1. Calculate the force of a 500 kg mass accelerating at 10 m/s^2. (Don't worry about units, they will be discussed in detail in the following chapters.)
2. If the mass remains constant, and the acceleration increases, then will the force increase or decrease?
3. How can you calculate acceleration if you know the force and mass of an object?

A **diagram** is a schematic drawing that illustrates and helps to quantify a real event or object. Figure 1.8 shows a diagram of a wave. The amplitude of the wave is the distance from the resting point of the wave to either a crest or trough. The wavelength is the peak to peak distance of the wave.

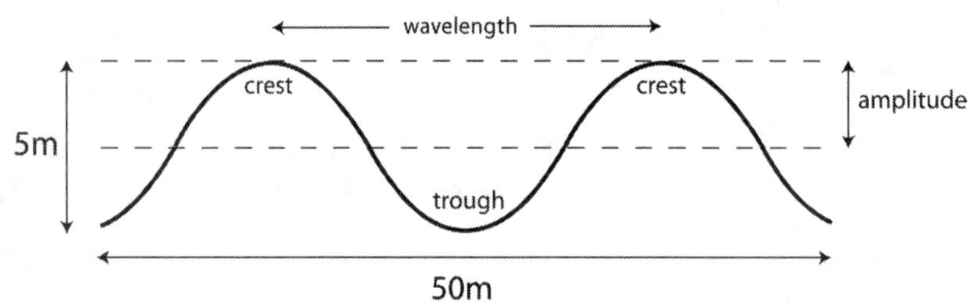

Figure 1.8 Diagram of a Wave

Practice Exercise 7: Diagrams

Use Figure 1.8 to answer questions 1 – 2.

1. What is the amplitude of the wave?
2. What is the wavelength?

A **cross section** shows an object or system as if it has been cut in half. Imagine that you have in your hand a whole apple. How do you know what is inside? You could eat it, but that would not represent the natural state of the apple. Now imagine cutting the apple in half — this allows you to view the cross section of the apple. Now you see that the apple has an internal structure that includes a core and seeds. Cross sections are often used to show us clearer pictures of biological entities like cells, as well as geological entities, like the earth's core.

Maps can be used to show how data change or are distributed over an area. You are probably familiar with topographical maps or globes, which show the elevation at different locations on Earth. Maps can be used to show either qualitative or quantitative data. Figure 1.9 at right shows quantitative data. In this case, the data is the change in speed and direction of air around an airfoil (such as an

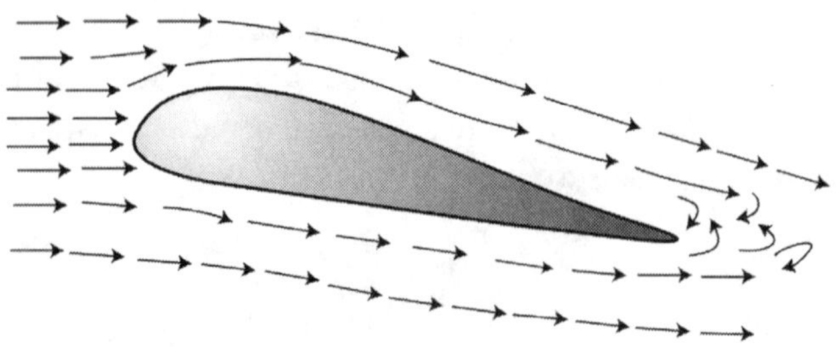

Figure 1.9 Air Flow Around an Airfoil

airplane wing). The speed is represented by an arrow. The longer the arrow, the higher the speed. The arrows point in the direction the air is flowing around the airfoil. A map of this data may allow a scientist to easily determine where areas of turbulence occur.

Practice Exercise 8: Maps

Use Figure 1.9 to answer the following questions.

1. Does the air flow faster above or below the airfoil?
2. Where is the air flow the most irregular?

DRAWING CONCLUSIONS

A conclusion is a judgement based on observation and experimentation. It is drawn from the results of the experiment. A conclusion may involve an inference. An inference is based on data, but not on direct observation. For instance, fossils of dinosaurs are valuable data. Explanations of how dinosaurs moved, what they ate, and how they became extinct are all inferences since no human alive today has witnessed a living dinosaur. The results are the end product of an experiment. The analytical method of investigation is an examination of parts of an experiment to seek reliable information that will support or reject the hypothesis. To really know the true outcome of an experiment, it must be performed many times. Through the result obtained from all the experiments, a summary or conclusion can be expressed. Think about Latisha's investigation into the freezing point of water, from the end of Section 1. What conclusions could she draw from her data? What improvements could be made to her experimental set up?

Practice Exercise 9:

Drawing Conclusions

Scientists have given us a diagram of the Big Dipper in the north sky as it looked 100,000 years ago. We know how it looks now. Scientists have also suggested what the Big Dipper will look like 100,000 years in the future. Study the diagrams to the right.

If the earth has not changed its position in the sky, then some change must have happened to explain the way the Big Dipper has changed its appearance. What conclusion can you draw about the apparent differences in the Big Dipper's shape over time?

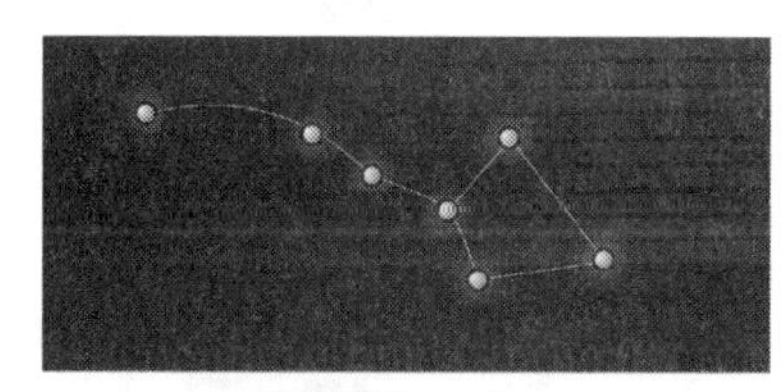

100,000 Years Ago

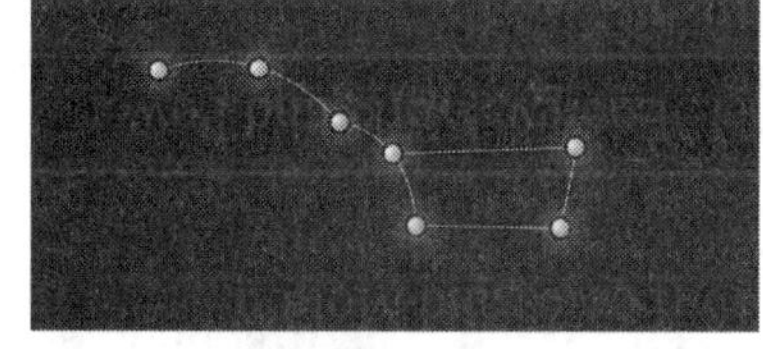

Present

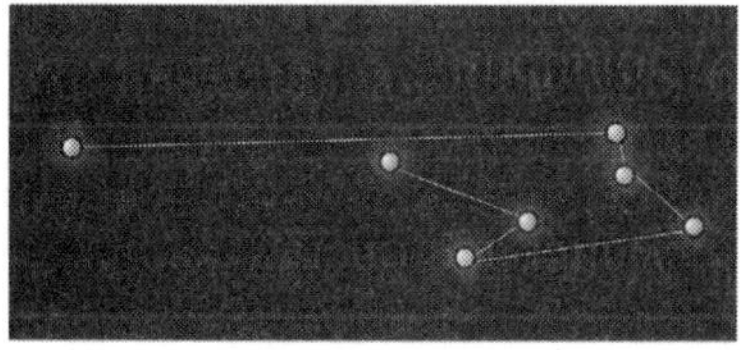

100,000 Years From Now

Observations vs. Inferences

An observation is knowledge gained through the use of one or more of the five senses. Observations are often a component of data collected during a scientific experiment. An **inference** is a possible interpretation of an event that is gained from facts, ideas, or previous experience. Observations are often used to draw inferences. Inferences can be useful in helping us understand natural phenomena, and they can be instrumental in projecting future outcomes. Inferences can contain words like possibly, usually, and perhaps. Inferring is often the process used in drawing conclusions from a scientific experiment.

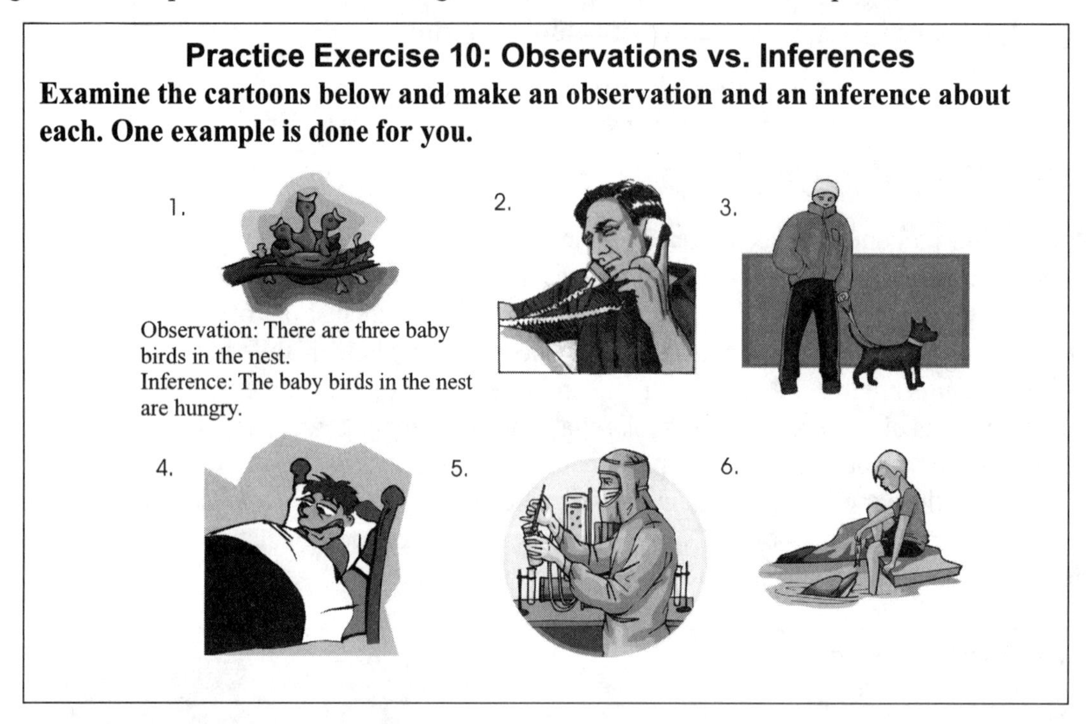

Making a Prediction

A **prediction** is a forecast of the possible results of events. Knowledge we have gained from observation and experimentation can help us to make predictions about seemingly unrelated events.

Try to make a prediction about the following situation. On a freezing cold wintry morning, Mrs. Allen got into her car in the driveway and left for work. Shortly afterward, her daughter, Ann, went out to the street to catch the bus for school. As she walked down the driveway, she noticed two small puddles of water where the car had been parked. The first puddle was frozen. The second puddle was not frozen. Why?

We already know that salt water freezes at lower temperatures than water without salt. We also know that it is common to put salt down on icy roads. Based on this information, we can make the inference that the water in the unfrozen puddle contained salt. We can predict that the puddle will remain unfrozen until it gets a few degrees colder.

Practice Exercise 11: Making a Prediction

Review the figures representing the Big Dipper in Practice Exercise 9. Predict what the Big Dipper might look like 500,000 years into the future. Explain how you arrived at this prediction.

Obtaining Unexpected Results

Unexpected results can occur during a scientific investigation. They can be a result of error, such as improper equipment usage, improper setup, or poor data collection. However, when these factors are carefully controlled, unexpected results can lead to better and more complete hypotheses.

Analysis of Error in Scientific Data

Any errors made during an experiment will result in a distortion of data. The following are examples of errors that can be made during an experiment.

- A student consistently reads the volume in a graduated cylinder at a slight angle instead of straight on. The recorded volume will be consistently high or low.
- Sometimes a spillage or other accident can result in an invalid data point.
- Pouring a liquid from one container to another and then recording the volume will usually result in some error since not all of the liquid will be transferred to the second container.
- Errors in data are commonly made simply by recording the data incorrectly.

Compare the data obtained by two students performing the same experiment. What is the relationship between the length of time and the degrees of temperature change? Student 1 might suggest that for every five minutes of time that elapsed, the temperature dropped five degrees. Student 2 has recorded data that suggests otherwise — the relationship between the length of time and the degrees of temperature change is quite different in his results. What can you infer?

Table 1.5 Time for Water to Freeze

Data from Student 1	0	5	10	15	20	25	35
Temperature of salt water	25°C	20°C	15°C	10° C	5°C	0°C	-10°C

Table 1.6 Time for Water to Freeze

Data from Student 2	0	5	10	15	20	25	35
Temperature of salt water	25°C	20°C	25°C	10°C	–5°C	0°C	–10°C

A careful analysis of the data would indicate that water does not suddenly warm 5 degrees or drop 15 degrees in temperature when placed in a freezer. An error must have been made either in the experiment or in recording the data. By repeating the experiment and carefully checking the data for any variations in the control, an analysis of the error can be made.

Practice Exercise 12: Looking at Error

Use Tables 1.5 and 1.6 to answer questions 1 – 3.

1. What kinds of errors could account for the students' very different data sets?
2. Which data set seems more likely to be accurate? Why?
3. What type of graph would best represent these data sets, allowing you to compare results?

Section Review 2: Presenting and Analyzing Data

A. Define the following terms.

table	circle graph	cross section	results
line graph	model	map	inference
multiple line graph	diagram	conclusion	prediction
bar graph			

B. Choose the best answer.

1. A judgement based on data gathered in an experiment is
   A. a skill.
   B. a conclusion.
   C. a hypothesis.
   D. an observation.

2. A forecast of possible or future events is a/an
   A. analysis.
   B. prediction.
   C. conclusion.
   D. observation.

3. The end products of your investigation or experiment are the
   A. results.
   B. guesses.
   C. predictions.
   D. questions.

4. Janet is interested in the growth rates of three different varieties of bean plants. She has ten plants of each of the three varieties, and she records the heights of each plant each day for a period of one month. How should Janet best present her data?
   A. on a data table with the days of the month along the top and the heights for each plant listed below
   B. on a multiple line graph that shows average plant height for each variety, each day for the month
   C. in a bar graph that shows the average height of each variety at the end of the experiment
   D. on a pie chart that shows the percentage of plants that had significant growth

The following information about the results of an experiment on goldfish is included in a research paper.

Goldfish in a 12"×24" tank were given the same amount of food as the goldfish in a 24"×60" tank. Equal numbers of goldfish were present in each tank. Results indicate that goldfish raised in the 24"×60"tank grew on average 1" longer than the goldfish in the 12"×24" tank.

5. Which of the following sentences is most likely to be in the conclusion of these results?
   A. Goldfish growth was not affected by the size of their habitat.
   B. Goldfish have a tendency to grow larger in a larger habitat.
   C. Diet is a deciding factor in goldfish growth.
   D. Goldfish prefer square tanks to round tanks.

6. Unexpected results are
   A. detrimental to an experiment.
   B. ignored by scientists.
   C. a result of a bad hypothesis.
   D. useful in obtaining a better hypothesis.

7. An explanation based on an assumption is also called a(n)
   A. inference.
   B. prediction.
   C. observation.
   D. scientific error.

C. Answer the following questions.

Refer to the following paragraph and pie chart to answer questions 1–4.

During a summer drought, a community asks the question of how it can better conserve water. Through research and questionnaires, the community assembled data on daily water usage in its neighborhood. The community graphed the data in a pie chart shown to the right.

Average Daily Water Use in the Home (based on 285 Liters/Day)

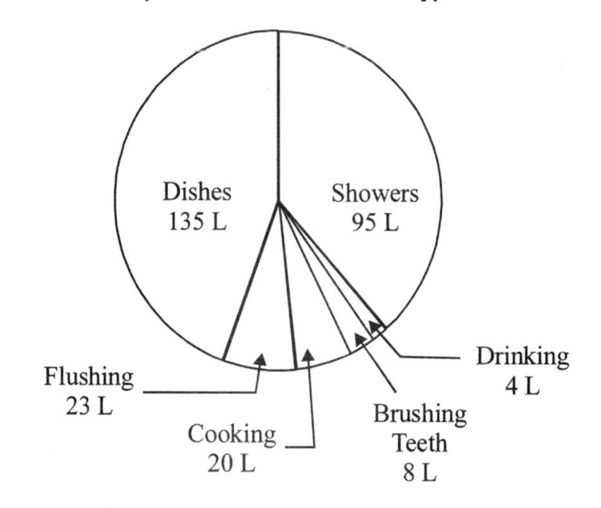

1. What area of water usage consumes the most water?
2. What area of water usage consumes the least amount of water?
3. Using this data, make a recommendation on how the community could best conserve water. Give some specific examples of how your recommendation could be implemented.
4. How would you investigate the water usage in your school? List the steps of your investigation and how you would gather data.

D. Read the following experiment, and Identify the steps in the Scientific Method.

Last year, Wendy planted seeds in a garden. She noticed that not all of the seeds became plants. This year, she asked herself, "On average, how many of the seeds in a package will grow?" She thought maybe 85% of the seeds in a package would not grow.

She bought three packages containing twenty seeds each. She planted each package of seeds in separate boxes, so she could keep careful count of the number of seeds that would grow. She drew a diagram of the boxes, indicating where each seed was planted. As the seeds sprouted, she put a green X on the place in her diagram where the seed was planted. If the seed didn't grow, she put a red X on the place in her diagram where the seed was planted.

At the end of her investigation, she noticed that four seeds in the first box did not grow. The second box had six seeds that did not grow. In the third box, only five of the seeds did not grow. Wendy concluded an average of five seeds in each package did not grow. For future gardens, Wendy assumed that 75% percent of the seeds in a package would grow.

1. What is the observation that led to the experiment?
2. State the question to be investigated.
3. What was Wendy's hypothesis?
4. Explain her experiment to test her hypothesis.
5. How did she collect her data?
6. What was Wendy's conclusion?
7. State the prediction she made.

Chapter 1 Review

A. Choose the best answer.

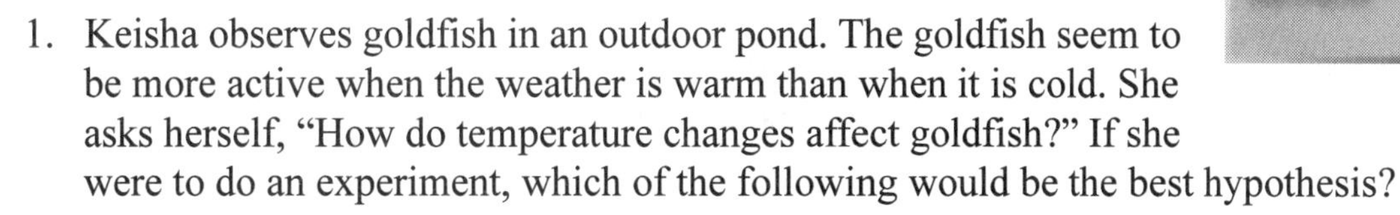

1. Keisha observes goldfish in an outdoor pond. The goldfish seem to be more active when the weather is warm than when it is cold. She asks herself, "How do temperature changes affect goldfish?" If she were to do an experiment, which of the following would be the best hypothesis?

   A. Goldfish prefer warm water to cold water.

   B. Goldfish are more active in warm water than in cold water.

   C. Goldfish live in warm and cold water.

   D. Temperature changes will kill goldfish.

2. For a science project, Steve conducted a taste test on 4 brands of regular cola. As part of his research, he determined the percentages of the main ingredients in each cola. He found that the percent of sugar in the colas strongly correlated with taste preference. He wanted to present the data showing the percentages of main ingredients in each cola as part of his project. Which of the following would be the best way to present this data?

   A. in separate pie charts
   C. in a bar graph
   B. in a multiple line graph
   D. in a data table

3. Four groups of rats are tested in a lab. Group 1 is given a special hormone for muscle growth. Group 2 is given 50 mg per day of the same growth hormone. Group 3 receives 100 mg per day of the hormone treatment. Group 4 does not receive any hormone treatment. Which of the following groups is the control group for this experiment?

   A. group 1 B. group 2 C. group 3 D. group 4

4. In a previous experiment, Josh determined that the growth of goldfish depends on the size of the container they are in. Now, Josh wants to know if the number of goldfish in a container affects their growth. To conduct his experiment, he placed 10 goldfish in a 20 gallon aquarium, 5 goldfish in a 10 gallon aquarium, and 2 goldfish in a 3 gallon aquarium. All the goldfish received the same food in equal amounts. He recorded goldfish growth over a 10-week period. From his data, he concluded that the more goldfish in an aquarium, the larger they grow. Why was his conclusion not valid?

   A. His conclusion was not valid because he should have put the 2 goldfish in a 4 gallon aquarium.

   B. His conclusion was not valid because it was based on old data.

   C. His conclusion was valid, but he should not make a conclusion based on just one experiment

   D. His conclusion was not valid because he did not use the same size aquarium for all the groups.

5. If Josh had placed the 2 goldfish in a 4 gallon aquarium, his data would have supported a conclusion about

   A. the effect of population density on the growth of goldfish.

   B. the effect of tank shape on the growth of goldfish.

   C. the effect of small aquariums on the growth of goldfish.

   D. the effect of large aquariums on the growth of goldfish.

Study the bar graphs below, and answer questions 6 – 8.

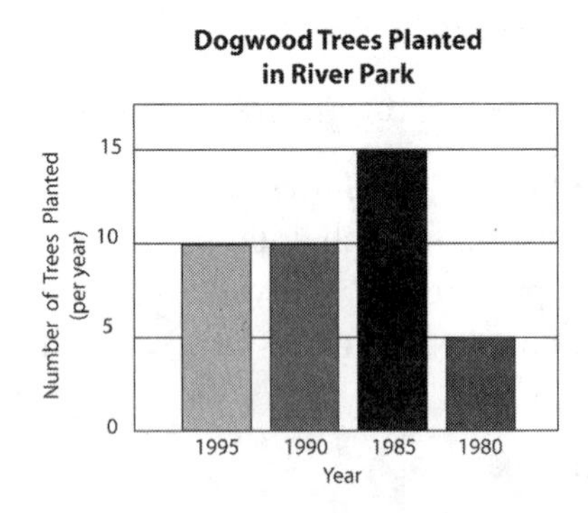

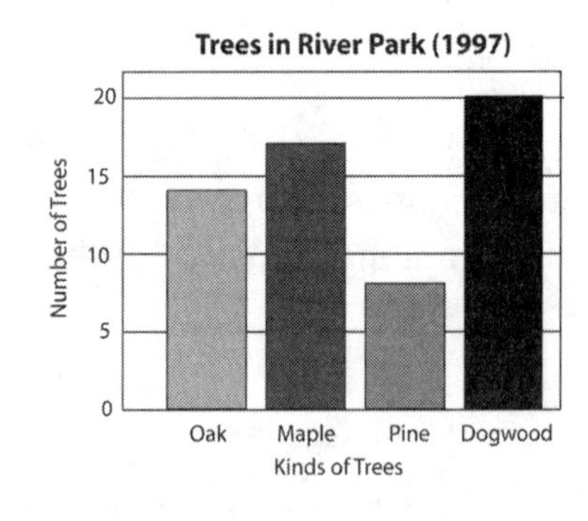

6. How many dogwood trees were in the park in 1997?

   A. 20 B. 15 C. 10 D. 6

7. What is the total number of dogwood trees planted between 1980 and 1995?

   A. 5 B. 15 C. 30 D. 40

8. What percent of the total number of dogwood trees planted remained alive in 1997? (assuming no new dogwood trees sprouted naturally and there were no dogwood trees in the park before 1980)

   A. 25% C. 75%

   B. 50% D. 100%

Study the table below, and answer questions 9 –11.

Incubation Temperature of Turtle Eggs Versus Sex of Hatchling				
Four Groups of 25 Eggs	Temperature	Number of Male	Number of Female	Eggs Not Hatched
Group 1	26°C	21	2	2
Group 2	28°C	13	11	1
Group 3	30°C	1	19	5
Group 4	32°C	1	20	4

9. At what temperature did most of the turtle eggs hatch?

   A. 26°C B. 28°C C. 30°C D. 32°C

10. What temperature produced the most females?

    A. 26°C B. 28°C C. 30°C D. 32°C

11. What temperature produced the most males?

    A. 26°C B. 28°C C. 30°C D. 32°C

12. The first step in presenting data is
    A. to record it with state offices.
    B. to check spelling errors.
    C. to pick a chart.
    D. to organize the data.

13. A conclusion is drawn based on
    A. a hypothesis.
    B. textbook statistics.
    C. analysis of data.
    D. an educated guess.

14. Identify the purpose of the control group in a scientific experiment.
    A. to clarify the hypothesis
    B. to reduce experimental error
    C. to determine the effect of no treatment
    D. to separate the independent and dependent variables

15. A hypothesis is checked by
    A. research.
    B. guessing.
    C. experimentation.
    D. researching on the Internet.

16. The statement, "An object fell 5 meters in 1 second" is
    A. a control.
    B. an observation.
    C. a conclusion.
    D. a theory.

17. Identify the best description of qualitative data.
    A. organized observations
    B. thoughtful explanations
    C. numerical measurements
    D. descriptions of characteristics

18. Identify the best description of quantitative data.
    A. organized observations
    B. thoughtful explanations
    C. numerical measurements
    D. descriptions of characteristics

Read the science experiment below, and then answer questions 19 – 21.

Observation: Some green bean plants grow in shaded areas while others grow in full sun.

Question: Do green bean plants grow better in full sunlight or do they grow better in shade?

Hypothesis: Green bean plants will grow better in full sunlight.

Method: Set up an experiment with green bean plants to record their growth over a seven week period.

Materials: 6 small green bean plants of approximately the same size.
6 Styrofoam cups filled with potting soil and with a hole in the bottom of each.
10 mL of water for each plant once a week.
A metric ruler.

Directions: Transfer the bean plants to the Styrofoam cups. Water each plant only once a week. Measure each plant and record the height. Place plants A and B in complete darkness. Place plants C and D in full sunlight. Place plant E and plant F in partial sunlight.

Observations: Closely measure the rate of growth of the control group plants and the experimental groups. Take notes about the conditions of each plant and record the measurements on tables.

Record Data: Make a graph to show the growth rate for each plant. Plot their growth on a multiple line graph showing height and time.

Conclusion: What conclusion can be drawn about the growth of the plants in relationship to the amount of light they each received? Does the investigation support the hypothesis? Can a theory be stated about the effect light had on the green bean plants?

Prediction: Based on the theory that is stated, make a prediction about the behavior of green bean plants in the presence of full sunlight.

19. What was the independent variable in the experiment with the green bean plants?
    A. the type of potting soil used
    B. the amount of sunlight
    C. the size of the container
    D. how frequently the plant was watered

20. Which plants were the control group plants?
    A. the plants kept in darkness
    B. the plants kept in partial sunlight
    C. the plants kept in full sunlight
    D. there was not a control group

21. Which of the following factors were kept constant in order for the data to be valid?
    A. the type of plant used
    B. the size of the container used to grow the plants
    C. the amount of water given to each plant
    D. all of the above are control variables

Chapter 2
Safety, Equipment, and Scientific Measurement

PHYSICAL SCIENCE STANDARDS COVERED IN THIS CHAPTER INCLUDE:

SCSh2	Students will use standard safety practices for all classroom, laboratory and field investigations
SCSh4	Students will use tools and instruments for observing, measuring and manipulating scientific equipment and materials.

SAFETY PROCEDURES IN THE LABORATORY

Safety procedures are set up to protect you and others from injury. The most important safety rule is to always follow your teacher's instructions. Before working in the laboratory, fully read all of the directions for the experiment. Laboratory accidents can be easily avoided if safety procedures are followed. Be sure that you wear appropriate clothing for the lab, and remove any dangling jewelry. Know where eyewash stations are located. Decide what personal protective equipment, like aprons, goggles, or gloves, are necessary. If there is an accident, spill, or breakage in the laboratory, report it to your instructor immediately.

Glassware Safety

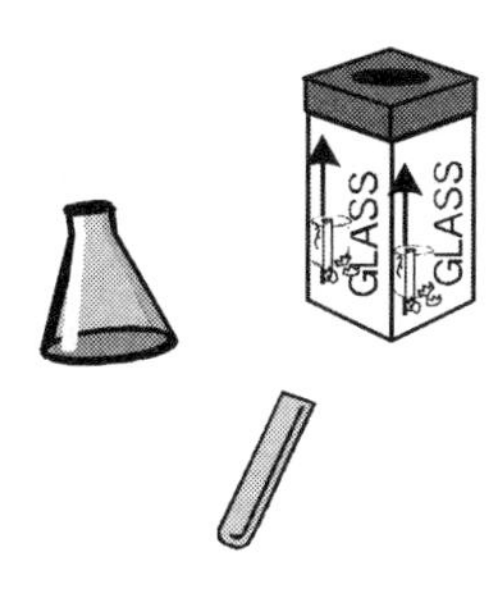

- Never use broken or chipped glassware. Dispose of broken or chipped glassware in a container specified by your teacher.
- Never heat glassware that is not thoroughly dry.
- Never pick up any glassware unless you are sure it is not hot. Remember, hot glass looks the same as cold glass.
- If glassware is hot, use heat-resistant gloves, or use tongs.
- Do not put hot glassware in cold water, or on any other cold surface.

Sharp Instrument Safety

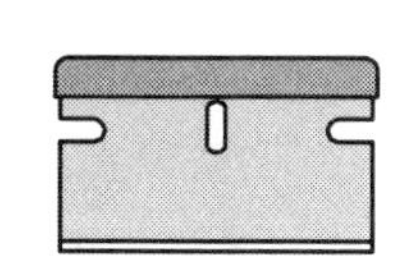

- Always use single edged razors.
- Handle any sharp instrument with extreme care.
- Never cut any material toward you. Always cut away from you. Immediately notify your teacher if your skin is cut.
- Dispose of used or ruined sharp instruments in a container specified by your teacher.

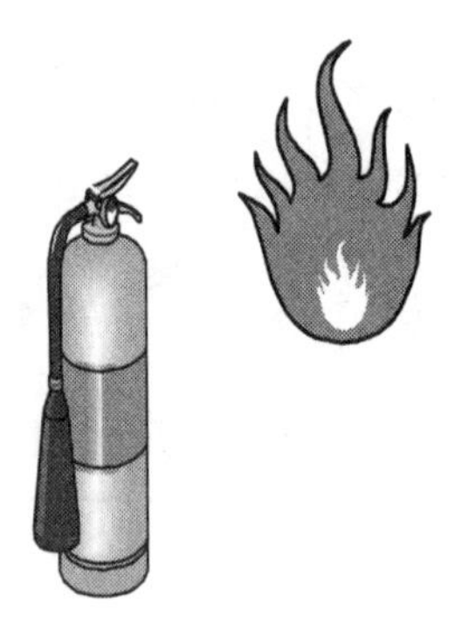

Fire and Heat Safety

- Never use an open flame without wearing safety goggles.
- Never heat anything (particularly chemicals) unless instructed to do so.
- Never heat anything in a closed container.
- When using a Bunsen burner to heat a substance in a test tube, move the test tube in and out of the flame. Never leave the test tube directly in the flame.
- Never reach across a flame.
- Always use a clamp, tongs, or heat-resistant gloves to handle hot objects.
- Fire extinguishers should be located in or near the lab. Do not tamper with the extinguishers in any way. Do not remove an extinguisher from its mounting unless instructed to by your teacher and/or during a fire.

Animal Safety

- Do not cause pain, discomfort, or injury to a live animal.
- Follow your teacher's directions when handling animals.
- Wash your hands thoroughly after handling animals or their cages.

Electrical Safety

- If an extension cord is needed to plug in an electrical device, use the shortest extension cord possible.
- Do not use socket multipliers to overload an electrical outlet.
- Never touch an electrical appliance or outlet with wet hands.

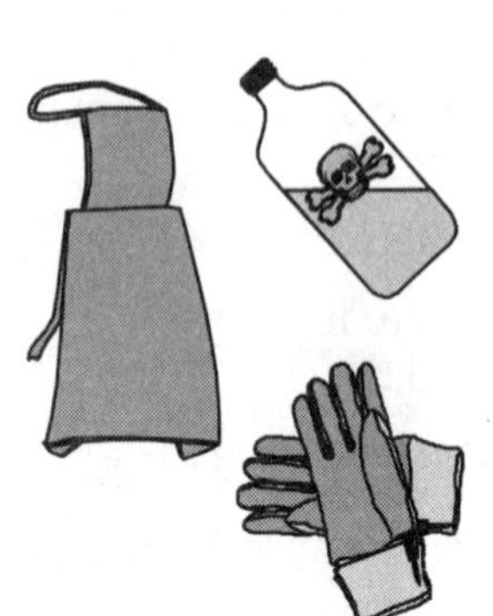

Chemical Safety

- Always wear a safety apron and protective gloves when handling chemicals to protect yourself from chemical spills. If a chemical contacts your skin, rinse immediately and notify your instructor, or seek emergency care.
- If instructed by your teacher to smell a chemical, never do so by sniffing directly from the container. Instead, hold the container away from your face and use your hand to waft some of the chemical odor toward your nose.
- Use proper ventilation in the lab through use of a chemical fume hood.
- Keep all lids closed when chemicals are not in use.
- Dispose of all chemicals as instructed by your teacher.

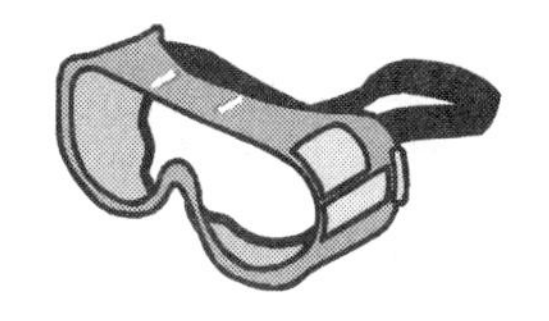

Eye and Face Safety

- Wear safety goggles when handling chemicals.
- When you are heating a test tube or bottle, always point it away from you and others.
- Remember, chemicals can splash or boil out of a heated test tube.
- If a chemical comes in contact with your eyes, use the eyewash fountain immediately, and seek emergency care.

Proper Dress

- Wear long-sleeved blouses, shirts, and pants rather than shorts.
- Tie back long hair to prevent it from coming into contact with chemicals or an open flame.
- Wear shoes without open ends.
- Remove or tie back any dangling jewelry or loose clothing to prevent them from getting caught on any equipment.

Official Safety Information

Product Name: Serenade ASO

Trade names/Synonyms:	QST 713 AS organic; QRD143AS
EPA Registration Number:	69592-12
Primary Hazards:	None known

Section 1: MATERIAL IDENTIFICATION

INGREDIENT 1

Common Name:	QST 713 strain of dried *Baccillus subtilis*
Percent:	1.34%

Section 2: PHYSICAL DATA

Boiling Point:	≥ 100° C
Density:	1.0 to 1.3 gram/cc

Section 3: REACTIVITY

Stability:	Material is non-reactive
Hazardous Polymerization:	Does not occur

Section 4: HEALTH HAZARDS

Corrosive:	Not Corrosive
Skin/Eye Irritation:	May be irritating to skin or eyes for some individuals

Section 5: FIRST AID

Emergency First Aid Procedures:

If in eyes:	Hold eye open and rinse slowly and gently with water for 15-20 minutes. Remove contact lenses, if present, after the first 5 minutes, then continue rinsing eye. Call a poison control center or doctor for treatment advice.
If on skin:	Take off contaminated clothing. Rinse skin immediately with plenty of water for 15-20 minutes. Call a poison control center or docor for treatment advice.

Source: http://www.agraquest.com/products/serenade/pdfs/Serenade_ASO-MSDS.pdf

Figure 2.1 Example of an MSDS

Through the Department of Labor, the United States Government runs the **Occupational Safety and Health Administration**, also called **OSHA**. The goal of OSHA is to protect the health and safety of America's workers. OSHA has many regulations and procedures that help maintain safe work environments. It also has a great deal of guidance for workers and employers from all industries. The national OSHA website is at www.osha.gov.

Manufacturers of chemicals are required to produce, update, and maintain a safety data sheet for each chemical they produce. This document is called a **material safety data sheet**, or **MSDS**. An MSDS lists information on chemical structure, chemical appearance, chemical properties, and personal safety. It also contains information on safe storage and disposal of chemicals.

Read and understand all MSDS sheets for chemicals used in the lab.

MSDS sheets can be found on the internet. Some websites of interest are: http://www.scorecard.org/chemical-profiles/ and http://msds.ehs.cornell.edu/msdssrch.asp.

Section Review 1: Safety Procedures in the Laboratory

A. List several rules for ensuring the various safety concerns below are met.

glassware safety
sharp instrument safety
fire and heat safety
animal safety
electrical safety
chemical safety
eye and face safety

B. Choose the best answer.

1. How should you pick up a piece of hot glassware?

   A. with bare hands
   B. with heat-resistant gloves
   C. with the sleeve of your shirt
   D. with a spatula

2. How should you hold a test tube containing a chemical?

   A. pointed away from your face
   B. pointed at your eye
   C. held right up to your nose
   D. very close to your partner's face

3. You should report a cut in your skin, glass breakage, or a chemical spill

   A. after the problem is handled.
   B. never.
   C. immediately.
   D. after you write down what happened.

4. Why shouldn't you wear dangling jewelry and baggy clothing to the laboratory?

   A. The baggier the clothes, the more chemical fumes are absorbed.
   B. The metal in the jewelry changes the expected reaction.
   C. Lab coats don't fit over baggy clothes.
   D. Jewelry and clothing could get caught on equipment, and clothes can catch on fire.

5. When you are done with an experiment, how should you dispose of any chemicals used?

   A. Mix them all up in a waste container and dump them in the trash.
   B. Pour them all down the sink.
   C. Follow the instructions given to properly dispose of the particular chemical(s).
   D. Mix the chemicals in a flask and heat the mixture until it evaporates into the air.

C. Answer the following questions.

1. Kara and Mitch are partners in physical science lab. Their class is conducting an experiment on supersaturated solutions. Kara and Mitch put on safety goggles and safety aprons and read the experiment's directions. The experiment directs them to add sugar to water in a beaker, heat the beaker with the solution on a hot plate, add more sugar to the solution, remove the beaker with the solution from the hot plate, and allow the solution to cool undisturbed. What additional safety equipment will Kara and Mitch need to use for this experiment?

2. Mr. Ohm's physical science class is making soap. The students will be using sodium hydroxide, which is a potent chemical. Which types of safety equipment will the class most likely need?

3. Name two pieces of safety equipment that should be used or made available when working with an open flame.

Equipment and Materials

Laboratory equipment and materials are tools or devices used in scientific investigations. Each article in the lab has a specific purpose. If you don't know the purpose of a piece of equipment, ask your teacher. Instruments are used to enhance the ability of our senses to observe.

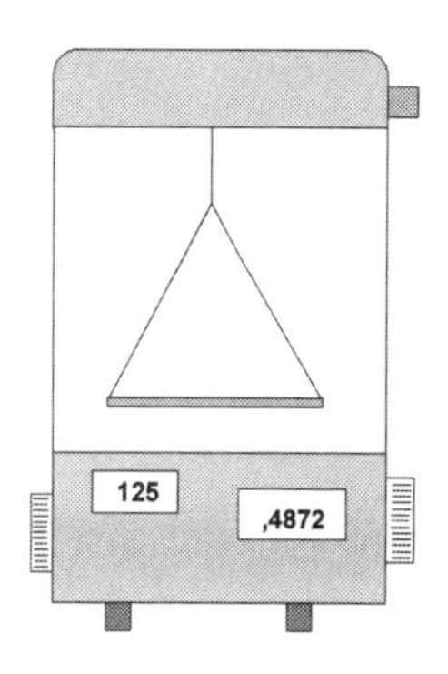

A **mechanical pan balance** is used to accurately determine mass to the nearest ten thousandth of a gram.

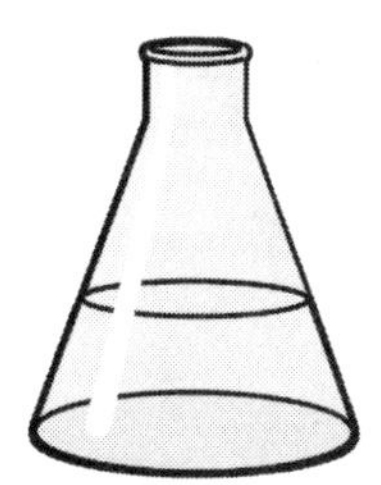

An **Erlenmeyer flask** is used to mix liquids; its narrow mouth prevents splashing and lessens the dispersion of noxious fumes. Although it is sometimes marked to make volume measurements, they are only approximate values.

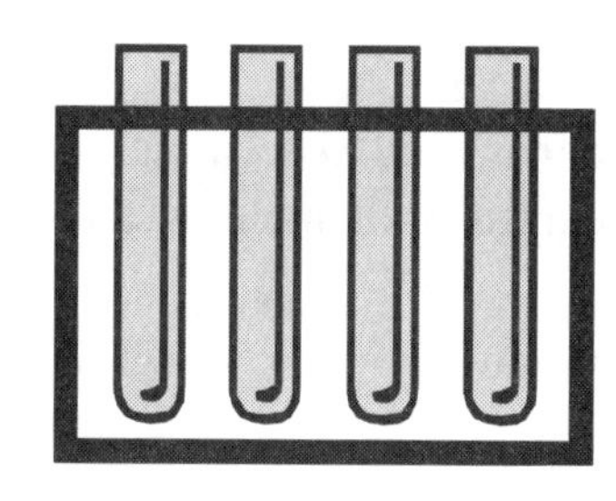

Test tubes are used to mix, measure, or heat liquids. Test tubes are not usually marked with measurements, so they are only used to make approximate measurements.

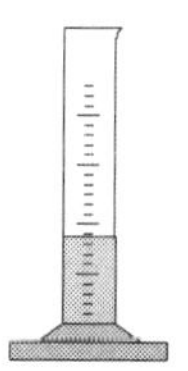

Graduated cylinders are "graduated" or marked with a scale for measurement. They are used to accurately measure liquid volume.

An **eyedropper** is used to dispense small measures of a liquid.

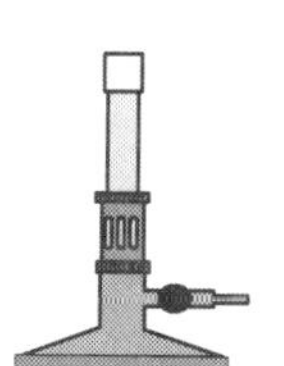

A **Bunsen burner** is a source of gas heat.

A **watch glass** is a holding container or covering device.

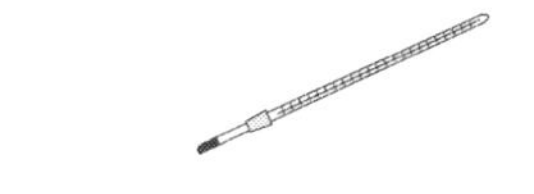

Thermometers measure temperature.

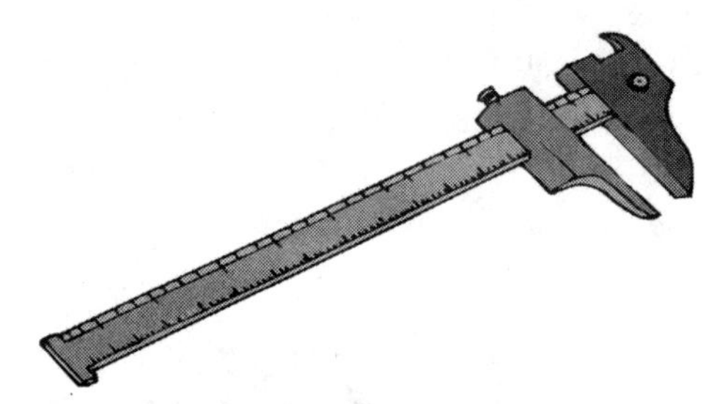

A **caliper** is used to accurately measure the thickness or diameter of an object.

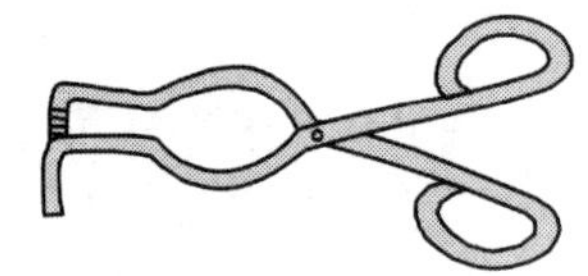

Tongs are used to grasp heated material.

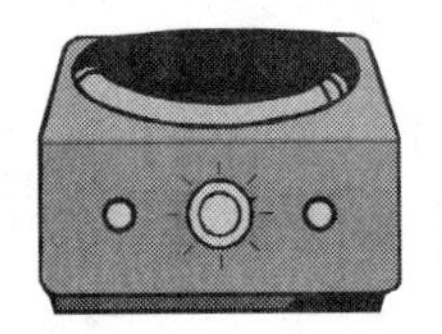

A **hot plate** is a source of electric heat.

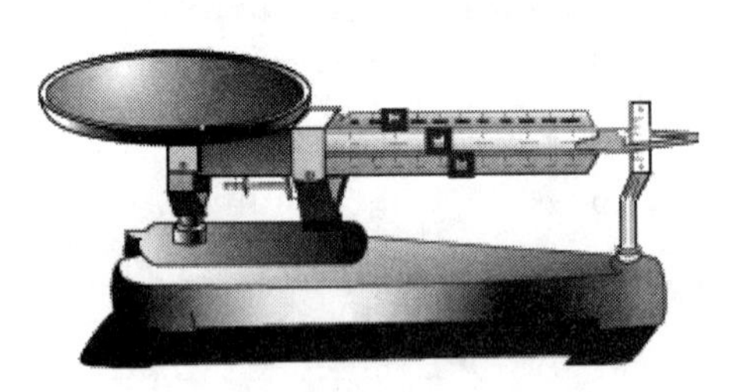

A **triple-beam balance** is used to determine the mass of heavier materials to the nearest gram.

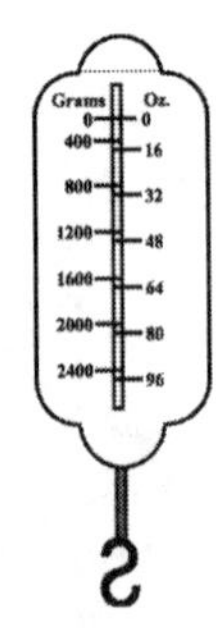

A spring scale is used to determine force.

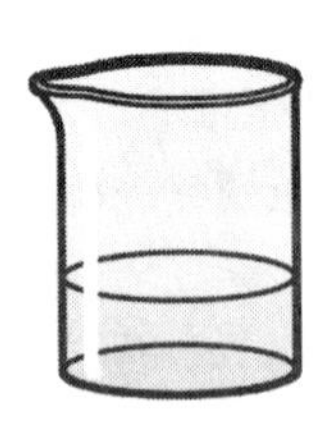

Beakers are used to mix and heat liquids. Like the Erlenmeyer flask, they are not intended for accurate volume measurement.

A **telescope** enables us to see objects too far away to be seen with the unaided eye.

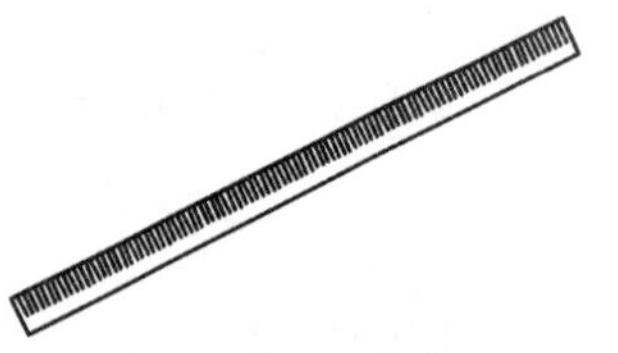

A **meter stick** measures length or width.

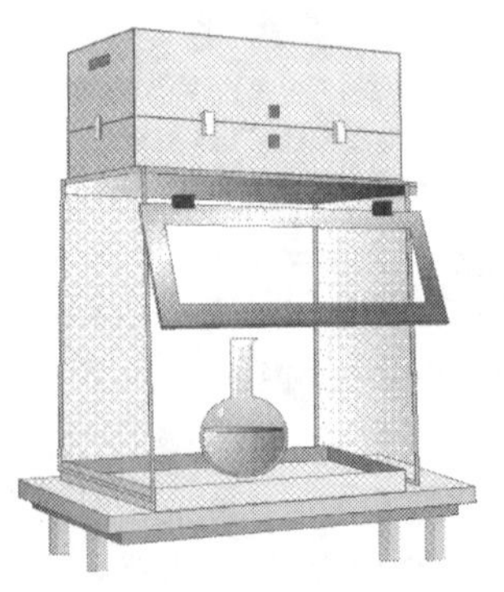

A chemical **fume hood** is used to contain and safely remove hazardous gases from the laboratory.

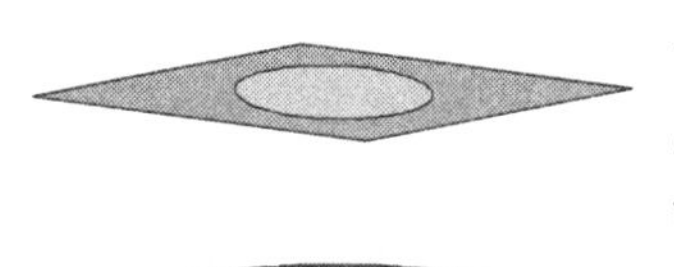

Wire gauze usually goes on top of a tripod to hold the glassware being heated.

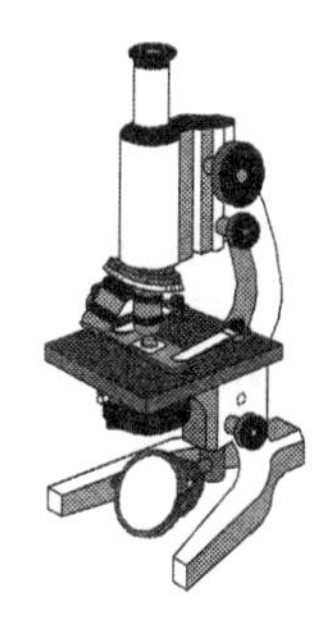

A **microscope** enables us to see very small objects or organisms too small to see with the unaided eye.

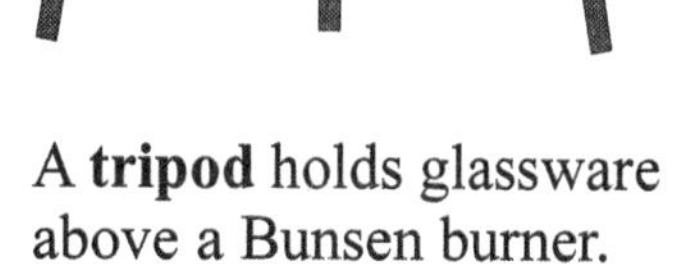

A **tripod** holds glassware above a Bunsen burner.

Lab Activity 1: Equipment

Visit a science lab and locate each piece of equipment listed on these past three pages. Do this frequently until you are able to find and name each piece.

Section Review 2: Equipment

1. Which of the following is used as a source of heat in the laboratory?
   A. thermometer B. Bunsen burner C. thermostat D. gasoline

2. Which has specific markings for measurement and is used to accurately measure liquid volume?
   A. test tubes B. beakers C. ruler D. graduated cylinders

3. Which of the following pieces of equipment is used to handle liquids but is not intended for accurate measurement?
   A. beaker B. test tube C. Erlenmeyer flask D. all of the above

4. If you were instructed to heat something on the Bunsen burner, you would need to set your container on a ______________ to hold your container over the burner.
   A. watch glass
   B. hotplate
   C. a piece of wire gauze held by a tripod
   D. Petri dish

Scientific Measurement

In order to describe the world around them, scientists often take measurements to quantify a certain phenomenon. However, there is uncertainty with every measurement. This uncertainty can arise from poor measuring devices or improper use of the measuring device (i.e., human error). Scientists can express the uncertainty of their measurements by using the proper amount of significant figures, which will be discussed in further detail in Chapter 4. Scientists must also include a unit with their measurements, otherwise the numbers have no meaning. The following sections discuss scientific units and precision in measurements in more detail.

Introduction to SI Units

The **SI units** of measurement are used throughout the world when performing calculations related to scientific investigations. It stands for *Le Système International d'Unites* and was established in France about 200 years ago. SI units were adapted from the metric system, and the base units are meter, gram, and second to measure length, mass, and time, respectively. In addition, volume, density, and temperature are measurements frequently used in the laboratory.

The **English system** of measurement, also called the **U.S. Customary System**, is used in the United States. In this system, the foot is the standard length, the pound is the standard weight, and the second is the standard for time. Although you are probably more familiar with the English system, the SI units are used in the scientific community, and throughout the world. Therefore, SI units will be the standard system of measurement used in this book. Some units and conversions are listed in Table 2.1 following.

Table 2.1 Metric - English Unit Conversions

English			Metric		
			Length		
12 inches	=	1 foot	1 inch	=	2.54 cm
3 feet	=	1 yard	3.281 feet	=	1 meter
5,280 feet	=	1 mile	3281 feet	=	1 kilometer
			Mass		
16 ounces	=	1 pound	0.035 ounces	=	1 gram
2000 pounds	=	1 ton	1 pound	=	0.453 kilograms
			Volume		
16 fluid ounces	=	1 pint	1 cm^3	=	1 mL
2 pints	=	1 quart	33.8 fluid ounces	=	1 liter
4 quarts	=	1 gallon	3.78 liters	=	1 gallon
			Weight		
1 pound on Earth	=	0.167 moon pounds	1 kilogram	=	2.2 pounds

Standard SI Measurements

The standard SI unit of measurement to determine **length** is **meter (m)**. To better visualize a meter, it is helpful to know that 1 meter is equal to 3.28 feet. Length measures the distance from one point to another and can be used to determine a person's height in meters, the distance between your home and your school in kilometers, the length of an almond in centimeters, or the thickness of a dime in millimeters. A ruler or **meter stick** is commonly used to measure length as seen in Figure 2.2 below.

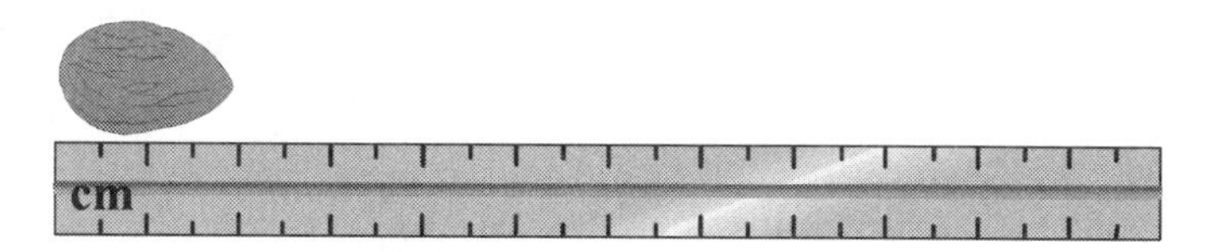

This almond measures about 2 cm in length.

Figure 2.2 Metric Ruler

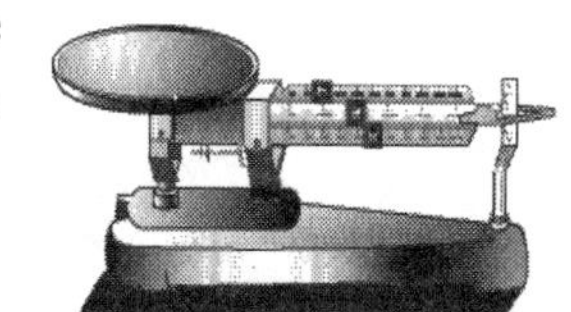

Mass is the measure of the amount of matter in an object. Its standard SI unit is the **kilogram (kg)**, and its tool of measurement is the balance.The mass of a needle, a rock and a person would be expressed in milligrams, grams and kilograms, respectively.

The terms mass and weight are often used interchangeably, but they are NOT the same. Mass is a property of an object. **Weight** is a measurement of the gravitational force that attracts an object to the earth. Therefore, the mass of an object on Earth is the same as it is on the moon, but the weight of an object will be different. Let's do a conversion. Let us say that Lois has a mass of 60 kg. She stands on a scale located on the earth and on the moon. What will the scale's reading be in the two locations?

$$60 \text{ kg} \times (2.2 \text{ pounds}/1\text{kg}) = 132 \text{ Earth pounds}$$

$$132 \text{ Earth pounds} \times (0.167 \text{ Moon pounds/Earth pounds}) = 22.04 \text{ Moon pounds}$$

So, Lois's weight will differ depending on the degree of gravitational force that is acting on her. Scientists prefer to use mass when reporting the results of an experiment because it does not vary.

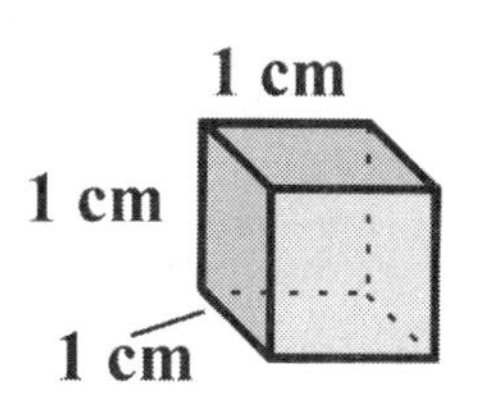

Volume = 1 cm^3 = 1 mL

Figure 2.3 Volume of a Cube

Volume is the amount of space occupied by an object. Volume is determined in different ways depending on the shape (i.e. cube, sphere, irregular) and state of matter (i.e. solid, liquid, gas) of the object. For a regularly shaped object, like a cube, volume is determined by multiplying the length times the height times the width of the object ($V = l \times h \times w$). The units used for volume are **cubic centimeters (cm^3** or **cc)** or **milliliters (mL)**. Figure 2.3 shows this relationship. One cubic centimeter is equal to one milliliter, which is equal to one thousandth of a liter.

A **graduated cylinder** is used to measure the volume of liquids. When liquids are placed in a graduated cylinder, a meniscus will form. A **meniscus** is the curve of liquid at its surface. The meniscus may curve down or up, depending on the liquid. To read the volume of the liquid, get eye level with the meniscus and measure from the bottom of the curve of the meniscus if it curves downward or the top of the meniscus if it curves upward. Figure 2.4 shows how to read a meniscus by getting eye level with it and measuring at the center point.

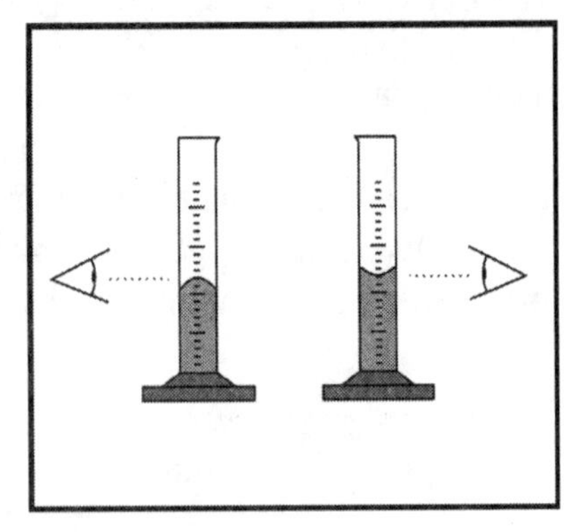

Figure 2.4 Reading a Meniscus

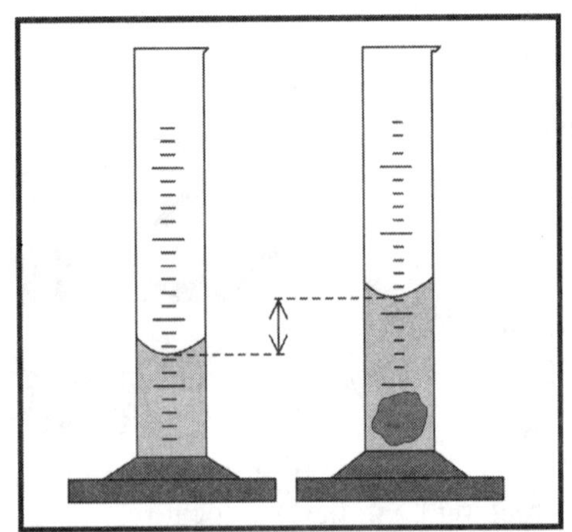

Figure 2.5 Water Displacement Method

To determine volume of irregularly shaped objects, the **water displacement** method is used. To measure volume by water displacement, place an amount of water in a container and determine its volume. Then add the object. When the object is placed in the container of water, the water level will rise. Measure the volume again. To determine the volume of the object, simply subtract the first measurement from the second.

Density, D, is the mass (m) of an object divided by its volume (V), $(D = m/V)$. The standard SI unit for density is **kg/m³**, but it is also commonly expressed in units of grams per cubic centimeter (g/cm^3). Density is a characteristic material property; thus the density of two objects of the same material is always the same even if the objects have different masses. For example, a gold ring and a gold brick both have the same density, because they are both made of gold. However, because the gold brick contains more matter than the gold ring, it has a greater mass.

Density explains why some things float, and other things sink. A rock at the bottom of a stream is denser than water, so it sinks to the bottom, whereas a piece of styrofoam is less dense than the water, so it floats. If an object sinks, it is denser than the liquid it is placed in. If an object floats, it is less dense than the liquid it is placed in.

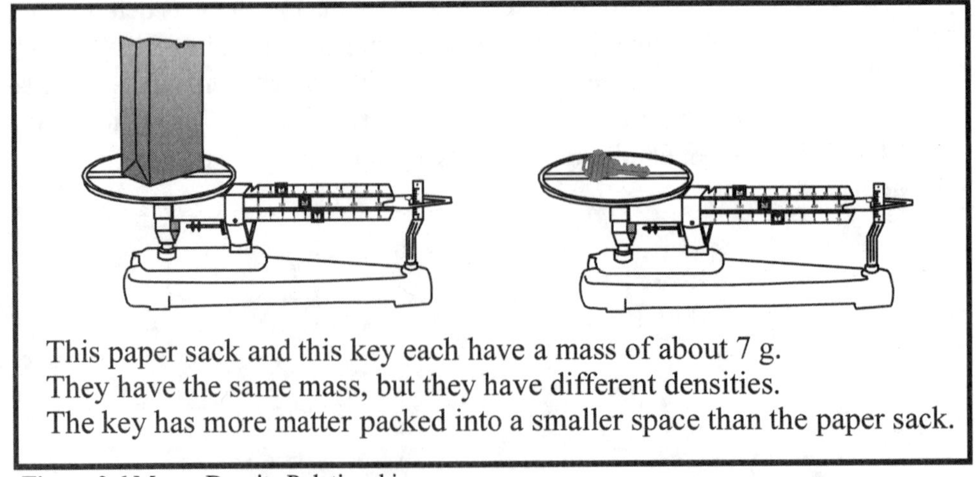

This paper sack and this key each have a mass of about 7 g.
They have the same mass, but they have different densities.
The key has more matter packed into a smaller space than the paper sack.

Figure 2.6 Mass - Density Relationship

Temperature measures how hot or cold something is. All measurements for temperature are taken in degrees. In the metric system, **Celsius** is used. The SI unit for temperature is **Kelvin**, and the English unit is **Fahrenheit**. To convert from one unit to another, use the following formulas:

$$C = \frac{(F - 32)}{1.8} \qquad F = 1.8\,C + 32 \qquad C = K - 273.15 \qquad K = C + 273.15$$

C is degrees Celsius; **F** is degrees Fahrenheit; **K** is Kelvin.

Figure 2.7 Temperature Conversion Formulas

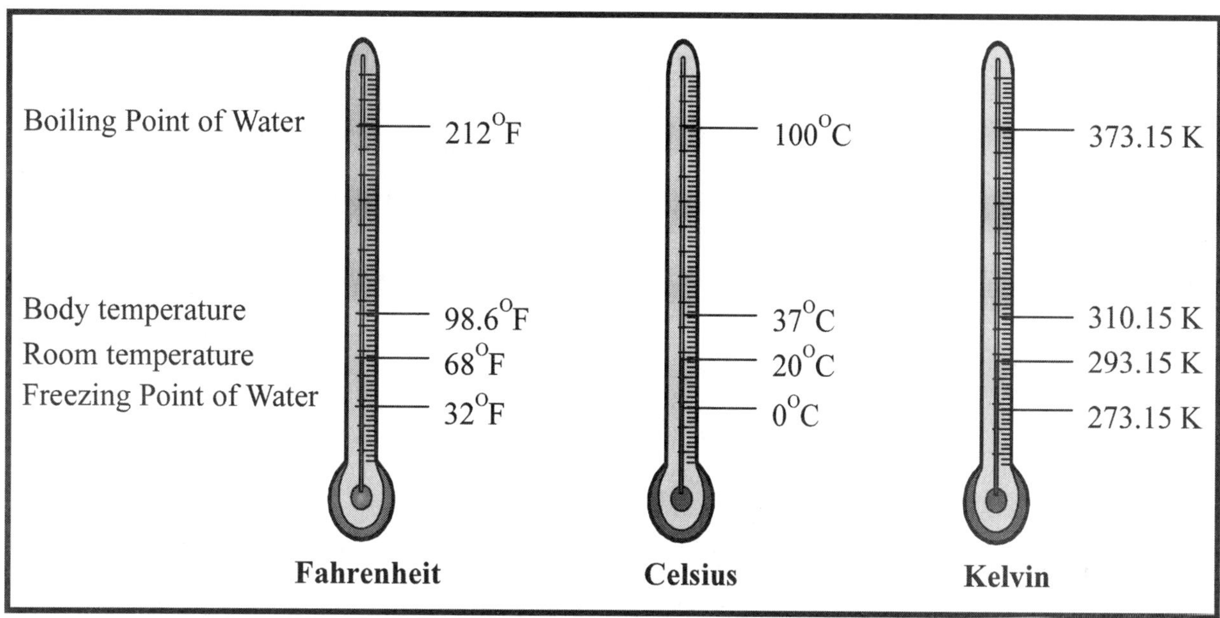

Figure 2.8 Common Temperatures in Different Temperature Scales

Metric Unit Conversions

The units in the metric system are defined in multiples of 10 from the standard unit. The metric prefixes indicate which multiple of 10 — 10, 100, or 1,000 — the standard unit should be multiplied or divided by. To convert from one unit to another in the metric system, you multiply and divide. Multiply when changing from a greater unit to a smaller one; divide when changing from a smaller unit to a larger one. The chart below is set up to help you know how far and which direction to move a decimal point when making conversions from one unit to another.

<u>Prefix</u>	kilo (k)	hecto (h)	deka (da)	unit (m, L, g)	deci (d)	centi (c)	milli (m)
<u>Meaning</u>	1000	100	10	1	0.1	0.01	0.001

Example 1:

2 L = 2000 mL

List or visualize the metric prefixes to figure out how many places to move the decimal point.

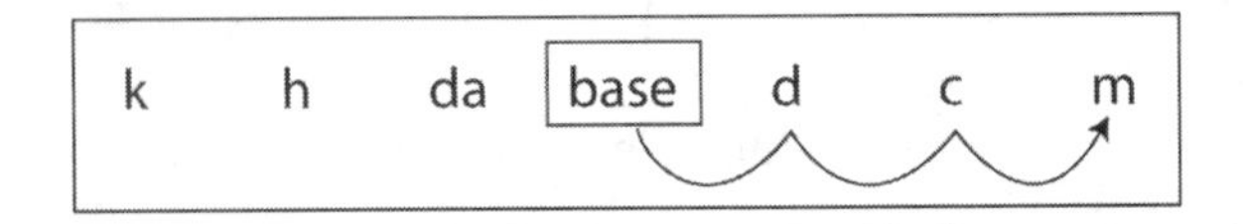

To convert the 2 L to mL, move the decimal point three places to the right, which is equivalent to multiplying by 1000. You will need to add three zeros as placeholders.

2.000 L = 2000 mL

Example 2:

5.25 cm = 0.0525 m

To convert from centimeters to meters, you need to move the decimal point two spaces to the left.

So, to convert 5.25 cm to m, move the decimal point two spaces to the left, which is equivalent to dividing by 100. Again, you need to add a zero.

005.25 cm = 0.0525 m

Practice Exercise 1: Unit Conversions

Convert the following measurements to the specified units.

1. 35 mg = _____ g	6. 32.1 mg = _____ kg	11. 72.3 cm = _____ m
2. 6 km = _____ m	7. 17.5 L = _____ mL	12. 0.003 kL = _____ L
3. 21.5 mL = _____ L	8. 4.2 g = _____ kg	13. 5.06 g = _____ mg
4. 4.9 mm = _____ cm	9. 0.417 kg = _____ cg	14. 1.058 mL = _____ cL
5. 5.35 kL = _____ mL	10. 2.057 m = _____ cm	15. 564.3 g = _____ kg

PRECISION AND ACCURACY

What is the difference between precision and accuracy? Many people use these terms interchangeably, but they are not the same. **Accuracy** refers to how "correct" a measurement is. The more "correct" a measurement is compared to the actual value, the more accurate it is. **Precision** refers to how small a scale is being used to make a measurement. The smaller the scale, the more precise the measurement. For example, a measurement made to the nearest eighth of an inch is more precise than a measurement that is rounded to the nearest inch.

The precision of a measurement is indicated by how many decimal points are used or how small a fraction is used. For example, a ruler marked in sixteenths of an inch will measure more precisely than a ruler marked in quarter inches. Or, a container marked in ounces will measure more precisely than a container marked in cups.

RELIABILITY

Reliability refers to how close repeated measurements of the same variable are to each other. For example, two lab groups have four members each. Each person times the same event. The times recorded are as follows:

First Group	Second Group
8.6s	7.20s
8.75s	11.25s
8.55s	10.85s
8.95s	6.50s

The first group's data are more reliable than the second group's data.
Look at the 2 thermometers below. Which one would give you a more precise temperature reading?

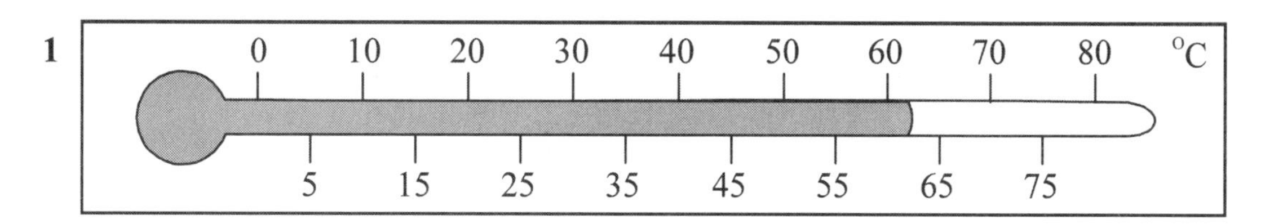

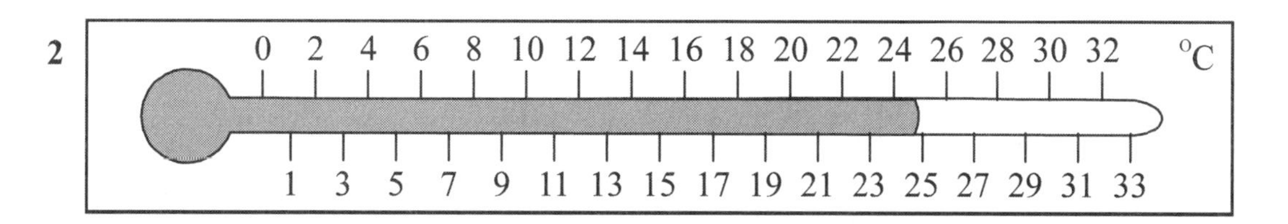

Answer: The second thermometer would give a more precise temperature reading because it measures to the nearest 1 degree. The first thermometer only measures to the nearest 5 degrees.

Example 3: Which of the following measurements is most precise?

A. 2.8623 cm C. 2.87 cm

B. 2.8 cm D. 2 cm

Answer: Choice A is the most precise because it is given to the nearest ten-thousandth. This choice indicates that the instrument used to take this measurement (probably not a ruler) could read to the nearest ten-thousandth of an inch.

Special Note: What if choice D were written as 2.0000 cm? Then the precision of choice D would be the same as the precision of choice A.

Example 3: Which of the following measurements is most precise?

A. $3\frac{1}{4}$ in.

B. $3\frac{1}{16}$ in.

C. 3 in.

D. $3\frac{1}{32}$ in.

Answer: Choice D is the most precise measurement because it is given in the smallest scale. This choice indicates that the ruler used to measure was marked in thirty-seconds of an inch. Remember, the smallest fraction indicates the greatest precision. Choice A indicates that the instrument used measured to the nearest one-fourth of an inch, choice B to the nearest sixteenth, and choice C to the nearest inch.

Special Note: What if choice A was written as $3\frac{8}{32}$ in? Then the precision of choice A, and the precision of choice D would be the same. They would both show that the ruler measured to the nearest thirty-second of an inch. When showing precision in a measurement, you do not reduce the fraction.

Practice Exercise 2: Precision

Which of the measurements below are the most precise?

	A	B	C	D	E
1.	A. $3\frac{1}{4}$ in.	B. $3\frac{1}{8}$ in.	C. $9\frac{1}{2}$ in.	D. $7\frac{7}{16}$ in.	E. $5\frac{3}{8}$ in.
2.	A. $6\frac{2}{8}$ cups	B. $4\frac{1}{4}$ cups	C. $6\frac{4}{16}$ cups	D. $7\frac{8}{32}$ cups	E. $4\frac{5}{16}$ cups
3.	A. 2.90 L	B. 7 L	C. 24 L	D. 24.3 L	E. 71.9 L
4.	A. 1 oz.	B. 0.0150 oz.	C. 0.015 oz.	D. 1.015 oz.	E. 1.5 oz.
5.	A. 42.6 m	B. 8.4 m	C. 2 cm	D. 21 cm	E. 9.8 cm
6.	A. 91 kg	B. 3 g	C. 37.2 g	D. 21.98 kg	E. 8.4 kg

7. Which would give you a more precise measurement: a graduated cylinder marked in mL or a graduated cylinder marked in dL?
8. Which would give you a more precise measurement: a pressure gauge that measured in pounds or a pressure gauge that measured in tenths of pounds?
9. Which would give you a more precise measurement: a ruler marked in centimeters or a ruler marked in millimeters?
10. Which would give you a more precise measurement: a machine that measures to the nearest tenth of an inch or one that measures to the nearest one hundredth of an inch?

Section Review 3: Scientific Measurement

A. Define the following terms.

SI Units	kilogram	graduated cylinder	Celsius
U.S. Customary System	weight	meniscus	Kelvin
length	volume	water displacement	Fahrenheit
meter	cubic centimeter	density	accuracy
mass	milliliter	temperature	precision

B. Choose the best answer.

1. Which unit is used to measure from one point to another?
   A. kilogram B. pound C. meter D. cubic centimeter

2. The volume of liquid held in an eyedropper is best measured in
   A. millimeters. B. milligrams. C. milliseconds. D. milliliters.

3. The amount of space occupied by an object is its
   A. volume. B. density. C. weight. D. mass.

4. The amount of matter in an object is its
   A. weight. B. mass. C. volume. D. density.

5. To determine the volume of irregularly shaped objects, a good method to use is
   A. multiply length, height, and width. C. water displacement.
   B. use a balance. D. use a thermometer.

6. The mass of an object divided by its volume, or stated another way, the amount of matter packed into a given space is its
   A. density. B. volume. C. mass. D. weight.

7. Density is a good way to identify an object because
   A. different masses of the same object will always have different densities.
   B. different masses of the same object will always have the same density.
   C. the same volume of the same object will always have different densities.
   D. the same mass of two different objects will have the same density.

8. What is the best way to read the volume of a liquid in a graduated cylinder?
   A. Read the level of the liquid going up the side of the cylinder.
   B. Make sure five people in the laboratory read it before you write it down.
   C. Glance at the cylinder and write down the first number you see.
   D. Read the level of liquid at the middle of the meniscus at eye level.

9. The measurement of how hot or cold something is indicates
   A. volume. B. mass. C. temperature. D. density.

10. The freezing point of water on the Celsius scale is
    A. 32 degrees. B. 0 degrees. C. 273.15 degrees. D. 293.15 degrees.

C. Answer the following questions.

1. Alan would like to determine how the solubility of salt in distilled water (measured in grams per liter) changes at different temperatures. List the pieces of lab equipment that he should use, and explain how each piece of equipment would be used.
2. Jamal found an unusual rock in the schoolyard. Give two examples of measurement he could take that might help him to identify the type of rock.
3. In her physics class, Geri needs to find the volume of one small BB (a metallic sphere). She does not trust her ability to take an accurate measurement of its diameter in order to calculate its volume. Suggest another way that Geri could experimentally find the volume of the small sphere. Be specific as to how she should carry out the experiment.
4. To calculate the density of a substance by using experimental data, name two measurements that should be taken.

Chapter 2 Review

CHAPTER REVIEW

Choose the best answer.

1. Identify the equipment used to measure mass.
   A. a spring scale.
   B. beam balance
   C. caliper
   D. graduated cylinder

2. Identify the correct statement about mass or weight.
   A. Weight is the quantity of matter in an object.
   B. The weight of a given object is a constant.
   C. The mass of a given object is a constant.
   D. Mass is a force measured in Newtons.

3. Kilograms are a unit of measurement for
   A. mass.
   B. height.
   C. volume.
   D. size.

4. Select the freezing point of water on the Kelvin temperature scale.
   A. –273.15 K.
   B. 0 K.
   C. 32 K.
   D. 273.15K

5. Identify the correct conversion of 4.2 grams (g) to kilograms (kg).
   A. 4.2 g = 0.0042 kg
   B. 4.2 g = 0.042 kg
   C. 4.2 g = 42 kg
   D. 4.2 g = 4,200 kg

6. Identify the correct conversion of 86°F to °C.
   A. 86°F = 16 °C
   B. 86°F = 30 °C
   C. 86°F = 97 °C
   D. 86°F = 123 °C

7. Identify the reason why scientists from countries around the world use the same measurement system.
   A. to simplify international copyright and patent laws
   B. to make it easier to understand and compare published results
   C. to make it easier for scientist to copy the work of other scientists
   D. to simplify monitoring of the code for the humane treatment of animals.

8. Graduated cylinders are marked in units of
   A. grams.
   B. meters.
   C. millimeters.
   D. milliliters.

9. Various safety rules apply in the laboratory. In order to protect your clothing, you should
   A. wear an apron.
   B. wear clothing treated with Teflon.
   C. wear clothing treated with Scotchgard™.
   D. wear as many layers as possible.

10. Reaching across a flame is
    A. never acceptable.
    B. always acceptable.
    C. sometimes acceptable.
    D. seldom acceptable.

11. Identify the material that looks the same as hot glass.
    A. hot coal
    B. cold stone
    C. hot metal.
    D. cold glass.

12. Select the piece of equipment that you would use to observe whether or not different solids float in a given liquid.
    A. funnel
    B. beaker
    C. Erlenmeyer flask
    D. Petri dish

13. Identify when you should report a chemical spill to your teacher.
    A. immediately.
    B. after you've cleaned up the spill.
    C. only if you think the spill is dangerous.
    D. after you've finished the experiment so your results are not ruined.

14. Identify the lab activity that should be conducted under a fume hood.
    A. measuring very high velocities
    B. using high voltage sources of laser light
    C. mixing chemicals that produce dangerous vapors
    D. massing a series of objects

Chapter 3
Scientific Inquiry and Communication

PHYSICAL SCIENCE STANDARDS COVERED IN THIS CHAPTER INCLUDE:

SCSh1	Students will evaluate the importance of curiosity, honesty, openness and skepticism in science.
SCSh7	Students will analyze how scientific knowledge is developed.
SCSh9	Enhancement of science reading ability..

DEVELOPMENT OF SCIENCE

There are several reasons for conducting scientific investigations. Our natural curiosity gives us the desire to know what is around us and how everything works. We gather information to explain all aspects of the universe and to attempt to solve the mysteries of life. Throughout history, great discoveries have been made, and we continue to learn more.

Scientific history stretches back to the earliest civilizations. The ancient Chinese, Arabs, and Greeks contributed greatly to the advancement of science. The Chinese made careful observations of powders and combustion. This led to the development of fireworks and gun powder. The Arabs were the first to use the number zero. The Greeks made many advances in the treatment of diseases, which led to the development of modern medicine. The contributions of these and other civilizations have led to our current understanding of the world around us.

Scientific understanding in the past was not what it is today. There have been many changes to the scientific ideas developed along the way. The very nature of science allows for shifts and changes in beliefs. Science is constantly changing due to newer and better ideas and technologies. Throughout history, many false ideas have been disproved and later abandoned.

THE DEVELOPMENT OF HELIOCENTRIC THEORY

Just because a belief is common does not mean it is correct. Our view of the world around us has changed tremendously over the course of human existence. Up until the 1500s, most people believed that Earth was the center of the universe and all solar bodies orbited around the earth. This is known as the **geocentric theory**, which means "earth centered." **Aristotle** (384 – 322 BC), the great Greek philosopher and scientist, originated this idea in the 4th century BC. Aristotle's model of the solar system consisted of perfectly circular orbits of the sun and the known planets around the earth. The orbits had to be circular because he considered the heavens to be more perfect than the earth. As such, he thought that the objects in the heavens were unchanging, and thus the orbits must be perfectly circular because they begin and end at the same point, and are, therefore, as close to unchanging as possible.

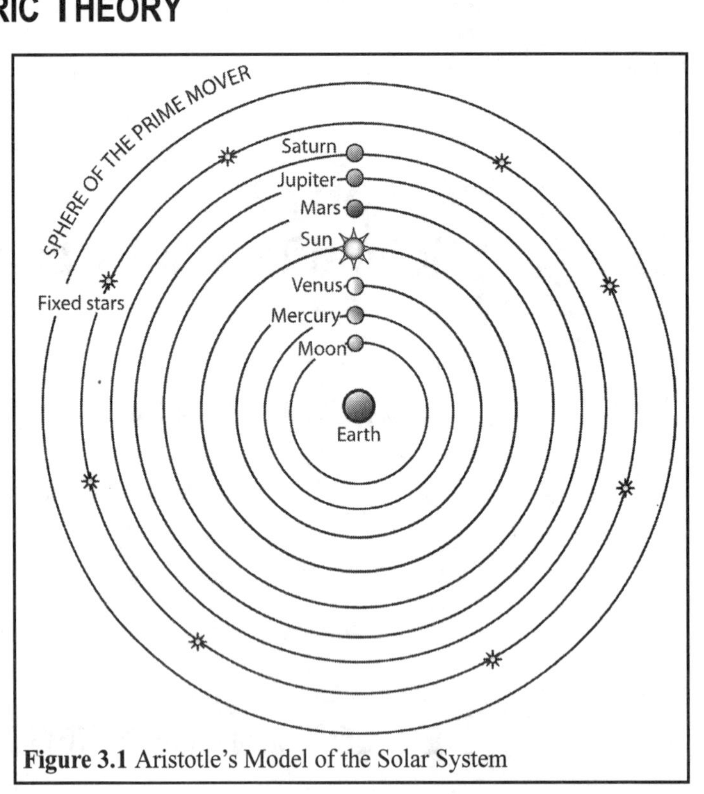

Figure 3.1 Aristotle's Model of the Solar System

Although Aristotle's theories of the solar system did not accurately explain several observed phenomena, his ideas held for almost 2000 years. Over time, his theories were adopted by the Roman Catholic Church and were considered scientific dogma. Challenging the geocentric theory was unacceptable because it was the same as challenging the beliefs of the church. Galileo was placed on house arrest for later challenging the geocentric theory. Copernicus waited until after his death to have his theories of the universe published because they challenged Aristotle's views.

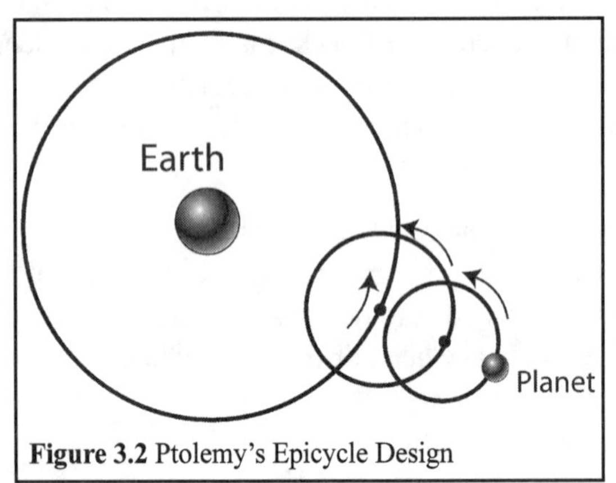

Figure 3.2 Ptolemy's Epicycle Design

Ptolemy (85 – 165 AD) (pronounced tah-lemy) made slight changes to Aristotle's views of the universe in the 2nd century AD. His changes helped to better describe the apparent backward or **retrograde motion** of the planets that was observed on occasion. He added what he called **epicycles** onto the circular orbit of each planet. The planet was thought to move around the epicycle, which in turn followed the circular orbit around the earth as shown in Figure 3.2 to the left. The idea of epicycles had already been proposed to explain some inconsistencies with Aristotle's model. Ptolemy's contribution was to add an additional epicycle onto the first epicycle.

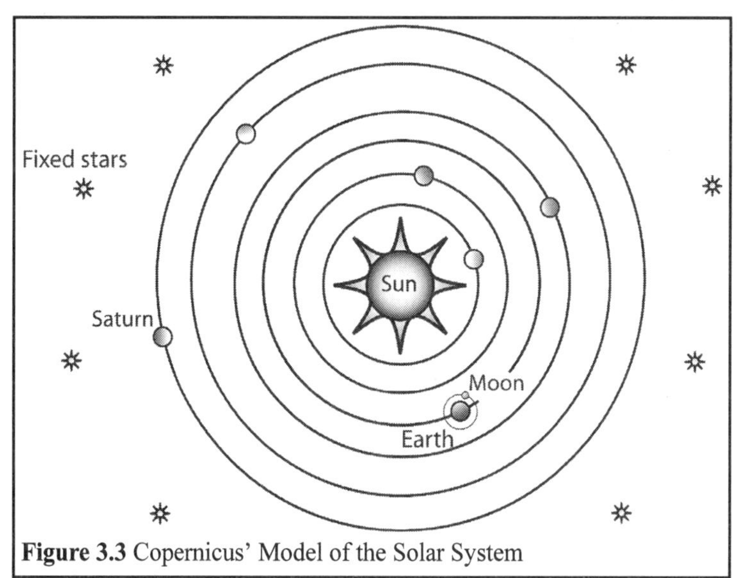

Figure 3.3 Copernicus' Model of the Solar System

It wasn't until the 16th century that **Nicolaus Copernicus** (1473 – 1543) proposed his "sun centered" or **heliocentric theory** of the solar system, which claimed that the earth and the other planets orbited the sun and the moon orbited the earth. At the time, scientists did not know of the existence of the three planets beyond Saturn (Uranus, Neptune, and Pluto), which explains why only the six inner planets are pictured in the Copernican model shown in Figure 3.3 to the left.

While Copernicus was correct in placing the sun as the center of the orbiting planets, he still held the belief that the planets moved around the sun in perfectly circular orbits. Thus, epicycles were still needed to explain all observed motion of the planets. However, the Copernican model required far fewer epicycles than the model of Ptolemy.

It was **Joahannes Kepler** (1571 –1630) who finally discovered that the planets actually orbit the sun in an **elliptical motion**. An ellipse is a flattened circle or egg shape. Kepler defined three laws that explain the relationship of an orbiting body to the body it orbits. Kepler defined these laws based on detailed observations of the planets and sun. He actually had no understanding of why the planets behaved the way they did. **Isaac Newton** (1642 –1727) was the one who discovered that the elliptical orbits of the planets and the near circular orbit of the moon were a result of gravity.

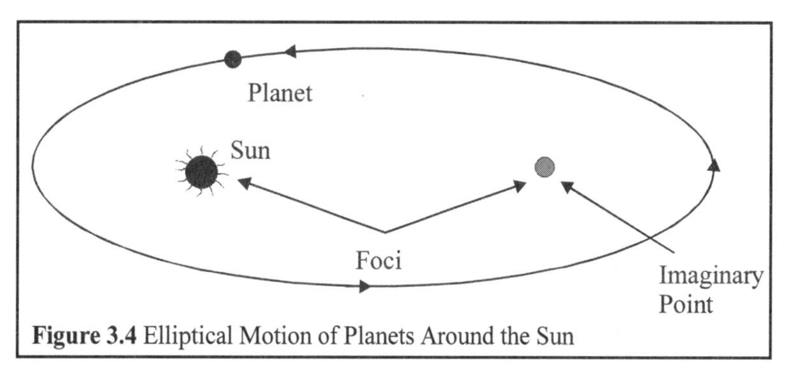

Figure 3.4 Elliptical Motion of Planets Around the Sun

Most experiments are conducted based on previous knowledge. Often scientists use existing theories to help them solve problems. A **scientific theory** is an explanation of how or why something happens based on research and testing that can be repeated and verified by all scientists in a particular field. Although a theory is often accepted after repeated testing, a theory cannot be absolutely proved due to uncertainty in measurement and the inability to test under all circumstances. Theories provide the basis for future investigations and help scientists solve problems. Often theories are made up of many hypotheses and supported by thousands of experiments over a long period of time. Theories are occasionally replaced by newer theories because they better predict a phenomenon or predict a larger range of phenomena. The history of several different scientific theories will be discussed throughout this text.

Section Review 1: Development of Science

A. Define the following terms.

geocentric theory	epicycles	elliptical motion
retrograde motion	heliocentric theory	scientific theory

B. Choose the best answer.

1. Basing investigations on previous knowledge
   A. inhibits new discoveries.
   B. allows scientists to test current thinking.
   C. is never done.
   D. is required by law.

2. When trying to solve a problem, a scientist can
   A. consider a relevant theory.
   B. read about similar problems in academic journals.
   C. perform experiments.
   D. do all of the above

Communication Within the Scientific Community

Scientists communicate to share ideas and to gain new perspectives. Communicating scientific discoveries allows other scientists to confirm or refute the work of their colleagues. It also provides news of recent discoveries to others in the scientific community so that other scientists may build on it and expand the original research. Expanded research can lead to new discoveries.

One method of communicating a scientific discovery is through the publication of a journal article. A **journal** is a specialized publication. There are many scientific journals focusing on various specialties or subject areas. Before a research article is published, the data, analysis, and conclusions presented undergo careful examination. The researchers revise their articles many times and often get input from colleagues. They must follow rules for authors outlined by the journal editors. Once the paper is submitted to the journal, it goes through the peer review process. **Peer review** is the scrutiny of an article by anonymous scientists in a similar field. They carefully look at every detail of the experiment to make sure acceptable guidelines were followed. Peer reviewers make sure the experiment was conducted using the scientific method. If the peer reviewers accept the paper, it usually gets published.

Conferences provide another way for scientists to communicate with each other. Scientific organizations such as the *American Association for the Advancement of Science* and the *National Academy of Sciences* meet to discuss new information. Research findings are presented at the conferences, and researchers have the opportunity to explain their methods and results in person. Conferences can be specialized just as journals are. Questions from the audience can often lead the researcher in new directions, or offer ideas for additional experimentation. These conference groups influence governmental policies related to science while also promoting science education.

Once the scientific community has shared information, the general public is next to learn about new discoveries. The media and health agencies provide the information to the public, and textbooks and teachers communicate the information to students. There are also journals published for the general public, such as *Scientific American, Discover, and Popular Science.*

LABORATORY REPORTS

Since new ideas often meet with resistance, it is important to follow the scientific method when conducting research. Use of the scientific method encourages objectivity. In addition, researchers must be prepared to defend all parts of their investigation. This is done by keeping a detailed laboratory notebook that describes all experiments. In professional research settings, each experiment or group of experiments is then translated into a laboratory report. The laboratory report is the basis for a journal article mentioned in the previous section. The main components of the lab report are listed below.

Introduction: Provide your reason for conducting the investigation. State the problem, ask a question, and make sure the hypothesis is clearly stated. Using previous experience and knowledge, suggest a possible and reasonable solution.

Materials and Methods: List the materials and methods used to perform the experiment. This section is very important. All scientists should be able to understand your materials and methods in order to repeat the experiment. Include everything that was done during the experiment.

Write this section as if you will forget what happened in the next minute. Include units of measurement and ingredients of solutions and other mixtures. This section is the main part of your report and should be a test of your hypothesis.

Results: Collect data from the experiment, and present it in a format that is easy to understand. Tables, charts, and graphs are great ways to present information in a clear way.

Conclusion: Write a conclusion based on the results of the experiment. What was learned? Was the hypothesis supported or refuted? Include any analysis of error in this section. Make a prediction about solutions to similar situations.

This basic format is used by researchers around the world. An accurate and thorough laboratory report is the best way to defend your scientific argument. Having references to show how background information was obtained and showing that the experiment was performed following the proper guidelines makes it easier for other scientists to accept the findings of the experiment.

Scientific Theory

A **scientific theory** is an explanation of how or why something happens, based on research and testing that can be repeated and verified by all scientists in a particular field. Although a theory is often accepted after repeated testing, a theory cannot be absolutely proved due to uncertainty in measurement and the inability to test under all circumstances. Theories provide the basis for future investigations and help scientists solve problems. Often theories are made up of many hypotheses and supported by thousands of experiments over a long period of time. Theories are occasionally replaced by newer theories because the new theories better predict a phenomenon or predict a larger range of phenomena. Often scientific theories help explain experimental data.

Critiquing Scientific Explanations

Newspapers, magazines, television, radio, the Internet, and books provide enormous amounts of information. People have access to more information today than ever before in the history of the world. Making informed decisions about new information is dependent on your being able to separate fact from fiction or opinions and to research relevant ideas.

Some topics are in the news every day and directly affect people's daily lives. For instance, claims about the effectiveness of natural herbs in preventing disease, stimulating brain activity, or increasing metabolic activity are prevalent in ads and articles in almost every form of media. The advertisements will have you believe that they are effective and safe. To be sure, you must evaluate the claims for yourself. Reading articles in scientific journals about studies done with the herbs will provide results from experiments and will offer less biased statements than an advertiser or manufacturer interested in a profit. Also, reading the labels on natural herb bottles will indicate whether what you will be ingesting is truly natural or contains other ingredients. If the products do contain other ingredients, you may research and become familiar with the extra ingredients and what effects they may have.

Many scientific explanations do not directly influence the products people buy, the medical care they receive, or the type of car they drive, but they are still important to the understanding of the world and the mechanisms that have led to the physical universe that we live in today. Examples include the extinction of the dinosaurs, global warming, the increase in asthma cases in the United States, the decrease in frog populations throughout the world, and the possibility of extinct microscopic life forms on Mars. Nobody denies the existence of the above named phenomena, but the reasons for their occurrences remain controversial. Scientists have varying explanations, most of which are supported by evidence or data.

Before making a decision as to which argument to adopt, read articles from current academic journals. Ensure the validity of the studies by identifying the steps of the scientific method. Make an informed decision based on research, and don't be afraid to choose a less popular view. If it weren't for individuals willing to contradict popular beliefs, we might still believe that the earth is the center of our solar system and that life spontaneously arises from nonliving matter. Remember, there are limitations to a scientific theory.

Practice Exercise 1: Reading a Scientific Article

Choose an article to read from a current science magazine, such as *Science*, *Science News*, *Nature*, or another peer-reviewed science journal, and answer the following questions.

1. What scientific explanation, discovery, or phenomenon does the article discuss?
2. Is the article related to any commonly accepted theory? If so, which one?
3. List the steps of the scientific method and discuss them as they pertain to the study or experiment discussed in the article.
4. Based on the results of the experiment, do you agree with the statements presented in the conclusion?
5. Where might you look for more information on the topic discussed in the article?

THINKING CRITICALLY ABOUT SCIENCE

It is important to note that, while science can do many things, it does have limitations. As discussed in the previous section, scientific thinking often changes as new theories and discoveries are made. In the case of spontaneous generation, scientists were limited by their ability to observe the organisms in question. Eventually the contributions of many different people, technologies, and ideas lead to the best possible explanation for natural phenomena.

LIMITATIONS TO SCIENTIFIC INQUIRY

It is important to mention what science cannot do. Scientific experiments are an attempt to answer a hypothesis or question. Science and scientific experiments cannot answer questions about value. For example, science cannot say that plastic is more valuable than concrete. Clearly plastic is utilized everywhere in our society, from milk cartons to medical equipment. Plastic is inexpensive, recyclable, and safe. However, it would be difficult to pave a road or build a building out of plastic.

Science and scientific experiments cannot answer questions about morality. Science cannot say that **cloning**, the synthesis of a being from the DNA of an existing being, is good or bad. It can only tell you the process by which animals are cloned.

Science and scientific experiments cannot answer questions about the supernatural (*super-* above; *natural-* nature). Questions about the supernatural are then, by definition, "above nature" and cannot be observed in the natural world. If a scientist cannot observe something with the five senses, he or she cannot design experiments to investigate it.

Section Review 2: Communication Within the Scientific Community

A. Define the following terms.

journal	peer review	conference
laboratory report	theory	

B. Choose the best answer.

1. Scientists communicate their findings
   A. because they are bored.
   B. to share results and offer new perspectives.
   C. because they were told to do so.
   D. but don't want their experiments repeated.

2. The scientific method is used as a guideline to conduct experiments to
   A. maintain objectivity.
   B. make it harder to do experiments.
   C. maintain subjectivity.
   D. stifle creativity.

3. Ways that researchers first communicate their discoveries to others are through
   A. journals and conferences.
   B. the Internet.
   C. television and radio.
   D. none of these

4. **I** "The atomic theory states that atoms are the basic building blocks of matter, and all material on Earth is composed of various combinations of atoms."
   II "I have a theory that everyone is late to the dinner party because of a traffic jam on the interstate."

Which of the above statements is a scientific theory and why?

   A. I, because scientific theories must always encompass both living and non-living things.
   B. II, because scientific theories usually involve one specific incident.
   C. I, because scientific theories are based on research and supported by experimental evidence.
   D. II because scientific theories are guesses.

5. The following four statements are headlines you read on various magazine covers while standing in the checkout line at the grocery store. Which one could be researched using scientific experimentation?
   A. "Diamond is more beautiful than ruby."
   B. "Metals are both helpful and harmful to the human body."
   C. "The earth was created in a miraculous event."
   D. "Poetry is the most artistic type of literature."

6. Which of the following should be included in a laboratory report?
   A. material and method
   B. results
   C. conclusions
   D. all of the above should be included
7. Scientific processes are limited because
   A. science cannot determine values.
   B. humans are always wrong.
   C. humans must be able to observe phenomena in order to collect data on it.
   D. A & C only

Science and Technology

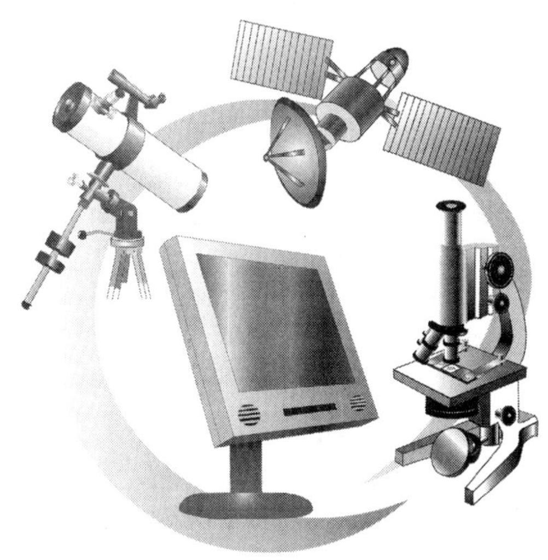

Technology is the application of science. Technology extends our natural abilities provided by our senses. With a microscope, a tool used to magnify tiny objects, we have sharper "eyes." With a computer, we have faster "brains." Measurement tools allow us to determine mass, volume, force, and density. Knowledge of how to use different types of radiation has allowed for the creation of X-rays and magnetic resonance imaging (MRI). These developments give doctors the ability to "see" internal living structures. Eyeglasses and hearing aids help us overcome physical limitations. **Biotechnology**, the application of biological processes, has allowed for the creation of new drugs, gene therapy, and rot-resistant plants. Smaller, clearer, and more environmentally-sound electronics have been created. Cars allow humans to travel farther and faster, forever changing our lives. All of these great technological advances come with a price, though. For every benefit, there is a drawback.

Science, technology, and people in the community are in constant interaction. Because of this interaction, a change in one area affects the other areas. Computers have changed the way scientists and societies manipulate data, store information, create documents, perform calculations, and communicate. Televisions have changed the way we obtain news and entertainment. Appliances, such as vacuums and dishwashers, have changed the way we take care of our homes.

Technology is usually developed in response to a problem or need. Some technologies simply increase the quality of life, whereas others are necessary to sustain life. For example, when human space exploration first began, Velcro was invented to keep objects from floating around in the absence of gravity. A more important technology that came from space exploration is the high temperature ceramic tiles placed on the bottom and sides of the space shuttle. These tiles are used to keep the shuttle from burning up during reentry through the earth's atmosphere. Without these tiles, the spacecraft could not withstand the high temperatures of reentry. In fact, it was damage to these tiles that led to the space shuttle Columbia disaster in February 2003. While Velcro is a useful invention, it is not needed to protect or sustain human life. It is merely a convenience item.

Technology is useful in helping humans to solve problems. Medical technology is developed to help cure diseases and treat conditions. Scientific technology is developed to help scientists design and understand better experiments. Societal technology is developed to help solve problems created by civilization, such as energy needs, pollution, hunger, and disease.

The effects of technology are not always local. Exhaust from a factory in the Midwest pollutes the air locally, but wind currents can carry the pollutants great distances and, therefore, contribute to acid rain in the Southeast. Communication satellites make global communication possible, allowing people in distant places to interact more frequently. The ability to view television broadcasts in other countries informs people about natural disasters and wars that they would not normally have knowledge of.

Ethical Practices

Scientific Ethics

Scientific ethics is defined as the standard for conduct as scientists carry out research, has been a concern throughout the history of the world. You already know about Galileo's imprisonment. He was placed on house arrest because his theory that the earth orbited the sun contradicted the Roman Catholic doctrine that the earth was the center of the universe and, thus, the sun must orbit the earth. The Manhattan Project revealed the mysteries of the atom. However, it also resulted in the most destructive and horrific warfare weapon ever made. The creation of the atom bomb during World War II created a huge ethical debate among scientists.

Scientific ethics is becoming a more frequent topic of conversation as we continue to unravel the mysteries of life. Ethical concerns today include the use of tissue from human embryos in stem cell research, cloning of animals and people, and the use of armor piercing bullets, just to name a few.

Ethics in science can be separated into two basic groups: the morality of the topics and the integrity of the methods.

Morality of the Topics

Ethicists, scientists, and layman often question whether science is good or bad, particularly in biological fields or fields in which testing is done on humans or animals. Ethical concerns are also introduced by religious groups when the results or processes of science violate their belief system.

Integrity of the Methods

The process of doing and reporting science also has an ethical component. In fact, many scientific journals have a code of ethics that authors must adhere to if they want to be published in that journal. Scientists must follow the scientific method and accurately report their data. For example, it would be highly unethical for a scientist to claim a new drug works when it really doesn't.

In conducting scientific research, the use of live animals or human test subjects often becomes necessary. Rules and regulations regarding live animal subjects are very strict and must be followed at all times. Failure to do so often results in criminal charges. Scientific research involving human test subjects is also very stringent. Humans must have **informed consent**, meaning that they must know all aspects of possible outcomes resulting from the experiment.

Section Review 3: Science and Technology

A. Define the following terms.

technology	biotechnology
informed consent	ethics

B. Choose the best answer.

1. Science and technology
   - A. interact a great deal.
   - B. have nothing to do with each other.
   - C. are life threatening.
   - D. create more work for people in society.

2. A negative effect of technology is that
   - A. it makes our lives easier.
   - B. it sometimes causes pollution.
   - C. it extends our natural abilities.
   - D. none of these

3. Which of the following is unethical behavior that violates the integrity of a scientists' results?
   - A. Reading the meniscus incorrectly on a graduated cylinder.
   - B. Falsely reporting the results of a properly observed measurement.
   - C. Hiring a lab assistant and asking him to improperly report results.
   - D. B and C

4. If you believe that a scientific activity is morally wrong, what is the BEST first step you could take toward correcting it?
   - A. Staging a protest at the scientist's laboratory.
   - B. Writing a newspaper editorial about the immorality.
   - C. Calling into a radio show to voice your protest.
   - D. Become as informed as possible about all aspects of the issue.

5. **Extended response.** Explore your thoughts. How do you think a consensus is reached on whether or not a line of scientific inquiry is moral or not? Do you think a consensus will ever be reached on whether animal testing or stem cell research is moral? What do you think about those issues?

Chapter 3 Review

A. Choose the best answer.

CHAPTER REVIEW

1. Identify the true statement about the development of science.
   A. Science started in the last decades of the 19th century.
   B. Science is no longer important in 21st century America.
   C. Several ancient civilizations made important contributions to science.
   D. It is unethical for scientists to contradict the findings of other scientists.

2. Identify the reason the heliocentric (sun-centered) theory replaced the geocentric (earth-centered) theory of the solar system.
   A. political support for the heliocentric theory
   B. religious support for the heliocentric theory
   C. powerful technologies that supported the heliocentric theory
   D. observations that were better explained by the heliocentric theory

3. Identify one reason scientists share their findings with other scientists.
   A. to prevent other scientists from conducting similar investigations
   B. to reduce the cost of conducting science investigations
   C. to encourage more young people to become students
   D. to allow others to build on and extend those findings

4. Select the statement that describes the most common reason graphs are included in many scientific research reports.
   A. to suggest a hypothesis
   B. to present results visually
   C. to explain the experimental method
   D. to suggest new questions raised by the findings

5. Identify the action that would be judged unethical by scientists.
   A. disagreeing publicly with the conclusions of a respected senior scientist
   B. publishing preliminary findings before final results are available
   C. developing a new hypothesis after collecting data
   D. changing data to fit an established theory

6. Select the source of science information that should be examined with the greatest care and skepticism.
   A. Internet blogs
   B. scientific journals
   C. recently published encyclopedias
   D. presentations at science conferences

7. Select the question about human stem cells that cannot be answered through application of the scientific process.

   A. Is human stem-cell research ethical and moral?

   B. Are stem cells present in the bone marrow of adults?

   C. Can stem cells be preserved for long periods by freezing them in liquid nitrogen?

   D. Do stem cells extracted from embryos differ from those extracted from placentas?

8. Identify the key characteristic of a scientific hypothesis.

   A. can be tested

   B. known to be true

   C. addresses a social need

   D. can be stated mathematically

9. Select the best description of the relationship between science and technology.

   A. Science is the product of technology.

   B. Technology progresses without scientific inquiry.

   C. Science inhibits technology and technology inhibits science.

   D. Technology enhances science and science enhances technology.

10. Identify the driving force behind the development of new technology.

    A. the desire to enhance religious faith

    B. the need to solve practical problems

    C. the need to develop new scientific theories

    D. the desire to limit the negative impacts of science

11. Select the statement that describes the reaction of science to new ideas.

    A. Science is fundamentally opposed to new ideas.

    B. Science immediately and uncritically accepts new ideas.

    C. Science evaluates new ideas before accepting or rejecting them.

    D. Science is based on the assumption that there is no such thing as a new idea.

12. When technology is applied to societal issues, the morality of that application must usually be determined. Which of the following situations has no moral component?

    A. Making AIDS medication freely available to infected Africans.

    B. Allowing North Korea to utilize nuclear reactor technology for energy needs.

    C. Computerizing car assembly equipment in a factory.

    D. All of these applications of technology have a moral component.

B. Answer the following questions.

13. Why did Aristotle's view of the universe stand for so long?
14. List four ways that scientific information is communicated to the general public.
15. List the basic parts of a laboratory report.
16. Why is it important to have an accurate, thorough laboratory report?
17. How do ethics apply to science?
18. Name ways that a scientist could be unethical in his methods.

Chapter 4
Math in Science

PHYSICAL SCIENCE STANDARDS COVERED IN THIS CHAPTER INCLUDE:

SCSh3	Students will identify and investigate problems scientifically.
SCSh5	Students will demonstrate the computation and estimation skills necessary for analyzing and developing reasonable scientific explanations.

MATH IN SCIENTIFIC MEASUREMENT

Math plays a large role in scientific measurement and the analysis of scientific data. In fact, many famous mathematicians also contributed greatly to the study of science. We use math to convert between units, to express uncertainty in measurements, to determine relationships among scientific data, and to determine unknown values based on measured data. The following sections describe the many ways that math is used in scientific investigations.

DIMENSIONAL ANALYSIS

Dimensional analysis is a structured method of helping you to convert units by using conversion factors. A **dimension** is a property that can be measured, such as length, time, mass, or temperature. It may also be calculated (or derived) by multiplying or dividing other dimensions. Some examples of derived dimensions include length/time (velocity), length^3 (volume), or mass/length^3 (density). Dimensions are **not** the same as units. The dimensions of a physical quantity can be measured in any appropriate unit. For instance, velocity can be measured in mph, m/s, etc., but it will always be a measure of length divided by time. Therefore, the dimensions of velocity are l/t.

A **conversion factor** is a fraction always equal to 1. For example, we know that there are 16 ounces in 1 pound, or 16 oz = 1 lb. We can write two conversion factors using this information:

$$\frac{16 \text{ ounces}}{1 \text{ pound}} \text{ or } \frac{1 \text{ pound}}{16 \text{ ounces}}$$

Ratios of equivalent values expressed in different units like these are known as conversion factors. To convert given quantities in one set of units to their equivalent values in another set of units, we set up **dimensional equations**. We will write our dimensional equations so that the old units cancel and we are left with only the new units. Thus, you always want to set up your dimensional equation so that the units you are trying to cancel appear in both the numerator **and** the denominator.

Example: How many inches are in four yards?

Step 1. **Begin by writing the term that needs to be converted: 4 yd**

Step 2. **Next, write the conversion formulas: 3 ft = 1 yd, 1 ft = 12 in**

Step 3. **Write the conversion formulas as fractions:** $\frac{3 \text{ ft}}{1 \text{ yd}}$ **and** $\frac{12 \text{ in}}{1 \text{ ft}}$

Step 4. **Multiply the term you are trying to convert by the conversion factors:**

$$4 \cancel{\text{yd}} \times \frac{3 \cancel{\text{ft}}}{1 \cancel{\text{yd}}} \times \frac{12 \text{ in}}{1 \cancel{\text{ft}}} = 144 \text{ in}$$

Notice that the units of "yards" and "feet" cancel, and you are left with inches.

WRONG: If you had chosen the wrong conversion factors, this is what you would have:

$$4 \text{ yd.} \times \frac{1 \text{ yd}}{3 \text{ ft}} \times \frac{1 \text{ ft}}{12 \text{ in}} = \frac{4 \text{ yd} \times \text{yd} \times \text{ft}}{3 \text{ ft} \times 12 \text{ in}} \text{ or } \frac{4 \text{ yd}^2 \times \text{ft}}{3 \text{ ft} \times 12 \text{ in}}$$

Notice, yards multiplied by yards are yards squared. None of the units cancel, so you know right away that this is wrong.

You can also write your dimensional formulas in a slightly different format that may help you keep better track of your units. This grid-like format is a different way to represent multiplying several conversion factors together. Each column contains a conversion factor that is needed to convert your units.

Example: How many seconds are there in the month of March?

60 seconds	60 minutes	24 hours	31 days
1 minute	1 hour	1 day	March

$$= \frac{(60) \times (60) \times (24) \times (31)}{(1) \times (1) \times (1)} = 2{,}678{,}400 \text{ seconds}$$

We are often asked to convert the units of area or volume, which are square or cubic terms, respectively. It is important to note that the numerical coefficient **must** also be squared or cubed when converting units that are squared or cubed.

Example: The volume of a barrel is 10 ft^3. How many in^3 of water will it hold?

10 feet³	(12)³ inches³
	1 foot³

$$= \frac{(10) \times (12) \times (12) \times (12)}{(1)} = 17{,}280 \text{ imches}$$

Example: How many kilograms per cubic meter (kg/m^3) are there in 3 grams per cubic centimeter (g/cm^3)?

$$\frac{3 \text{ grams}}{\text{cm}^3} \times \frac{1 \text{ kilogram}}{1000 \text{ grams}} \times \frac{(100)^3 \text{ cm}^3}{1 \text{ m}^3} = \frac{(3)\times(1)\times(100)\times(100)\times(100)}{(1000)} = 3{,}000 \ \frac{\text{kg}}{\text{m}^3}$$

This is an example of converting units of density, which is a measure in mass per unit volume.

Another useful scientific term that often needs converting is velocity (speed), measured in length per unit time.

Example: What is the highway speed limit of 65 miles per hour in feet per second?

$$\frac{65 \text{ miles}}{1 \text{ hour}} \times \frac{5{,}280 \text{ feet}}{1 \text{ mile}} \times \frac{1 \text{ hour}}{60 \text{ minutes}} \times \frac{1 \text{ minute}}{60 \text{ seconds}} = \frac{(65)\times(5{,}280)\times(1)\times(1)}{(1)\times(1)\times(60)\times(60)} = 95.3 \ \frac{\text{feet}}{\text{second}}$$

Practice Exercise 1: Dimensional Analysis

1. Jared can work 54 math problems in one hour. How many problems can he work in 10 minutes?
2. Leah rides 22 feet per second on her bicycle. How many miles per hour does she ride?
3. A jar of honey has a density of 14 kg per m^3. What is its density in g/mm^3?
4. If a pitcher will hold 2 ft^3 of lemonade, how many in^3 of lemonade will it hold?
5. Juan's car gets an average of 24 miles per gallon of gas. How far can Juan go on 1 quart of gas?
6. From question 5, how many gallons of gas will it take Juan to travel 528 miles?
7. How many cubic centimeters (cm^3) of water are in 1 m^3?
8. How many cubic feet are in a hole that is 3 feet deep, 4 feet wide, and 6 feet long?
9. How many yards are in 3 miles?
10. John Smoltz throws his fastball 99 miles per hour. If he starts 60 feet, 6 inches from home plate, how many seconds does it take the ball to get to the plate?

SIGNIFICANT FIGURES

All measurements are approximations because no measuring device can give perfect measurements and no human can take a measurement without a small degree of **uncertainty**. In taking a measurement, all the digits known for certain are recorded along with the last digit, which is an **estimation**. Each digit in the measurement is called a **significant figure**.

Consider the metric ruler shown below. You are asked to give the measurement at the point designated by the arrow.

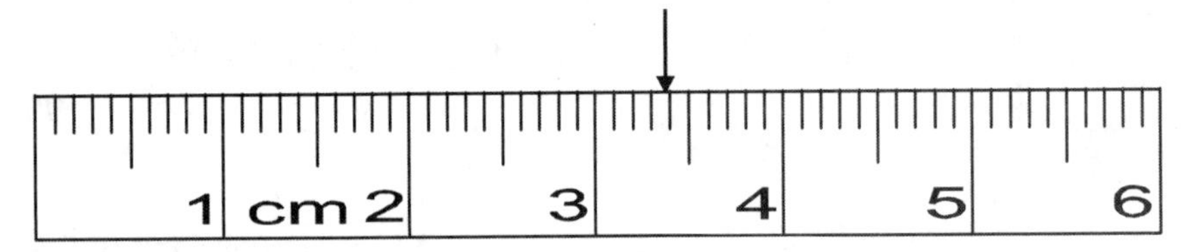

Figure 4.1 Metric Ruler

Step 1. You know that the reading is between 3 cm and 4 cm.
So, the first digit in the measurement is 3.

Step 2. Looking at the ruler, you also know the reading is between 3.3 cm and 3.4 cm.
So, the first two digits are 3.3.

Step 3. The first two digits are known for certain. The third digit, however, is going to be an estimation. You know that the measurement is between 3.3 cm and 3.4 cm. The third digit will tell if the reading is closer to 3.3 cm or closer to 3.4 cm. Imagine ten tiny lines between 3.3 and 3.4. To which one is the arrow pointing? 7, 8, or 9 would all be good estimates. Let's use 8 in this example.

The measurement shown above, then, is 3.38 cm and contains three significant figures.

Measurements made with flasks, balances, graduated cylinders, thermometers, etc. should be done using the same procedures outlined above. Any given measurement should contain no more than one estimated digit.

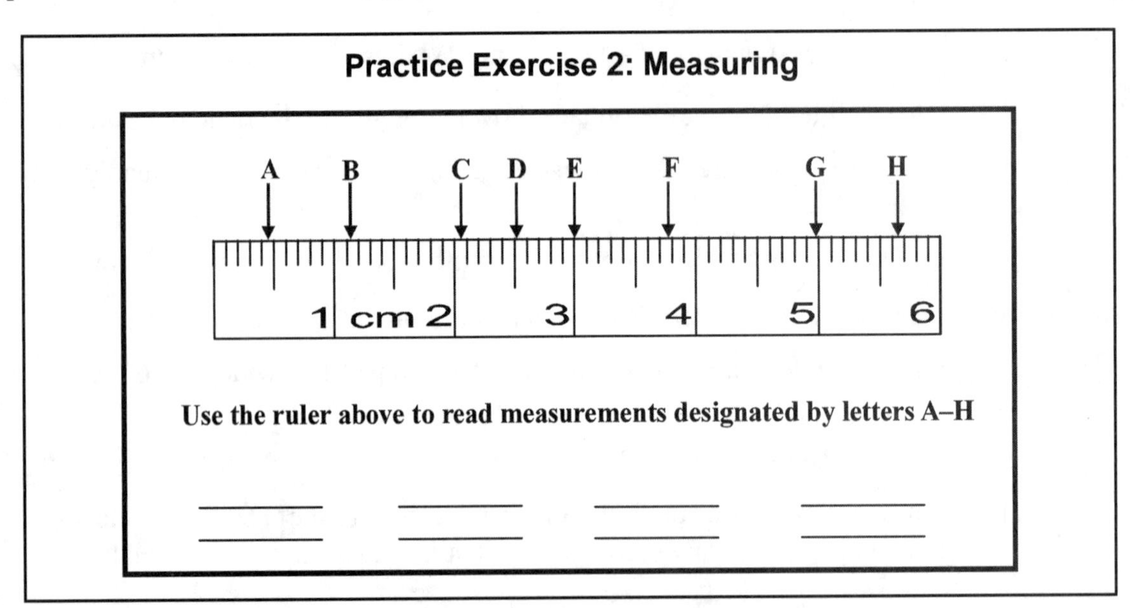

Rules for Determining the Number of Significant Figures

The number of **significant figures** in a measurement is the number of digits that are known with some degree of reliability. Sometimes a measurement is given, and you must determine how many significant figures it contains. Following are the rules for determining how many significant figures are contained in a measurement.

Rules	Examples: Measurement	Examples: # of Significant Figures
1. All nonzero digits are significant.	3.456	4
	7.9	2
2. Zeroes between nonzero digits are significant.	3.0006	5
	6.05	3
3. Zeroes to the left of the 1st nonzero digit are not significant. The zeroes act as placeholders that indicate the position of the decimal point.	0.0003	1
	0.072	2
	0.001	1
4. Zeroes to the right of a decimal point occurring after a nonzero digit are significant.	0.450	3
5. When a number ends in zeroes that are not to the right of a decimal point, significance of the zeroes can be shown in 2 ways:	190	2
A. Writing in the decimal point	190.	3
	1.90×10^2	3
B. Using scientific notation (for a detailed explanation of this topic, see pages 84– 87)	50,600	3
	5.06×10^4	3
	5.060×10^4	4
	5.0600×10^4	5
6. Exact numbers are known with complete certainty and can be considered to have an infinite number of significant figures. Examples include counts of objects and conversion factors.	100 people	infinite
	12 in. = 1 ft	infinite
	60 min. = 1 hr	infinite

Practice Exercise 3: Determining the Number of Significant Figures

How many significant figures are in each of the following measurements?

1. 3.006
2. 8×10^3
3. 0.00307
4. 0.9008
5. 7,980
6. 7.40×10^5
7. 6.300
8. 20.
9. 50,670
10. 0.00101
11. 6.02×10^7
12. 0.7089
13. 0.00231
14. 65,641
15. 1,010

Calculating Using Significant Figures

Maintaining the appropriate number of significant figures preserves the accuracy of results even after the measurements have been used in calculations. When doing calculations with measurements, the accuracy of the calculated result is dependent on the least accurate measurement.

Addition and Subtraction

When adding or subtracting, the result is rounded off to the digit farthest to the right in all components. In other words, the number of decimal places in the answer should equal the smallest number of decimal places in any term.

Example 1:

```
  3345.23  ◄—— This number is accurate to the hundredths place.
+    1.178 ◄—— This number is accurate to the thousandths place.
  ---------
  3346.408
```

The least accurate measurement in this problem has its final digit in the hundredths place. Therefore, the answer must also be rounded to the hundredths place. The 8 will round the zero up to 1. The correct answer, then, is 3346.41.

Example 2:

```
   83.5    ◄—— This number is accurate to the tenths place.
 – 45.012  ◄—— This number is accurate to the thousandths place.
  --------
   38.488
```

The least accurate measurement in this problem has its final digit in the tenths place. Therefore, the answer must also be rounded to the tenths place. The 8 in the hundredths place will round the 4 up to 5. The correct answer, then, is 38.5.

Practice Exercise 4: Adding and Subtracting with Significant Figures

Do the following calculations. Round your answers to the correct number of significant figures.

1. 0.30 + 2.678
2. 67.54 − 1.093
3. 1,089 + 34.023
4. 3,500 + 6,713
5. 76.10 − 0.05
6. 101 − 7.32
7. 900.01 + 68.020
8. 897.76 − 0.06
9. 10,657 + 13,000

Multiplication and Division

When multiplying and dividing, the result is rounded off so that it has the same number of significant figures as the factor with the least number of significant figures.

Example 1:

```
     70.54 ◄—— This number has 4 significant figures.
×      406 ◄—— This number has 3 significant figures.
  ---------
 28,639.24
```

The factor that has the least number of significant figures is 406, which has 3 significant figures. The result, then, must also have 3 significant figures. The final significant digit will be in the hundredths place. The number 3 does not round 6 up. Therefore, the correct answer is 28,600.

Example 2:

$$\begin{array}{r} 100.02 \\ \underline{\div\ 0.012} \\ 8{,}335 \end{array}$$

100.02 ← This number has 5 significant figures.
÷ 0.012 ← This number has 2 significant figures.

The factor 0.012 has only 2 significant figures. Therefore, the result can have only 2 significant figures. The final significant digit will be in the hundredths place. The number 3 does not round 3 up. The correct answer, then, is 8300.

Practice Exercise 5: Multiplying and Dividing with Significant Figures

Do the following calculations. Round your answers to the correct number of significant figures.

1. 456.1×0.032
2. $7{,}400 \times 0.09$
3. 703×0.290
4. $20.03 \div 7.13$
5. $813.0 \div 1.005$
6. $5.02 \div 0.02$
7. 700.03×1.40
8. $0.0004 \div 0.79$
9. $34{,}567 \div 11{,}000$

SCIENTIFIC NOTATION

Scientists often express very large or very small numbers in "powers of ten" or "exponential" notation. For example, 546,000 as 5.46×10^5 or 0.0017 as 1.7×10^{-3}.

Scientific notation expresses a number as the base number times ten raised to a power or exponent, which is expressed as:

$$\text{base} \times 10^{\text{power}}$$

where the **exponent** equals the number of decimal places in the original number. The base number contains only one digit before the decimal and the remaining nonzero digits after the decimal point. One benefit of scientific notation is that the number of significant figures is clearly defined. We don't know if the zeros in 546,000 are significant. With scientific notation the number of significant figures is explicit. For example, 5.46×10^5 has 3 significant figures and 5.460×10^5 has 4 significant figures.

Using Scientific Notation for Large Numbers

Scientific notation simplifies very large numbers that have many zeros. For example, Pluto averages a distance of 5,900,000,000 kilometers from the sun. In scientific notation, a decimal is inserted after the first digit (5.); the rest of the digits are copied except for the zeros at the end (5.9), and the result is multiplied by 10^9. The exponent equals the total number of digits in the original number minus 1, or the number of spaces the decimal point moved to the left.

$$5{,}900{,}000{,}000 = 5.9\times10^9$$

The following are more examples:

Example: $32{,}560{,}000{,}000 = 3.256\times10^{10}$ $\quad$ $5{,}060{,}000 = 5.06\times10^6$

Practice Exercise 6: Converting from Conventional Numbers to Scientific Notation

Convert the following numbers to scientific notation.

1. 4,230,000,000
2. 64,300,000
3. 951,000,000,000
4. 12,300
5. 20,350,000,000
6. 9,000
7. 450,000,000,000
8. 6,200
9. 87,000,000
10. 105,000,000
11. 1,083,000,000,000
12. 304,000

Example: To convert a number written in scientific notation back to conventional form, reverse the steps as described above.

Example: $4.02\times10^5 = 4.02000 = 402{,}000$ Move the decimal 5 spaces to the right and add zeros.

Practice Exercise 7: Converting from Scientific Notation to Conventional Numbers

Convert the following numbers from scientific notation to conventional numbers.

1. 6.85×10^8
2. 1.3×10^{10}
3. 4.908×10^4
4. 7.102×10^6
5. 2.5×10^3
6. 9.114×10^5
7. 5.87×10^7
8. 8.047×10^8
9. 3.81×10^5
10. 9.5×10^{12}
11. 1.504×10^6
12. 7.3×10^9

Using Scientific Notation for Small Numbers

Scientific notation also simplifies very small numbers that have many zeros. For example, the diameter of a helium atom is 0.000000000244 meters. It can be written in scientific notation as 2.44×10^{-10}. The first number is always greater than 0, and the first number is always followed by a decimal point. The negative exponent indicates how many digits the decimal point moved to the right. The exponent is negative when the original number is less than 1. To convert small numbers to scientific notation, follow the examples below.

Example: $0.00058 = 5.8 \times 10^{-4}$ $\quad$ $0.00003059 = 3.059 \times 10^{-5}$

Practice Exercise 8: Converting from Conventional Numbers to Scientific Notation

Convert the following numbers to scientific notation.

1. 0.00000254
2. 0.00000000508
3. 0.000008004
4. 0.00047
5. 0.000000005478
6. 0.00000059
7. 0.00000004712
8. 0.00025
9. 0.0000000501
10. 0.0000006
11. 0.0000000000875
12. 0.00004

Now convert small numbers written in scientific notation back to conventional form.

Example: $3.08 \times 10^{-5} = 00003.08 = 0.0000308$ Move the decimal 5 spaces to the left, and add zeros.

Practice Exercise 9: Converting from Scientific Notation to Conventional Numbers

Convert the following numbers from scientific notation to conventional numbers.

1. 1.18×10^{-7}
2. 2.3×10^{-5}
3. 6.205×10^{-9}
4. 4.1×10^{-6}
5. 7.632×10^{-4}
6. 5.48×10^{-10}
7. 2.75×10^{-8}
8. 4.07×10^{-7}
9. 5.2×10^{-3}
10. 7.01×10^{-6}
11. 4.4×10^{-5}
12. 3.43×10^{-2}

Performing Calculations Using Scientific Notation

Multiplication

Follow the steps below to multiply numbers written in scientific notation.

Step 1. **Multiply the coefficients**

Step 2. **Add the exponents.**

Step 3. **Adjust the coefficient to have one digit before the decimal by raising or lowering the exponent. Round to the correct number of significant figures. The base will remain 10.**

Example: Multiply $(3.15 \times 10^7) \times (6.75 \times 10^5)$

To solve:

- Rewrite: $(3.15 \times 6.75) \times (10^{7+5})$
- Multiply coefficients, add exponents: 21.2625×10^{12}
- Change to correct scientific notation and significant figures: 2.13×10^{13}

Practice Exercise 10: Multiplying Numbers Written in Scientific Notation

Perform the following calculations without converting the numbers to conventional form.

1. $(9.98 \times 10^5) \times (7.26 \times 10^2)$	7. $(8.7 \times 10^2) \times (9.3 \times 10^5)$
2. $(5.639 \times 10^{-3}) \times (6.369 \times 10^2)$	8. $(8.56 \times 10^9) \times (5.17 \times 10^{22})$
3. $(2.75 \times 10^6) \times (5.3 \times 10^{-1})$	9. $(5.56 \times 10^{-11}) \times (3.14 \times 10^{-6})$
4. $(5.4 \times 10^{-3}) \times (2.8 \times 10^{-7})$	10. $(1.002 \times 10^{-3}) \times (8.635 \times 10^{14})$
5. $(3.968 \times 10^{12}) \times (2.349 \times 10^{-3})$	11. $(3.26 \times 10^{-5}) \times (6.57 \times 10^5)$
6. $(3.964 \times 10^8) \times (4.52 \times 10^{-13})$	12. $(2.86 \times 10^{23}) \times (3.06 \times 10^5)$

Division

Follow the steps below to divide numbers written in scientific notation.

Step 1. **Divide the coefficients**

Step 2. **Subtract the exponents.**

Step 3. **Adjust the coefficient to have one digit before the decimal by raising or lowering the exponent. Round to the correct number of significant figures. The base will remain 10.**

Example: Divide 4.5×10^8 by 5.2×10^4

To solve:

- Rewrite: $(4.5 \div 5.2) \times (10^{8-4})$
- Divide coefficients, subtract exponents: 0.8654×10^4
- Change to correct scientific notation and significant figures: 8.65×10^3

Practice Exercise 11: Dividing Numbers Written in Scientific Notation

Perform the following calculations without converting the numbers to conventional form.

1. $(9.28 \times 10^3) / (7.83 \times 10^{-6})$
2. $(4.939 \times 10^9) / (7.364 \times 10^3)$
3. $(9.85 \times 10^{-8}) / (8.3 \times 10^{-8})$
4. $(5.6 \times 10^7) / (2.2 \times 10^1)$
5. $(8.63 \times 10^4) / (2.468 \times 10^{-9})$
6. $(1.28 \times 10^{-8}) / (7.52 \times 10^{-16})$
7. $(9.1 \times 10^{-3}) / (8.3 \times 10^8)$
8. $(1.006 \times 10^{-9}) / (5.42 \times 10^{-11})$
9. $(6.34 \times 10^{12}) / (6.96 \times 10^{-3})$
10. $(9.087 \times 10^8) / (8.659 \times 10^5)$
11. $(4.56 \times 10^6) / (3.05 \times 10^9)$
12. $(2.79 \times 10^{19}) / (2.45 \times 10^{15})$

Addition and Subtraction

Follow the steps below to add or subtract numbers written in scientific notation.

Step 1. Express the numbers as the same power of 10.

Step 2. Add or subtract the coefficient.

Step 3. Express the answer in correct scientific notation, with correct significant figures.

Example: Add 5.56×10^4 and 6.2×10^2

To solve:

- Move the decimal to express in same power of 10: 6.2×10^2 to 0.062×10^4
- Add the coefficients; keep the base and exponent the same: $5.56 + 0.062 = 5.622 \times 10^4$
- Express in the correct number of significant figures: 5.622×10^4

Example: Subtract $(4.2 \times 10^5) - (9.3 \times 10^4)$

To solve:

- Move the decimal to express in same power of 10: 9.3×10^4 to 0.93×10^5
- Subtract the coefficients, keep base and exponent the same: $4.2 - 0.93 = 3.27 \times 10^5$
- Express in the correct number of significant figures: 3.27×10^5

Practice Exercise 12: Adding and Subtracting Numbers Written in Scientific Notation Perform the following calculations without converting the numbers to conventional form.	
Addition	**Subtraction**
1. $(6.28 \times 10^{4}) + (7.94 \times 10^{-5})$	13. $(9.98 \times 10^{5}) - (7.33 \times 10^{-7})$
2. $(4.939 \times 10^{9}) + (7.364 \times 10^{2})$	14. $(3.934 \times 10^{9}) - (8.864 \times 10^{4})$
3. $(9.75 \times 10^{-8}) + (8.2 \times 10^{-8})$	15. $(9.45 \times 10^{-9}) - (8.2 \times 10^{-9})$
4. $(5.5 \times 10^{7}) + (2.4 \times 10^{2})$	16. $(5.3 \times 10^{7}) - (2.4 \times 10^{1})$
5. $(8.53 \times 10^{4}) + (2.368 \times 10^{-7})$	17. $(8.54 \times 10^{9}) - (2.567 \times 10^{-7})$
6. $(1.68 \times 10^{-9}) + (7.54 \times 10^{-14})$	18. $(1.22 \times 10^{-7}) - (7.54 \times 10^{-16})$
7. $(9.4 \times 10^{-3}) + (8.7 \times 10^{8})$	19. $(9.5 \times 10^{-3}) - (8.3 \times 10^{9})$
8. $(1.118 \times 10^{-9}) + (5.44 \times 10^{-13})$	20. $(1.002 \times 10^{-2}) - (2.12 \times 10^{-11})$
9. $(6.38 \times 10^{22}) + (6.99 \times 10^{-3})$	21. $(6.64 \times 10^{12}) - (4.47 \times 10^{-6})$
10. $(9.067 \times 10^{8}) + (9.959 \times 10^{4})$	22. $(9.387 \times 10^{5}) - (2.647 \times 10^{2})$
11. $(2.52 \times 10^{9}) + (3.205 \times 10^{9})$	23. $(2.01 \times 10^{4}) - (3.97 \times 10^{4})$
12. $(2.785 \times 10^{11}) + (2.525 \times 10^{13})$	24. $(7.74 \times 10^{17}) - (2.15 \times 10^{6})$

Section Review 1: Math in Scientific Measurement

A. Define the following terms.

dimensional analysis
dimension
conversion factor
dimensional equation
significant figure
scientific notation

B. Choose the best answer.

1. The fastest runners in the world can run at a speed of around 10 m/s. How fast is this in km/h?
   A. 12 km/h B. 36 km/h C. 360 km/h D. 24 km/h

2. There are about 3.06×10^{21} atoms in 1 g of gold. How many atoms are present in 1.70×10^{-2} g of gold?
   A. 4.76×10^{19} B. 5.20×10^{19} C. 5.20×10^{41} D. 4.76×10^{-41}

Using Mathematics in Scientific Experiments

Understanding Slope

The **slope** of a line refers to how steep a line is. Slope is also defined as the **rate of change**. When we graph a line using ordered pairs of the form (x_1, y_1) and (x_2, y_2), we can easily determine the slope. Slope is often represented by the letter ***m***.

The formula for slope of a line is: $m = \frac{\text{rise}}{\text{run}} = \frac{y_2 - y_1}{x_2 - x_1}$

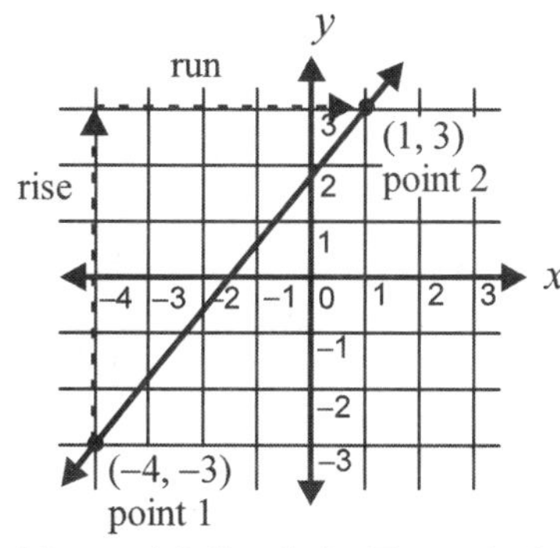

Figure 4.2 Graph for Example 1

Example 1: What is the slope of the line shown in Figure 4.2 that passes through the ordered pairs (-4, -3) and (1, 3)?

y_2 is 3, the y-coordinate of point 2.

y_1 is –3, the y-coordinate of point 1.

x_2 is 1, the x-coordinate of point 2.

x_1 is – 4, the x-coordinate of point 1.

Use the formula for slope given above:

$$m = \frac{3-(-3)}{1-(-4)} = \frac{6}{5}$$

The slope is $\frac{6}{5}$. This shows us that we can go up 6 (rise) and over 5 to the right (run) to find another point on the line. This relationship is shown in Figure 4.3.

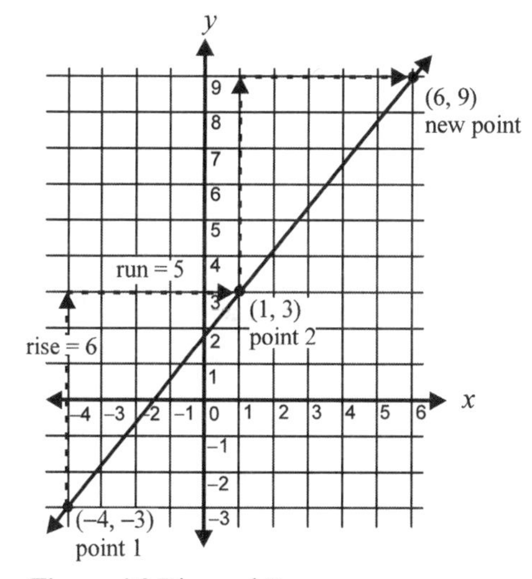

Figure 4.3 Rise and Run

Example 2: Find the slope of a line through the points (– 2, 3) and (1, –2). It doesn't matter which pair we choose for point 1 and point 2. The answer is the same.

let point 1 be (–2, 3)

let point 2 be (1, –2)

$$\text{slope} = \frac{(y_2 - y_1)}{(x_2 - x_1)} = \frac{-2-3}{1-(-2)} = -\frac{5}{3}$$

When the slope is negative, the line will slant left. For this example, the line will go down 5 units and then over 3 to the right. This is shown in Figure 4.4.

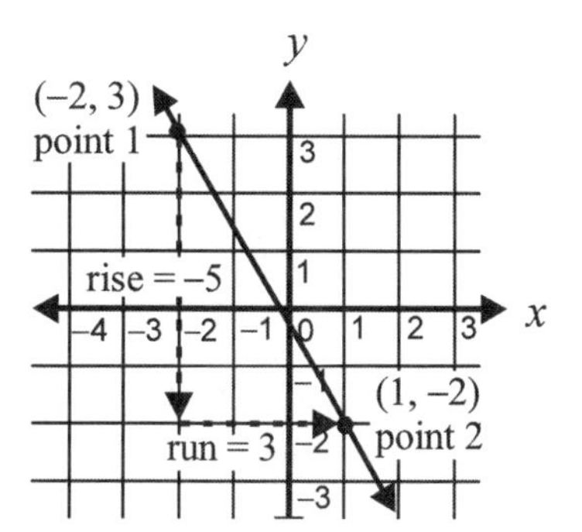

Figure 4.4 Graph for Example 2

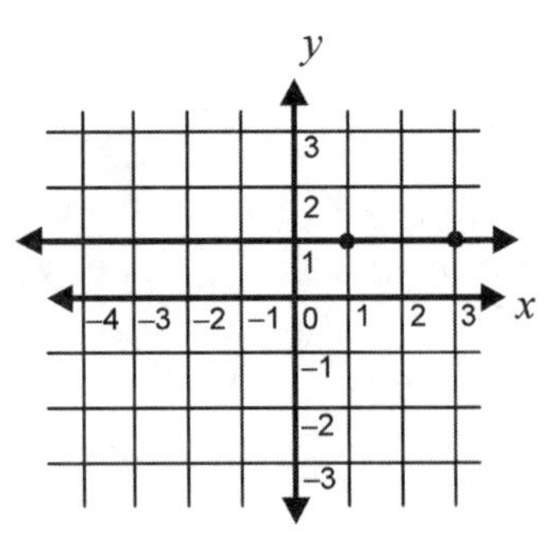

Figure 4.5 Graph for Example 3

Example 3: What is the slope of a line that passes through (1, 1) and (3, 1)? slope $= \frac{1-1}{3-1} = \frac{0}{2} = 0$

When $y_2 - y_1 = 0$, the slope will equal 0, and the line will be horizontal.

Example 4: What is the slope of a line that passes through (2, 1) and (2, –3)?

$$\text{slope} = \frac{-3-1}{2-2} = \frac{4}{0} = \text{undefined}$$

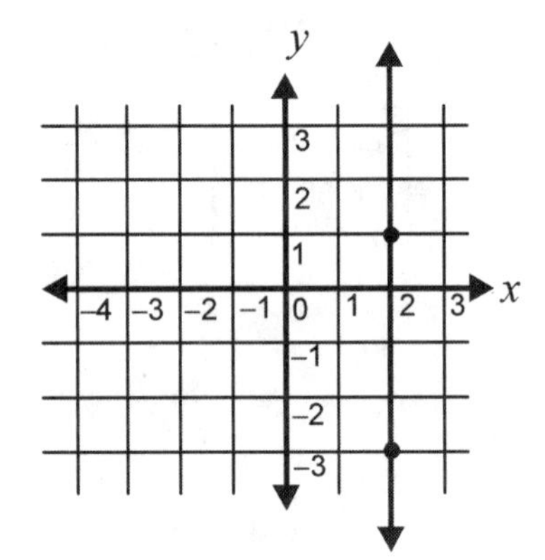

Figure 4.6 Graph for Example 4

When $x_2 - x_1 = 0$, the slope is undefined, and the line will be vertical.

The lines in Figure 4.7 below summarize what we know about slope.

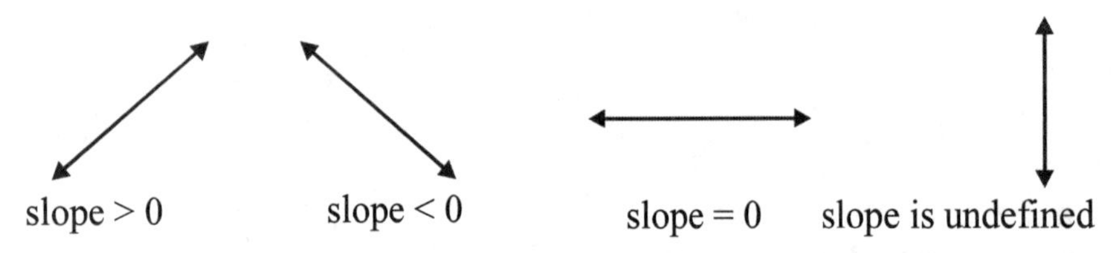

Figure 4.7 Results of Different Slope Values

Practice Exercise 13: Understanding Slope

Find the slope of the line that goes through the following pairs of points. Use the formula .

$$\text{slope} = \frac{y_2 - y_1}{x_2 - x_1}$$

Then, using graph paper, graph the line through the two points, and label the rise and the run. (See Examples 1 and 2 on page 89.)

1. (2, 3) (4, 5)
2. (1, 3) (2, 5)
3. (–1, 2) (4, 1)
4. (1, –2) (4, –2)
5. (3, 0) (3, 4)
6. (3, 2) (–1, 8)
7. (4, 3) (2, 4)
8. (2, 2) (1, 5)
9. (3, 4) (1, 2)
10. (3, 2) (3, 6)
11. (6, –2) (3, –2)
12. (1, 2) (3, 4)
13. (–2, 1) (–4, 3)
14. (5, 2) (4, –1)
15. (1, –3) (–2, 4)
16. (2, –1) (3, 5)
17. (2, 4) (5, 3)
18. (5, 2) (2, 5)
19. (4, 5) (6, 6)
20. (2, 1) (–1, –3)

WRITING AN EQUATION FROM DATA

Data is often written in a columnar format, with two columns being the simplest. The data for the independent (x) and dependent (y) variables can be written as **ordered pairs** in the form of (x, y). If the increases or decreases in the ordered pairs are at a constant rate, then a **linear equation** for the data can be found.

Example: Dan set his car on cruise control and noted the distance he went every 5 minutes. Write an equation for the data set shown in Table 4.1.

Table 4.1 Data for Above Example

Minutes in Operation (x)	Odometer Reading (y)
5	28,490 miles
10	28, 494 miles

Step 1. Write two ordered pairs in the form (minutes, distance) for Dan's driving, (5, 28490), and (10, 28494), and find **slope**. Don't forget to include units in your slope calculation. A review of slope is included in the Appendix.

$$m = \frac{y_2 - y_1}{x_2 - x_1} = \frac{28{,}494 - 28{,}490 \text{ miles}}{10 - 5 \text{ min}} = \frac{4}{5} \text{ miles/min}$$

Step 2. Use the ordered pairs to write the equation in the **slope-intercept** form $y = mx + b$. Place the slope, m, that you found in Step 1 and one of the ordered pairs of points as x_1 and y_1 in the following **point-slope** formula:

$$y - y_1 = m(x - x_1)$$
$$y - 28{,}490 = \tfrac{4}{5}(x - 5)$$
$$y - 28{,}490 + 28{,}490 = \tfrac{4}{5}x - 4 + 28{,}490$$
$$y = \tfrac{4}{5}x + 28{,}486$$

It doesn't matter which pair of points you use, the answer will be the same. Units were not included in the above equation for the sake of simplicity, but if they were, it would look like this:

$$y \text{ (miles)} = \frac{4 \text{ miles}}{5 \text{ min}} x \text{ (min)} + 28{,}486.$$

If you are ever unsure of what the units should be, use dimensional analysis to figure it out.

Practice Exercise 14: Writing an Equation from Data

Write an equation for each of the following sets of data, assuming the relationship is linear.

1. Doug's Doughnut Shop		**2. Jim's Depreciation on His Jet Ski**	
Years in Business	Total Sales	Years	Value
1	$55,000	1	$4,500
4	$85,000	6	$2,500
3. Gwen's Green Beans		**4. Stepping on the Breaks**	
Days Growing	Height in Inches	Seconds	MPH
2	5	2	51
6	12	5	18
5. At the Gas Pump		**6. Stepping on the Accelerator**	
Gallons Purchased	Total Cost	Seconds	MPH
5	$6.00	4	35
7	$8.40	7	62

GRAPHING LINEAR DATA

Many types of data are related by a constant ratio. This type of data is **linear**. The slope of the line described by linear data is the ratio between the data. Plotting linear data with a constant ratio can be helpful in finding additional values.

Example: A department store prices socks per pair. Each pair of socks costs $0.75. Plot pairs of socks versus price on a Cartesian plane.

A **Cartesian plane** allows you to graph points with two values. Recall that the Cartesian coordinate plane is made up of two number lines, the horizontal x-axis and the vertical y-axis. Each point graphed on the plane is designated by an ordered pair of coordinates.

Remember: The first number always tells you how far to go right or left from 0, and the second number tells you how far to go up or down from 0.

Step 1. Since the price of the socks is constant, you know that one pair of socks costs \$0.75, 2 pairs of socks cost \$1.50, 3 pairs of socks cost \$2.25, and so on. Make a list of a few points.

Table 4.2 Data for Example 1

Pair(s) x	Price (\$) y
1	0.75
2	1.50
3	2.25

Step 2. Plot these points on a Cartesian plane, and draw a straight line through the points.

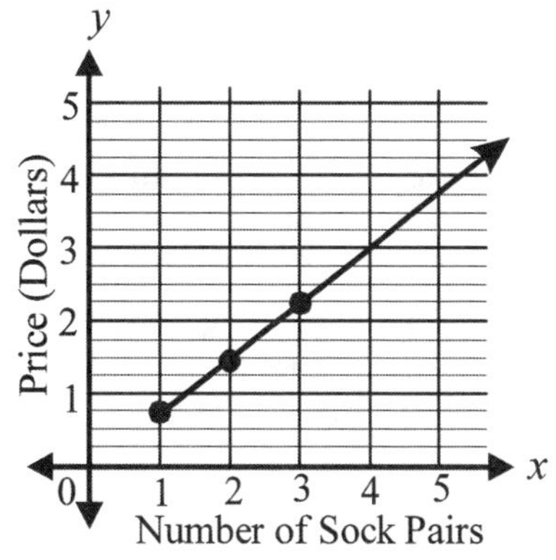

Figure 4.8 Graph for Example

Question: What is the slope of the data? What does the slope describe?

Solution: You can determine the slope either by the graph or by the data points. For this data, the slope is \$0.75/pair. Remember, slope is rise/run. For every \$0.75 going up the y-axis, you go across one pair of socks on the x-axis. The slope describes the price per pair of socks.

Question: Use the graph in Figure 4.8 to answer the following questions. How much would 5 pairs of socks cost? How many pairs of socks could you purchase for \$3.00? Extending the line gives useful information about the price of additional pairs of socks.

Solution 1: The line that represents 5 pairs of socks intersects the data line at \$3.75 on the y-axis. Therefore, 5 pairs of socks would cost \$3.75.

Solution 2: The line representing the value of \$3.00 on the y-axis intersects the data line at 4 on the x-axis. Therefore, \$3.00 will buy exactly 4 pairs of socks.

Practice Exercise 15: Graphing Linear Data

Use the information given to make a line graph for each set of data, and answer the questions related to each graph.

1. The diameter of a circle versus the circumference of a circle is a constant ratio. Use the data given to the right to graph a line to fit the data. Extend the line, and use the graph to answer the next question.
2. Using the graph of the data in question 1, estimate the circumference of a circle that has a diameter of 3 inches.
3. If the circumference of a circle is 3 inches, about how long is the diameter?
4. What is the slope of the line you graphed in question 1?
5. What does the slope of the line in question 4 describe?

Circle

Diameter	Circumference
4 in	12.56 in
5 in	15.70 in

6. The length of a side on a square and the perimeter of a square are a constant ratio. Use the data given to the right to graph this relationship.
7. Using the graph from question 6, what is the perimeter of a square with a side that measures 4 inches?
8. What is the slope of the line graphed in question 6?

Square

Length of Side	Perimeter
2 in	8 in
3 in	12 in

9. Conversions are often constant ratios. For example, converting from pounds to ounces follows a constant ratio. Use the data given to the right to graph a line that can be used to convert pounds to ounces.
10. Use the graph from question 9 to convert 40 ounces to pounds.
11. What does the slope of the line graphed for question 9 represent?

Measurement Conversion

Pounds	Ounces
2	32
4	64

12. Graph the data to the right, and create a line that shows the conversion of weeks to days.
13. About how many days are in 21 weeks?

Time

Weeks	Days
1	7
2	14

14. Graph a data line that converts feet to inches.
15. Using the graph in question 14, how many inches are in 4.5 feet?
16. What is the slope of the line converting feet to inches?

17. An electronics store sells DVDs for $25 each. Graph a data line showing total cost versus the number of DVDs purchased.
18. Using the graph in question 17, how many DVDs could be purchased for $150?

Scatter Plots

A **scatter plot** is a graph of ordered pairs involving two sets of data. These plots are used to detect whether two sets of data, or variables, are truly related.

In order to determine the relationship between income and education, twenty people aged 25 years and older were interviewed. The number of years spent in school and the yearly income of each interviewee were recorded on the scatter plot in figure 4.9.

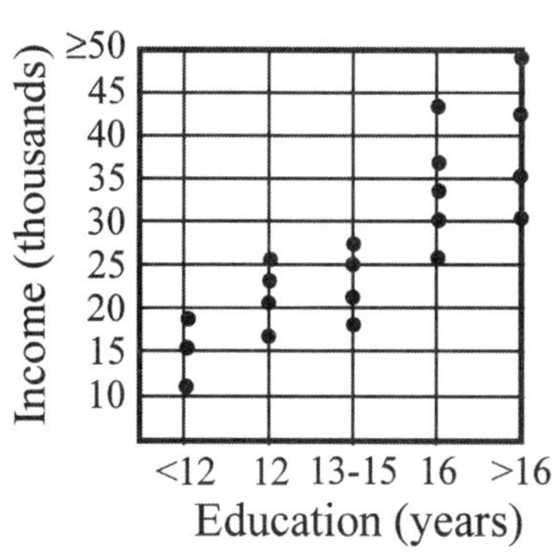

Figure 4.9 Data for Scatter Plot

Although the data does not have an exactly linear relationship, there does appear to be a general upward trend. Draw a line on the scatter plot so that half of the points are above the line and half are below it. That is called a **trendline**, or line of best fit; it is a linear approximation of the relationship between the data points. You will recognize the general equation of the trendline as: $y = mx + b$

Here we can add on to the concept of the independent and dependent variables discussed in Chapter 1. The independent variable is always plotted on the x-axis scale, called the **abscissa.** The dependent variable is always plotted on the y-axis scale, called the **ordinate**. The y values indicate the response to a change in the x value. This relationship is shown graphically by the trendline. A positive relationship means that an increase in the independent variable causes a corresponding increase in the dependent variable. The trendline that you drew into Figure 4.9 should indicate a **positive relationship**: an *increase* in the amount of education causes a *corresponding increase* in the amount of income earned.

Figure4.10 Data

Let's try it again. The scatter plot in Figure 4.10 shows the results of interviews of 15 girls aged 2 – 12. Each girl was asked how many stuffed animals she owned. The interviewer was trying to determine the relationship between the age of the girl and the number of stuffed animals she owned. Here the independent variable (age) is plotted on the abscissa and the dependent variable (the number of stuffed animals owned) is plotted on the ordinate. Draw a trendline with half of the point above the line and half below. Your trendline should indicate a **negative relationship**: that is, an *increase* in age causes a *corresponding decrease* in the number of stuffed animals owned.

Finally, look at the plot in Figure 4.11. Rita wanted to determine whether or not there was a relationship between the temperature of a classroom in which she was taking a test and her score on that test. She plotted the temperature of the classroom on the abscissa and the test grade that she received on the ordinate. Try to draw a trendline for her results. There doesn't seem to be a trend to draw a line through, does there? Rita's grades are very unreliable, but the fluctuation does not appear to be in response to temperature. Therefore, we can conclude there is **no relationship** between Rita's test grades and the temperature of the classroom.

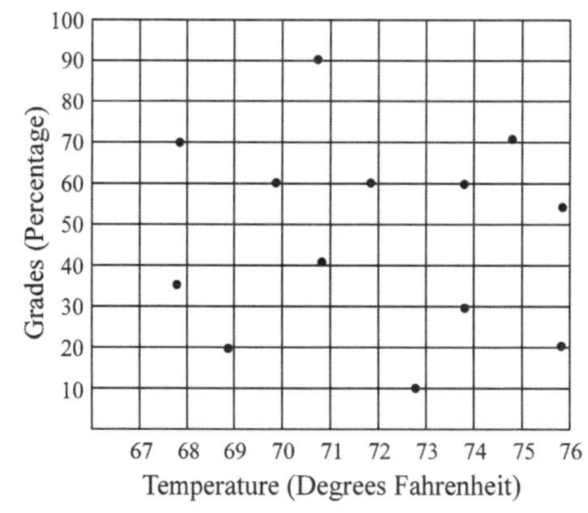

Figure 4.11 Data Showing No Relationship

Practice Exercise 16: Scatter Plots

Examine each of the scatter plots on the following page. Determine which type of relationship is shown between the two variables: positive, negative, or no relationship

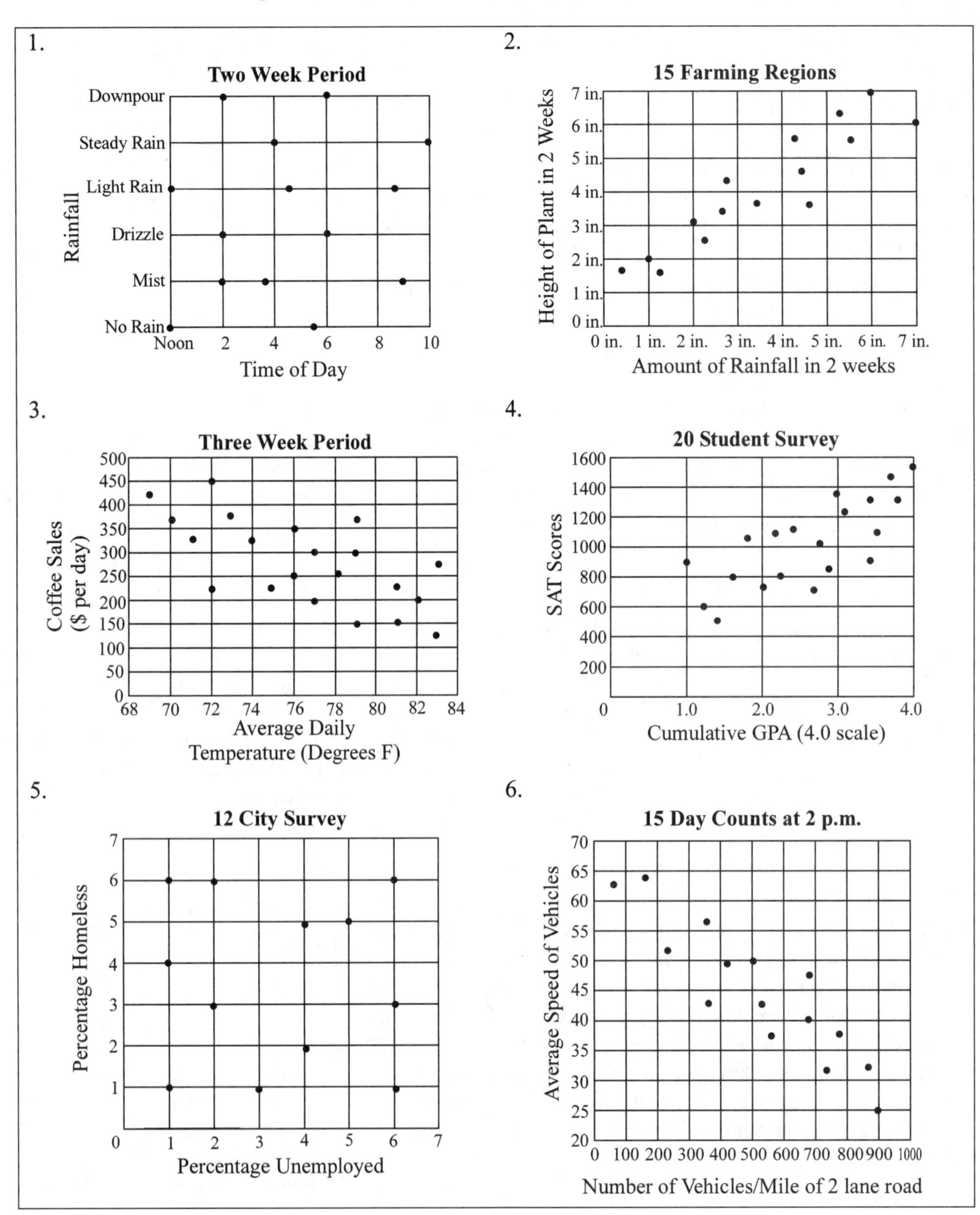

THE LINE OF BEST FIT

At this point, you now understand how to plot points on a Cartesian plane. You also understand how to find the data trend on a Cartesian plane. These skills are necessary to accomplish the next task, determining the line of best fit, or trendline.

In order to find a **line of best fit**, you must first draw a **scatter plot** of all data points. Once this is accomplished, draw an oval around all of the points plotted. Draw a line through the points in such a way that the line separates half the points from one another. You may now use this line to answer questions.

Example: The following data set contains the heights of children between 5 and 13 years old.

Make a scatter plot and draw the line of best fit to represent the trend. Using the graph, determine the height for a 14 year-old child.

Table 4.3 Data Set for Above Example

Age 5: 4'6", 4'4", 4' 5"	Age 8: 4'8", 4'6", 4'7"	Age 11: 5'0", 4'10"
Age 6: 4'7", 4'5", 4'6"	Age 9: 4'9", 4'7", 4'10"	Age 12: 5'1", 4'11", 5'0", 5'3"
Age 7: 4'9", 4'7", 4'6", 4'8"	Age 10: 4'9", 4'8", 4' 10"	Age 13: 5'3", 5'2", 5'0"

In this example, the data points lay in a positive sloping direction. To determine the line of best fit, all data points were circled, then a line of best fit was drawn. Half of the points lay below, half above the line of best fit drawn bisecting the narrow length of the oval.

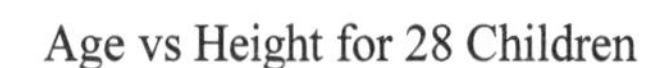

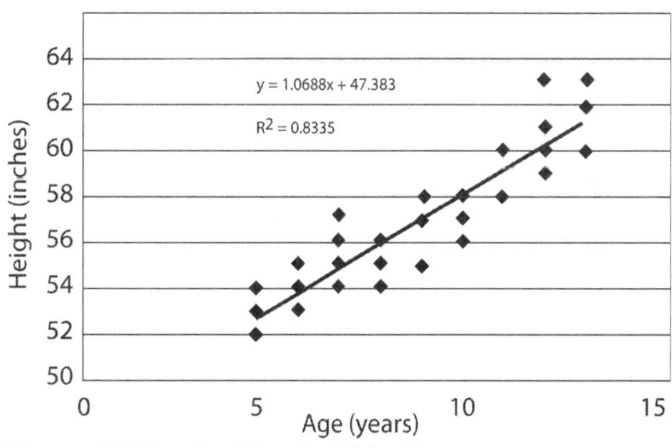

Figure 4.12 Graph of Data in Table 4.3

Let us determine the height a 14 year old would be by looking at the graph. To do this, we simply extend the trendline out to the value of 14 on the abscissa. This is called **extrapolation**, and it allows us to use the line of best fit to estimate the value of a dependent variable that you did not measure. Figure 4.12 tells us visually that, based on the data collected, a 14 year old should be 62 inches tall. Convert that from height in inches to the more familiar units of height: feet. Does that seem right?

There are ways to see if this predicted value has any statistical usefulness. Computer programs like Microsoft® Excel© have tools that help fit a trendline to your data, and give you an equation that describes the trendline. Let's say that you took the data shown in Table 4.3 and used Excel© to plot it in a scatter plot. You would get a plot similar to that shown in Figure 4.12. After plotting the data, you would click the command "Add Trendline" in the pulldown menu under "Chart." This would generate the equation shown on the chart, as well as the R^2 value. The R^2 value helps you determine the "goodness of fit;" the closer R^2 is to one, the better the fit. An exactly linear relationship is defined by an R^2 value of exactly one. Any R^2 value that is less than one indicates that the relationship is not exactly linear.

Practice Exercise 17: The Line of Best Fit

Plot the data sets below, then draw the line of best fit. Next, use the line to estimate the value of the next measurement.

1. Selected values of the Sleekster Brand Light Compact Vehicles:
   New Vehicle: $13,000; 1 year old: $12,000, $11,000, $12,500;
   2 year old: $9,000, $10,500, $9,500; 3 year old: $8,500, $8,000, $9,000;
   4 year old: $7,500, $6,500, $6,000

   What is the value of a five year-old vehicle?

2. The relationship between string length and kite height for the following kites:
   (L = 500 ft, H = 400 ft); (L = 250 ft, H = 150 ft); (L = 100 ft, H = 75 ft);
   (L = 500 ft, H = 350 ft); (L = 250 ft, H = 200 ft); (L = 100 ft, H = 50 ft)

   What is the height of a kite with a string length of 600 feet?

3. Relationship between Household Incomes (HI) and Household Property Values (HPV):
   (HI = $30,000, HPV = $100,000); (HI = $45,000, HPV = $120,000); (HI = $60,000, HPV = $135,000); (HI = $50,000, HPV = $115,000); (HI = $35,000, HPV = $105,000); (HI = $65,000, HPV = $155,000)

   What is the property value for a household making $90,000?

INTERPOLATION AND EXTRAPOLATION

Interpolation is a method for determining new data points that lie between two known values. **Linear interpolation** assumes that the two known values can be joined by a line, or the rate of change is constant. If you know the equation for the line joining the known points then you can predict the value of any point on that line. The data in Table 4.4 below was recorded and graphed in Figure 4.13 to show the distance it takes a vehicle to stop at various speeds.

Table 4.4 Data for Stopping Distance

Speed (mph)	Stopping Distance (ft.)
30	78
40	142
50	208
60	284
70	375
80	475

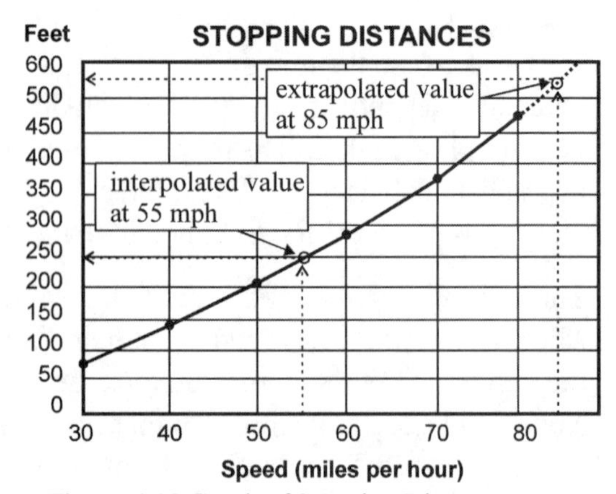

Figure 4.13 Graph of Stopping Distance

Example: What is the stopping distance for a vehicle traveling 55 mph?

Table 4.4 does not give the value for stopping distance at 55 mph, but you can interpolate the value using the graph in Figure 4.13. At 55 mph, the stopping distance of a vehicle is approximately 250 feet. The interpolated value is between the known stopping distance values at 50 mph and 60 mph.

You can also calculate interpolated data points by calculating the slope and line equation between 2 adjacent points. The value we are interested in lies between data points (50, 208) and (60, 284). The first step in calculating the interpolated data point is finding the slope of the line connecting the two data points surrounding our point of interest.

$$m = \frac{284 - 208 \text{ (ft)}}{60 - 50 \text{ (mph)}} = \frac{76 \text{ ft}}{10 \text{ mph}} = 7.6 \text{ ft/mph}$$

Therefore, the stopping distance of the car increases by 7.6 feet for every additional 1 mph of speed. Now that we know the slope, we can calculate the equation of the line by substituting our known values into the point-slope equation for a line.

$$y - 208 = 7.6 \cdot (x - 50) = 7.6x - 380$$
$$y = 7.6x - 380 + 208 = 7.6x - 172$$

Finally, we can calculate our interpolated value by substituting 55 into the x in the above equation.

$$y = (7.6 \cdot 55) - 172 = 246 \text{ feet}$$

Notice that the calculated value for stopping distance at 55 mph is different than the estimation from the graph. This illustrates an important point: *Interpolation is simply an approximation.* In this example, we assumed that the rate of change remains constant, when in fact, the rate of change is slowly increasing. To improve the accuracy of our interpolated point it is important to use the right relationship between data points. Therefore, you should choose a trendline that best fits your data in order to minimize error in your data analysis. Maybe a higher order relationship would have given us a more accurate value for stopping distance at 55 mph.

Extrapolation is a method for determining a quantity that is beyond those that are known. Extrapolation assumes that the data follows the same trend, or changes at the same rate, as the data that is known.

Example: What is the stopping distance for a vehicle traveling 85 mph?

The table and graph do not give values for a vehicle traveling 85 mph. However, if the graph is extended by assuming the values beyond 80 mph increase at a constant rate, you can predict what the stopping distance at 85 mph might be. At 85 mph, the stopping distance should be about 545 feet. You could also calculate the extrapolated point by using the line equation calculated earlier, which will give you a value of 474 feet. This value is very different than the value estimated from the graph. Let's recalculate the line by using the last two data points in table 3.4: (70, 375) and (80, 475).

$$y - 475 = 10 \cdot (x - 80) = 10x - 800$$
$$y = 10x - 800 + 475 = 10x - 325$$

Substituting 85 into this equation gives us a stopping distance of 525 feet, which is much closer to the value estimated from the graph.

Practice Exercise 18: Interpolation and Extrapolation

Use the graph in Figure 4.13 to answer the following questions.

1. Would you interpolate or extrapolate to find the stopping distance at 62 mph?
2. Would you interpolate or extrapolate to find the stopping distance at 20 mph?
3. According to the graph, what would be the stopping distance at 44 mph? Did you interpolate or extrapolate to find the answer? Compare to the calculated value at 44 mph.
4. Predict what the stopping distance might be for a vehicle traveling at 25 mph.

Challenge Activity:

Look again at figure 4.12. Use the trendline equation generated in Microsoft® Excel© and plug in 14 as the *x* value. This allows you to estimate the height of a 14-year old. What result do you get? How many significant figures should the answer that you generate have? Given the R^2 value, what can you say about the accuracy of the extrapolation? This equation indicates that a 50-year old will be about 26 inches taller than a 25-year old; is that generally true? If not, what does that say about the broad applicability of the equation?

Section Review 2: Using Mathematics in Scientific Experiments

A. Define the following terms.

slope	Cartesian plane	negative relationship	interpolation
ordered pairs	scatter plot	no relationship	linear interpolation
linear equation	positive relationship	line of best fit	extrapolation
abscissa	ordinate	trendline	

Chapter 4 Review

Choose the best answer.

1. Identify the conversion factor that would convert gram per milliliter (g/mL) to kilograms per liter (kg/L).

   A. g/mL × (1 kg/1000 g) × (1000 ml / 1 L)
   B. g/mL × (1 kg/1000 g) × (1 L /1000mL)
   C. g/mL × (1000 g/ 1 kg) × (1000 mL / 1 L)
   D. g/mL × (1000 / 1 kg) × (1 L / 1000mL)

2. A water bottle holds 500 cm^3. How many cubic meters does the water bottle hold?

   A. 0.0005 m^3
   B. 5 m^3
   5,000 m^3
   500,000,000 m^3

3. How many significant figures does 0.01250 have?

   A. 3
   B. 4
   C. 5
   D. 6

4. Identify the correct way to write 0.00542 in scientific notation.

   A. 5.42×10^{-3}
   B. 5.42×10^{-4}
   C. 5.42×10^{3}
   D. 5.42×10^{4}

5. Identify the product of $(3.21 \times 10^{2)} \times (2.035 \times 10^{5})$ that is stated with the correct precision.

   A. 6×10^{7}
   B. 6.5×10^{7}
   C. 6.53×10^{7}
   D. 6.532×10^{7}

6. Which of the following is the graph of the line which has a slope of –2?

A.

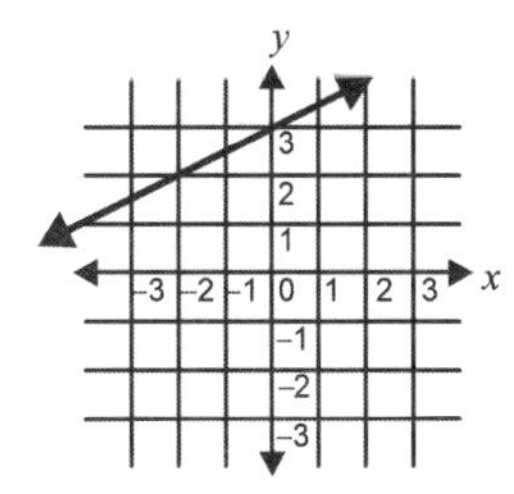

B.

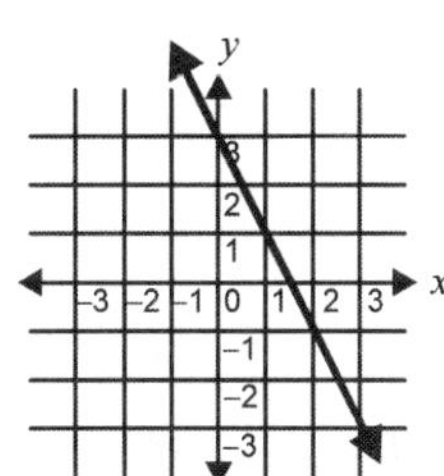

C.

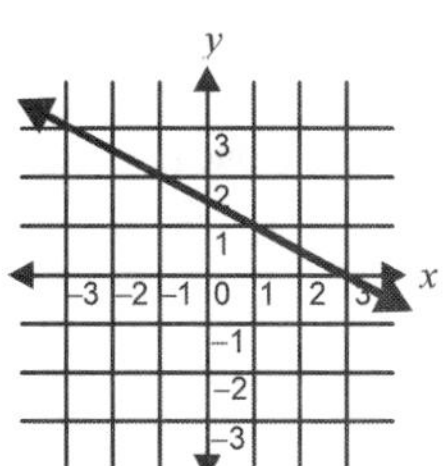

D.

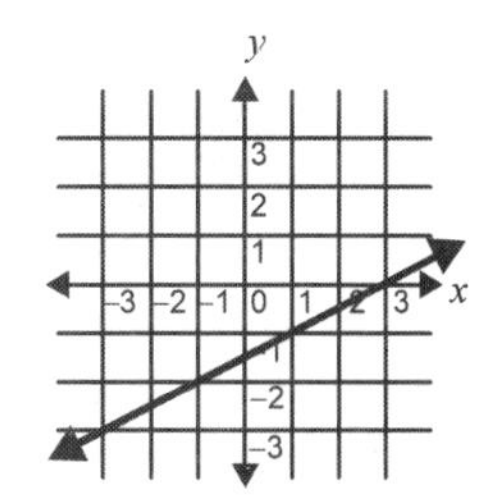

7. Paulo turned on the oven to preheat it. He noted the temperature after one minute and after two minutes. His data are presented in the table below.

Time (minutes)	Temperature (°F)
1	200°
2	325°

Identify the equation that fits Paulo's data.

   A. $y = 200x$
   B. $y = 100x + 100$
   C. $y = 250x - 50$
   D. $y = 125x + 75$

The data table and graph below show the relationship between miles per hour (mi/h) and kilometers per hour (km/h)..

mi/h	km/h
30	48
55	88

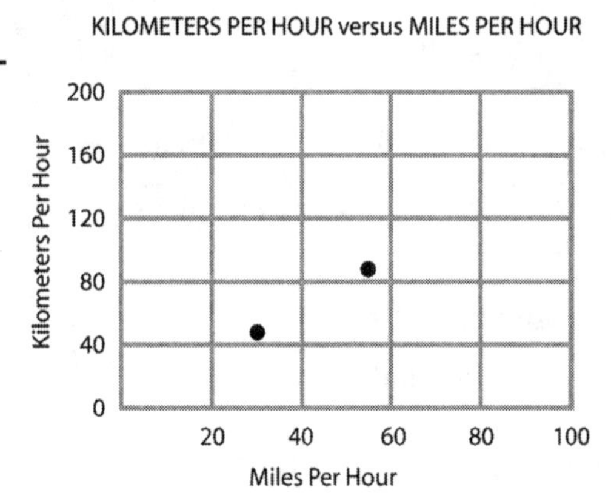

8. Identify the best approximation of the conversion of 45 mi/h to km/h
   A. 45 mi/h = 50 km/h
   B. 45 mi/h = 60 km/h
   C. 45 mi/h = 70 km/h
   D. 45 mi/h = 80 km/h

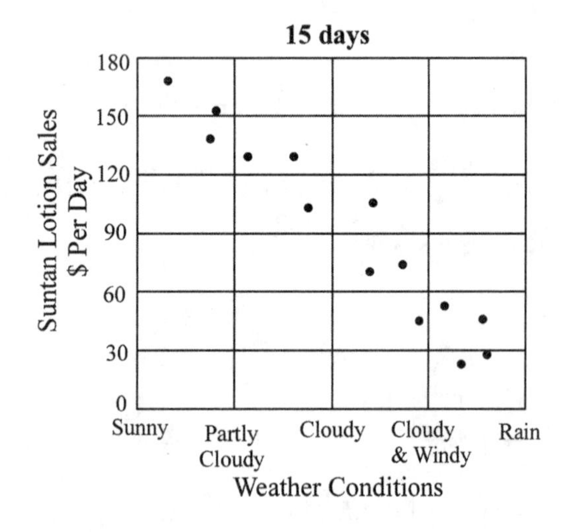

9. The scatter plot has what type of relationship?
   A. no relationship
   B. positive
   C. negative
   D. cannot be determined

10. What method would you use to determine a new data that lies between two known values?
    A. extrapolation
    B. intrapolation
    C. inductive reasoning
    D. deductive reasoning

Chapter 5
Structure, Properties, and Bonding of Elements

PHYSICAL SCIENCE STANDARDS COVERED IN THIS CHAPTER INCLUDE:

SPS1	Students will investigate our current understanding of the atom.
SPS4	Students will investigate the arrangement of the periodic table.

THE STRUCTURE OF ATOMS

MODERN ATOMIC THEORY

The **atomic theory** states that all matter is made up of tiny particles called atoms. **Matter** is defined as anything that has mass and takes up space. **Atoms** are the basic unit of all things. Everything that has matter is made up of smaller particles called atoms.

ATOMIC STRUCTURE

Atoms are made up of **subatomic particles**. These particles include protons, neutrons, and electrons. **Protons** have a positive charge, and **electrons** have a negative charge. **Neutrons** have neither a positive nor negative charge; they are neutral. Figure 5.1 shows a simplified model of a helium atom. The **nucleus** of the atom consists of protons and neutrons. The electrons are outside the nucleus. The overall charge of the atom is neutral because the number of protons equals the number of electrons.

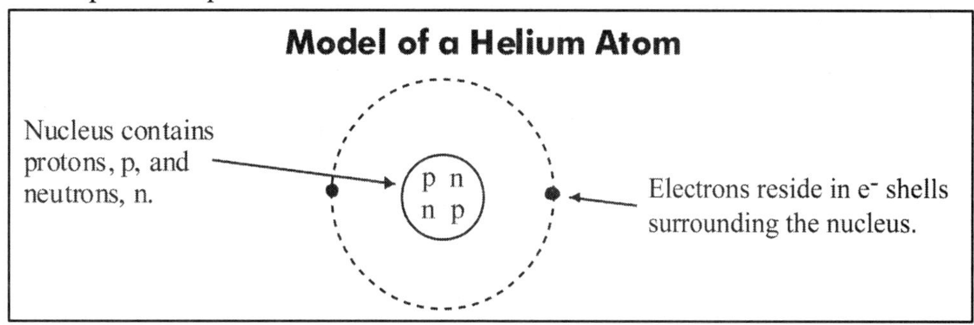

Figure 5.1 Model of a Helium Atom

The negatively charged electrons are electrically attracted to their oppositely charged counterparts, the protons. This attraction holds them in orbit around the nucleus. The area that they occupy is called an **orbital**. These orbitals are designated s, p, d and f, each of which has a different shape. When associated with their quantum mechanical **energy level** (1, 2, 3…), these orbitals define the electron distribution of an atom. The s orbital is found closest to the nucleus; it can be imagined as a sphere around the nucleus. The smallest atom, hydrogen (H), consists of 1 proton and 1 electron. The proton is in the nucleus (in fact, for hydrogen, the proton essentially is the nucleus) and the electron orbits the nucleus in the 1s orbital. The next largest atom is helium (He), which consists of 2 protons, 2 neutrons and 2 electrons. As shown in Figure 5.1, the 4 protons and neutrons of the helium atom are held together in the nucleus, while the 2 electrons orbit the nucleus in the 1s orbital.

An s orbital will only hold two electrons, however. The next element in the periodic table is lithium (Li), which has 3 protons, 3 neutrons and 3 electrons. Two of lithium's electrons will go into the 1s orbital, and the third will go into the 2s orbital, as shown in Figure 5.2. While having the same orbital designation (s) as the 1s orbital, the 2s orbital has a higher quantum number (2), and is thus a higher energy orbital, located farther from the nucleus. The greater the distance between two charged particles, the weaker the electrical force that holds them together. Therefore, since the lithium electron in the 2s orbital spends most of its time farther away from the nucleus than the atom's 1s electrons, it is less tightly bound to the nucleus.

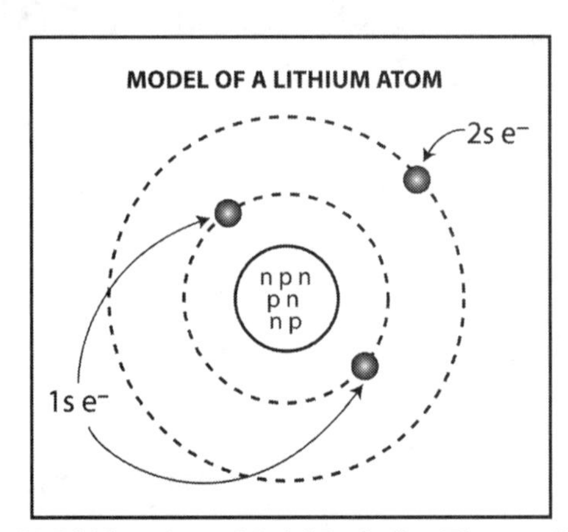

Figure 5.2 Lithium Atom Model Including Electron Shells

Even without explicitly addressing the quantum mechanical considerations necessary to examine larger atoms, we can still understand a bit about how atoms are put together. The quantum number and orbital designation of each electron can be accounted for in an atom's **electron configuration**. This is the organizational concept behind the periodic table, and will be discussed in the next section. For now, it is sufficient to recognize that electrons sequentially fill various quantum energy levels (1,2,3, etc…), and the various shells within those energy levels. As each energy level becomes full, electrons begin to fill the next highest level. The highest energy level orbital containing electrons is the atom's **valence shell**; it contains the electrons that exist farthest away from, and thus the least tightly bound to, the nucleus of the atom. These "outer electrons" are called valence electrons, and are free to participate in bonding with other atoms. Table 5.1 shows the electron configuration for sulfur, which has 16 electrons. It has four valence electrons in the 3p orbital. Note that the p orbital consists of three sub-orbitals — p_x, p_y and p_z — each of with holds a maximum of two electrons, for a total of six.

Number of Electrons to fill the Energy Level

$1s_2$

$2s_2$, $2p_{x2}$, $2p_{y2}$, $2p_{z2}$

$3s_2$, $3p_{x2}$, $3p_{y1}$, $3p_{z1}$

Table 5.1 Electron configuration of sulfur

Section Review 1: The Structure of Atoms

A. Define the following terms.

atomic theory	subatomic particles	electron	neutron	orbital
matter	proton	atom	nucleus	electron configuration
valence shell				energy level

B. Choose the best answer.

1. Which of the following parts of an atom has a positive charge?
   A. protons B. neutrons C. electrons D. electron shells

2. Which of the following parts of an atom has a negative charge?
   A. protons B. neutrons C. electrons D. the nucleus

3. Which of the following parts of an atom has no charge?
   A. protons B. neutrons C. electrons D. the nucleus

4. Which subatomic particles are found in the nucleus of an atom?
   A. protons and electrons
   B. electrons and neutrons
   C. protons and neutrons
   D. protons, neutrons, and electrons

5. What is the maximum number of electrons that are contained in an s orbital?
   A. 1 B. 2 C. 3 D. 4

C. Answer the following questions.

1. Compare and contrast protons, neutrons, and electrons.
2. What is a valence electron?

ORGANIZATION OF THE PERIODIC TABLE

ATOMIC NUMBER AND ATOMIC MASS

Elements are substances that cannot be further broken down by simple chemical means. An element is composed of atoms with the same number of protons. Each element has its own symbol, atomic number, atomic mass, and electron shell arrangement (electron configuration). The atomic number represents the number of protons found in a given atom.

The mass of an atom, referred to as **atomic mass**, is related to the number of protons, electrons, and neutrons in the atom. Protons and neutrons account for the majority the atom's mass. The unit of atomic mass, as expressed in the periodic table, is called an **atomic mass unit (amu)**. One atomic mass unit is defined as a mass equal to one-twelfth the mass of one atom of carbon-12. The amu is also known as the **dalton (Da)**.

To find the number of neutrons most commonly found in an element, subtract the atomic number from the atomic mass, and round to the nearest whole number.

Carbon is represented in the periodic table as follows:

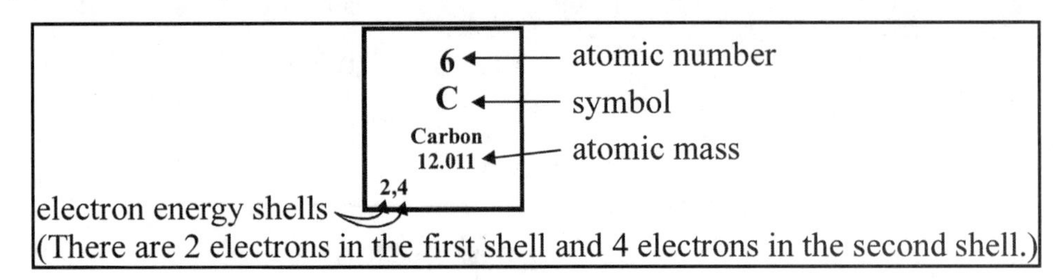

Figure 5.3 Periodic Table Information for Carbon

To find the neutrons most commonly found in an atom of carbon, subtract the atomic number, 6, from the atomic mass, 12.011, to get 6.011. Round to the nearest whole number to get 6. Carbon atoms most often have 6 neutrons in their nuclei.

PERIODIC TABLE

Look at Figure 5.4 on the next page. As you already know this is called the **periodic table**. The periodic table arranges all known elements by atomic number, starting with atomic number 1 (hydrogen) and "ending" with atomic number 111 (roentgenium). However, if the periodic table were only organized by atomic number, it would not be very useful. It also organizes the elements by their properties. The horizontal rows of the table are called **periods**. By moving across the periods from left to right, one can determine two things; how many valence electrons a given element has and the order in which their orbitals fill (called the **electron configuration**). The vertical columns of the periodic table are called **groups** or **families**; all members of any vertical group have the same number of valence electrons in the same orbital. An element's placement in the rows and columns of the table has meaning, and enables the observer to understand many properties of that element.

Figure 5.4

THE PERIODIC TABLE OF THE ELEMENTS

Atomic Number → 36
Symbol → Kr
Name → Krypton
Atomic Mass → 83.80

Noble Gases: 18 VIIIA

Period	1 IA	2 IIA	3 IIIB	4 IVB	5 VB	6 VIB	7 VIIB	8 VIIIB	9 VIIIB	10 VIIIB	11 IB	12 IIB	13 IIIA	14 IVA	15 VA	16 VIA	17 VIIA	18 VIIIA
1	1 H Hydrogen 1.007g																	2 He Helium 4.0026
2	3 Li Lithium 6.941g	4 Be Beryllium 9.01218											5 B Boron 10.81	6 C Carbon 12.011	7 N Nitrogen 14.0067	8 O Oxygen 15.9994	9 F Fluorine 18.998403	10 Ne Neon 20.179
3	11 Na Sodium 22.9898	12 Mg Magnesium 24.305											13 Al Aluminum 26.98154	14 Si Silicon 28.0855	15 P Phosphorus 30.97376	16 S Sulfur 32.06	17 Cl Chlorine 35.453	18 Ar Argon 39.948
4	19 K Potassium 39.0983	20 Ca Calcium 40.08	21 Sc Scandium 44.9559	22 Ti Titanium 47.90	23 V Vanadium 50.9415	24 Cr Chromium 51.996	25 Mn Manganese 54.9381	26 Fe Iron 55.847	27 Co Cobalt 58.9332	28 Ni Nickel 58.69	29 Cu Copper 63.546	30 Zn Zinc 65.38	31 Ga Gallium 69.723	32 Ge Germanium 72.61	33 As Arsenic 74.9216	34 Se Selenium 78.96	35 Br Bromine 79.904	36 Kr Krypton 83.80
5	37 Rb Rubidium 85.4678	38 Sr Strontium 87.62	39 Y Yttrium 88.9059	40 Zr Zirconium 91.22	41 Nb Niobium 92.9064	42 Mo Molybdenum 95.94	43 Tc Technetium 97.91	44 Ru Ruthenium 101.07	45 Rh Rhodium 102.9055	46 Pd Palladium 106.4	47 Ag Silver 107.868	48 Cd Cadmium 112.41	49 In Indium 114.82	50 Sn Tin 118.71	51 Sb Antimony 121.75	52 Te Tellurium 127.60	53 I Iodine 126.9045	54 Xe Xenon 131.30
6	55 Cs Cesium 132.9054	56 Ba Barium 137.33		72 Hf Hafnium 178.49	73 Ta Tantalum 180.9479	74 W Tungsten 183.84	75 Re Rhenium 186.2	76 Os Osmium 190.2	77 Ir Iridium 192.22	78 Pt Platinum 195.09	79 Au Gold 196.9665	80 Hg Mercury 200.59	81 Tl Thallium 204.383	82 Pb Lead 207.2	83 Bi Bismuth 208.9808	84 Po Polonium 208.98244	85 At Astatine 209.98704	86 Rn Radon 222.02
7	87 Fr Francium 223.01976	88 Ra Radium 226.0254		104 Rf Rutherfordium 261.1	105 Db Dubnium 262.11	106 Sg Seaborgium 263.12	107 Bh Bohrium 262.12	108 Hs Hassium 264.13	109 Mt Meitnerium 266.14	110 Ds Darmstadtium 271	111 Rg Roentgenium 272	*112*	*113*	*114*	*115*	*116*	*117*	*118*

Series															
Lanthanide Series →	57 La Lanthanum 138.9055	58 Ce Cerium 140.12	59 Pr Praseodymium 140.9077	60 Nd Neodymium 144.24	61 Pm Promethium 144.91279	62 Sm Samarium 150.4	63 Eu Europium 151.96	64 Gd Gadolinium 157.25	65 Tb Terbium 158.9254	66 Dy Dysprosium 162.50	67 Ho Holmium 164.9304	68 Er Erbium 167.26	69 Tm Thulium 168.9342	70 Yb Ytterbium 173.04	71 Lu Lutetium 174.967
Actinide Series →	89 Ac Actinium 227.02779	90 Th Thorium 232.0381	91 Pa Protactinium 231.0359	92 U Uranium 238.029	93 Np Neptunium 234.0482	94 Pu Plutonium 244.06424	95 Am Americium 243.06139	96 Cm Curium 247.07035	97 Bk Berkelium 247.07030	98 Cf Californium 251.0796	99 Es Einsteinium 252.08	100 Fm Fermium 257.09515	101 Md Mendelevium 258.1	102 No Nobelium 259.100	103 Lr Lawrencium 262.11

Figure 5.5 shows a portion of the periodic table. Notice the pattern of electrons in the outer shells of elements in the same family versus elements in the same group.

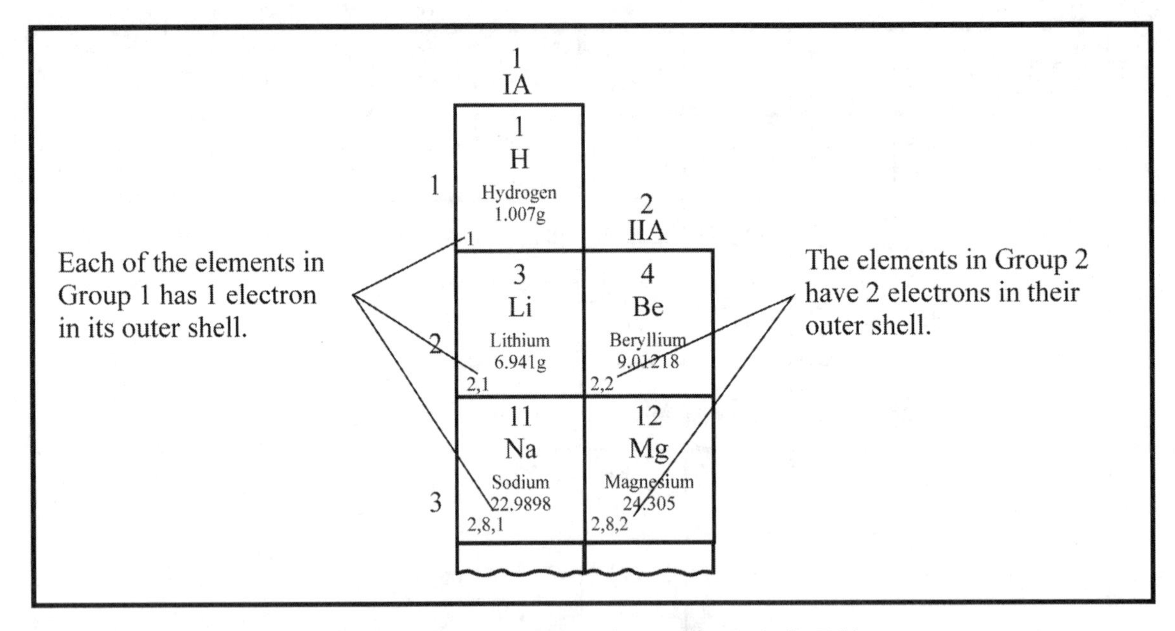

Figure 5.5 Groups IA and IIA of the Periodic Table

Challenge Activity

Look at figure 5.5. The arrows point to the number of valence electrons in each group: 1 valence electron in each Group IA element and 2 in each Group IIA element. Can you figure out what the other numbers are? Hint: look back at Table 5.1.

ELEMENTAL CLASSIFICATION

Elements can all be classified as metals, nonmetals, metalloids, or noble gases depending on where they are located in the periodic table.

Metals make up the majority of the periodic table and are located on the left side. Hydrogen is the only exception. Although it is in the IA group/family, it is a gas at room temperature and not considered a metal. Most metals are solids at room temperature. They are hard, they have luster (are shiny), and they conduct heat and electricity. The metals in the periodic table are located to the left of the bold line. Metals are more likely to give up an electron because they only have a few electrons in the outermost shell.

Nonmetals are on the right side of the periodic table. Nonmetals are usually gases or dull, brittle solids at room temperature. The nonmetals are shown to the right of the bold line and include elements such as hydrogen, helium, carbon, nitrogen, oxygen, fluorine, and neon. Nonmetals have a tendency to gain electrons in order to completely fill their outermost shell.

The elements diagonally between the metals and the nonmetals are called **metalloids**. These elements have properties of both metals and nonmetals. Metalloids are frequently used in computer chip applications. The metalloids are located along the bold line and include boron, silicon, germanium, arsenic, antimony, tellurium, and astantine. Some metalloids naturally act as semiconductor materials that can conduct electricity at elevated temperatures

The **noble gases**, or rare gases, are in Group 18, the far right column on the periodic table. They are nonmetals that do not react readily with any other elements.

Historical Development of the Periodic Table

The trends explained by the periodic table as we know it today began to be recognized in the 19th century, as scientists discovered more and more information about the known elements. As we have seen, they began to recognize that certain elements exhibited similar properties, but the reasons were not yet clear. There were several attempts to arrange the elements to fit scientific observations, but each of these arrangements had problems.

Then, in 1869, the Russian chemist **Dmitri Mendeleev** (1834 – 1907) constructed his periodic table. In this table, the elements were arranged in order of increasing atomic mass. Mendeleev found that, in the table, similar properties were seen at regularly spaced, periodic intervals. Some of the elements were not known in Mendeleev's time; therefore, he had to leave blank spots in his table in order to group elements with similar properties into the same column. He predicted the properties and atomic masses of elements yet to be discovered that fit into these blank spots. These elements have since been discovered, and Mendeleev's predictions have been found to be very accurate.

There were some inconsistencies in Mendeleev's periodic table. For example, he had potassium listed before argon in his table because the atomic mass of argon is greater than that of potassium. However, the inert gas argon obviously did not belong in a group with such highly reactive elements as lithium and sodium. It was not until 1913 that scientists were able to determine the atomic number of elements. They found that with few exceptions, the order of increasing atomic number is the same as the order of increasing atomic mass. Argon and potassium are one of those exceptions. The atomic number of argon is less than the atomic number of potassium. With this information, argon was correctly placed in a group with the other inert gases. The modern periodic table chemists use today appears very similar to that of Mendeleev. However, the modern periodic table is organized in order of increasing atomic *number*, instead of mass.

The last modification of the periodic table occurred in the middle of the 20th century, when **Glenn Seaborg** (1912 –1999) discovered plutonium and the other transuranic elements between 94 and 102. He modified the table by placing the actinide series below the lanthanide series. Although the basic structure of the periodic table now appears to be correct, new elements called the transactinides are being created at labs across the world. After each element is created and fully characterized, it is submitted to the **International Union of Pure and Applied Chemistry (IUPAC)** for confirmation and naming.

Section Review 2: Organization of the Periodic Table

A. Define the following terms.

atomic mass	period	metal	metalloid
periodic table	Group	nonmetal	noble gas

B. Choose the best answer.

1. Which element has two electrons in its 2p orbital?
   A. He B. C C. Be D. O

2. Lithium and sodium are in the same group of elements in the periodic table. Which of the following statements is true regarding these two elements?
   A. They have the same number of electrons in their valance shell.
   B. They have the same number of protons in their nucleus.
   C. They are both noble gases.
   D. They have different chemical properties.

3. Which of the following is ***not*** a property of most metals?
   A. solid at room temperature
   B. have luster
   C. conduct heat and electricity well
   D. do not react readily with any other elements

4. Where might you find a metalloid element used?
   A. computer motherboard
   B. kitchen potholder/oven mitt
   C. electrical power lines
   D. atmospheric gas mixture

C. Answer the following questions.

1. What do elements in the same group have in common?
2. Name two physical properties of nonmetals.
3. Name two physical properties of metals.
4. Which Group of elements is very stable and does not react readily?

Reactivity of Elements in the Periodic Table

In general, an element is most stable when its valance shell is full. Recall that the valance shell is the highest energy level containing electrons. Period 1 elements (hydrogen and helium) have the valence shell 1s, which can only contain 2 electrons. Period 2 elements (lithium through neon) may have a 2s or 2s, 2p valence shell configuration that can hold up to 8 electrons. The 2s orbital can contain up to 2 electrons, and the 2p orbital can contain up to 6 (two in each of the sub-orbitals p_x, p_y and p_z) for a total of 8. Look at the periodic table in Figure 5.4 and count this out.

You will learn more about electron configurations and energy levels in AP Chemistry or college chemistry, but for now it is important to realize that bonding between elements occurs primarily because of the placement of electrons in the valence shell, particularly the unfilled orbital of the valence shell. Remember this: *Bonding is all about energy!*

For instance, the energy needed to remove an electron from an atom is called the **ionization energy.** Another term is directly related to ionization energy: **electronegativity**. The electronegativity of an atom is a description of the atom's energetic "need" for another electron. An atom with a high electronegativity "wants" another electron; it would be very difficult to remove an electron from an atom that already wants another electron. Therefore, the ionization energy of that atom would also be high. Elements in the same group tend to have similar chemical reactivity based on their willingness to lose or gain electrons. We will look at some of these trends in the following section.

ELEMENTAL FAMILIES

Group 1(IA) elements, with the exception of hydrogen, are called the **alkali metals**. All the elements in Group 1 are very reactive. Since they only have one electron in their valance shell, they will give up that one electron to another element in order to become more **stable**.

When an element loses or gains an electron, it forms an **ion**. An ion is an atom that has lost or gained electrons. Ions have either a positive or a negative charge. When the elements in Group 1 give up the one electron in their valance shell, they form positive ions (or **cations**) with a +1 charge. The positive +1 charge comes from having one more proton than electron. The alkali metals become more reactive as you move down the periodic table because the lone electron in the valence shell is further from the positive charge of the nucleus, and thus the electrical attraction is less. ***Group 1 elements form ions with a +1 charge.***

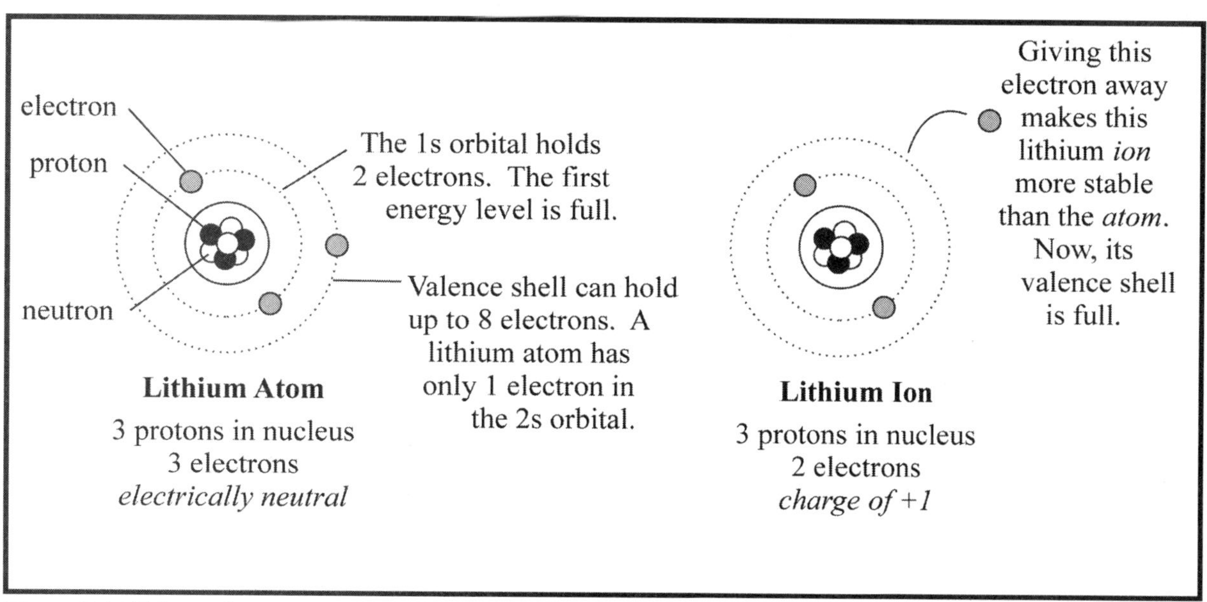

Figure 5.6 Lithium - Group 1

Group 2 (IIA) elements are called the **alkaline earth metals**. They have 2 electrons out of a possible 8 in their valence shell. These metals are less reactive than the alkali metals but are still very reactive. The alkaline earth metals will give away both of their electrons in their valance shell in order to be more stable. Therefore, they form positive ions with a +2 charge. The +2 charge comes from having two more protons than electrons. The alkaline earth metals also become more reactive as you move down the periodic table. ***Group 2 (IIA) elements form ions with a charge of +2.***

Groups 3 – 12(IIIB-IIB) elements in the middle of the periodic table are called **transition metals**. In general, the reactivity of these metals increases as you go down the periodic table and from right to left.

Group 13 – 15 (IIIA, IVA and VA) contain both metals and non-metals. In Group 13, boron is a metalloid; going down the column, all other elements are metals. Group 13 elements form oxides with the general formula R_2O_3. Group 14 is headed by carbon, a prominent nonmetal; going down the column, there are both metalloids (silicon and germanium) and metals. Group 14 elements form oxides with the general formula RO_2. Having four valence electrons (a half-full valence shell) lends these elements a special stability. Group 15 also shows the variation from nonmetal (nitrogen and phosphorous) to metalloid (arsenic and antimony) to metal (bismuth). These elements generally form oxides of the formula R_2O_3 or R_2O_5. ***Group 13 elements form +3 cations and Group 15 elements form negatively charged, –3 ions (or anions). Group 14 elements are generally too stable to ionize.***

Group 16 (VIA) elements have 6 out of a possible 8 electrons in their valence shell. These elements want to gain two electrons to fill the valence shell. Said another way, Group 16 elements have a high electron affinity, particularly oxygen. ***Group 16 elements form anions with a –2 charge.***

Group 17 (VIIA) elements are called the halogens. They have seven electrons in their valence shell, and only require one more to achieve a full valence shell. They have a very high electron affinity and are the most reactive nonmetal elements. They are generally designated with the symbol X and exist in the form X_2, as in Cl_2 gas. They also react with hydrogen, as in HC1. ***Group 17 elements form ions with a –1 charge.***

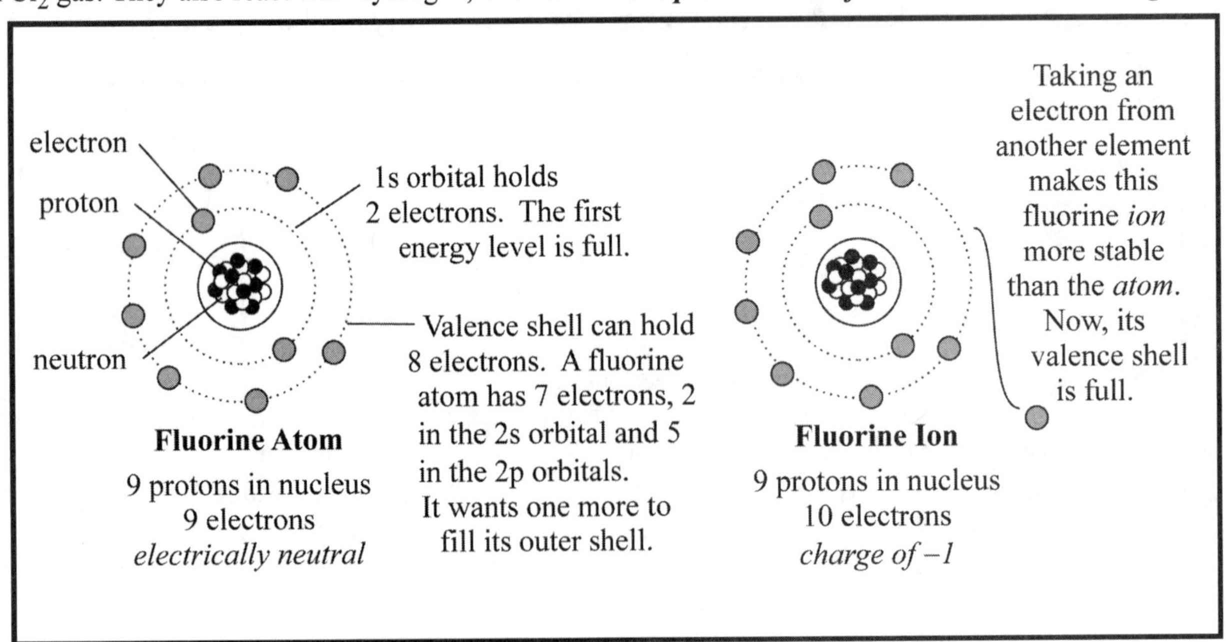

Figure 5.7 Fluorine - Group 17

Group 18 (VIIIA) are the **noble gases**. The noble gases have 8 electrons in their outer shells with the exception of helium, which fills its first energy level with only 2. (Remember, the first energy level will only hold 2 electrons.) The noble gases are very stable elements because their outermost electron shell is completely filled. They will not react readily with any other elements.

SUMMARY OF PERIODIC TRENDS OF ELEMENTS

- Reactivity of metals increases down the periodic table.
- Reactivity of non-metals increases up the periodic table.
- In general, atomic radius increases down the periodic table.
- Atomic radius decreases left to right across the periodic table. This trend may seem opposite from what you would guess. Since the atoms increase in number of protons, neutrons, and electrons, you may think the atomic radius would also get larger. However, the opposite is true. Since the atoms have an increasing number of protons, the positive charge in the nucleus increases. The greater the positive charge in the nucleus, the closer the electrons are held to the nucleus due to the electrical force between them. So, in general, the atomic radius decreases from left to right on the periodic table.
- In general, ionization energies increase left to right across the periodic table and decrease down the periodic table. Ionization energy is a measure of how tightly an electron is bound to an atom.
- In general, electronegativity increases from left to right, and decreases going down the periodic table. Fluorine has the highest electronegativity.

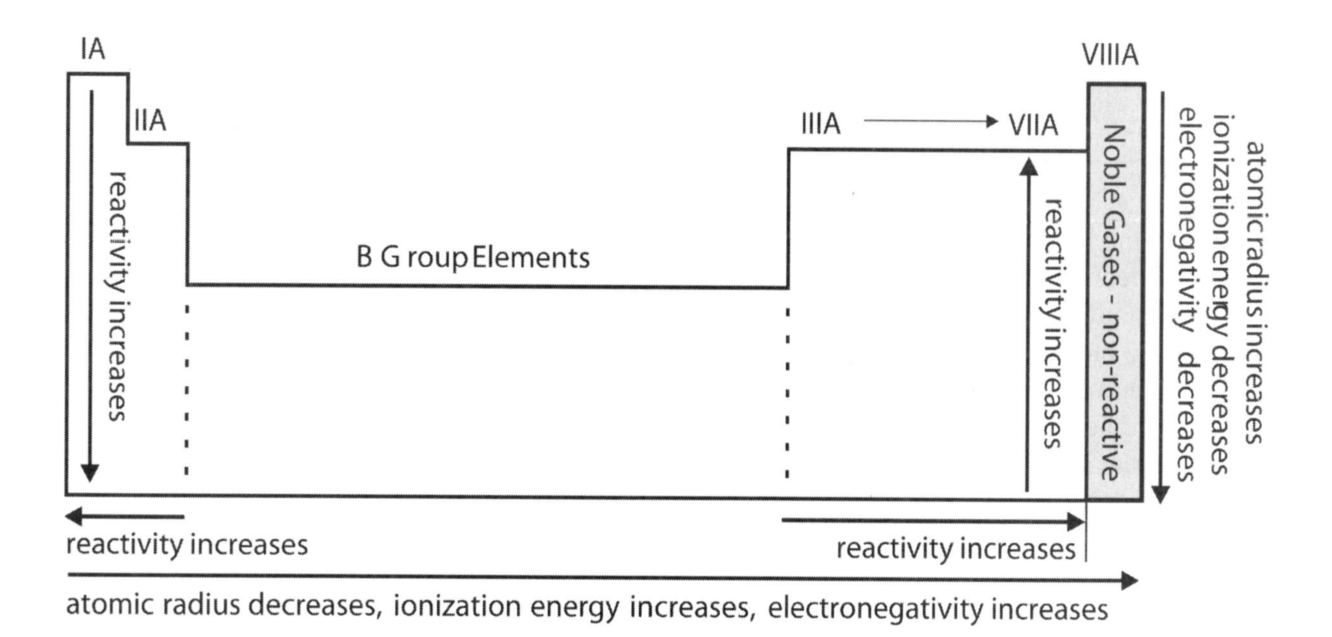

Figure 5.8 Trends of the Elements in the Periodic Table

Section Review 3: Reactivity of Elements in the Periodic Table

A. Define the following terms.

valence electron	alkali metals	transition metals
ionization energy	ion	anion
electron affinity	cation	halogens
chemical reactivity	alkaline earth metals	

B. Choose the best answer.

1. Choose the valence shell configuration of sulfur (S).

   A. $3s_2$, $3px_2$, $3py_1$, $3pz_1$ $3s_2$, $3px_1$, $3py_1$, $3pz_1$

   B. $3s_2$, $3px_2$, $3py_2$, $3pz_1$ $3s_2$, $3px_2$, $3py_2$, $3pz_2$

2. The element of oxygen appears in the periodic table as shown below. An oxygen ion would most likely have what charge?

   8
   O
   Oxygen
   15.9994
   2,6

   A. +1
   B. +2
   C. –1
   D. –2

3. Given the following set of elements as found in the periodic table, which 2 elements would have the most similar chemical properties?

3 Li Lithium 6.941g 2,1	4 Be Beryllium 9.01218 2,2
11 Na Sodium 22.9898 2,8,1	12 Mg Magnesium 24.305 2,8,2

   A. lithium and beryllium
   B. lithium and sodium
   C. sodium and beryllium
   D. sodium and magnesium

4. Which of the following statements is ***not*** true of noble gases?

   A. Except for helium, they have 8 electrons in their outer subshell.
   B. They do not react readily with other elements.
   C. They usually exist as ions.
   D. They are in Group 18 (VIIIA).

C. Answer the following questions.

1. What does an atom become when it gains or loses an electron?
2. Which group/family of elements is the most stable? Why?
3. Look at the following block of atoms as found in the periodic table. Which of the elements is most reactive and why?

7	8	9	10
N	O	F	Ne
Nitrogen	Oxygen	Fluorine	Neon
14.0067	15.9994	18.998403	20.179
2,5	2,6	2,7	2,8

4. If an atom gains two electrons, what is the charge of the resulting ion?
5. What chemical characteristic do elements in a group/family share?
6. Why do alkali metals become more reactive as you move down the periodic table?

Bonding of Atoms

A **compound** is a substance composed of two or more atoms joined together. A molecule is the smallest particle of a chemical compound that retains the properties of that compound. Atoms of different elements can combine chemically to form molecules by sharing or by transferring **valence electrons**. Valence electrons are either lost, gained, or shared when forming compounds.

Ionic Bonds

An **ion** is an atom with a charge. It is formed by the *transfer* of electrons. When one atom "takes" electrons from another atom both are left with a charge. The atom that took electrons has a negative charge. (Recall that electrons have a negative charge); the atom that "gave" electrons has a positive charge. The bond formed by this transfer is called an **ionic bond**. Ionic bonds are very strong. Ionic compounds have high melting points and high boiling points. These compounds tend to have ordered crystal structures and are usually solids at room temperature. Ionic compounds will usually dissolve in water, and they have the ability to conduct electricity in an aqueous (dissolved in water) or a molten state.

Aluminum oxide is an example of a compound with an ionic bond. In aluminum oxide, two atoms of aluminum react with three atoms of oxygen. The two aluminum atoms give up three electrons each to form positive ions with +3 charges. The three oxygen atoms gain two each of the six electrons given up by the two aluminum atoms to form negative ions with –2 charges. Figure 5.9 illustrates this electron transfer. Note that the orbital shape (circular) has been simplified for clarity.

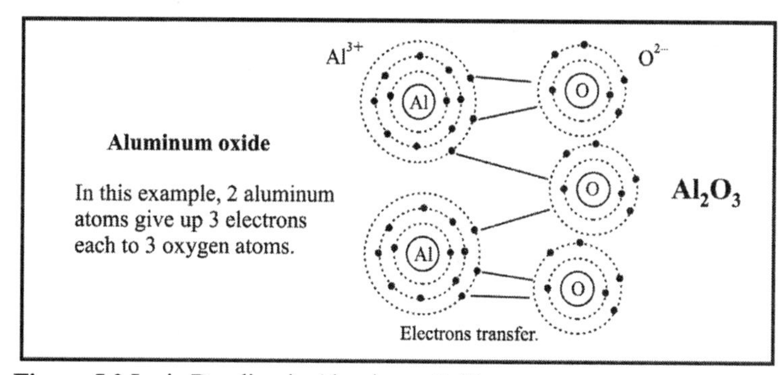

Figure 5.9 Ionic Bonding in Aluminum Oxide

COVALENT BONDS

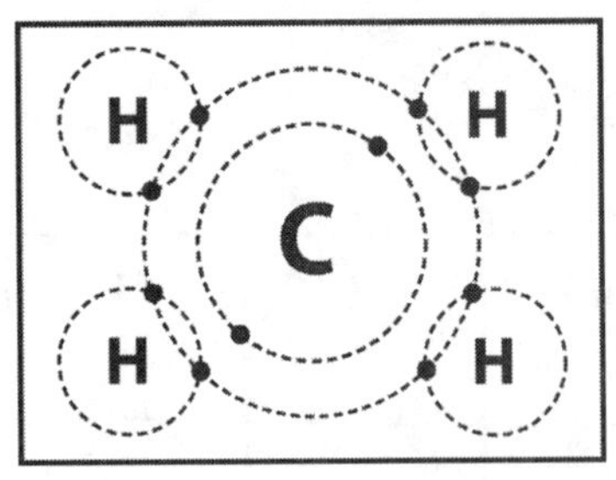

Figure 5.10 Methane Molecule (CH_4)

Covalent bonds are formed when two or more elements *share* valence electrons to create a more stable outer electron structure. Covalent bonds are weaker than ionic bonds. Covalent bonds are usually present in compounds formed by one or more non-metal atoms. Covalent compounds have low melting points and low boiling points. They tend to be brittle and do not conduct electricity well.

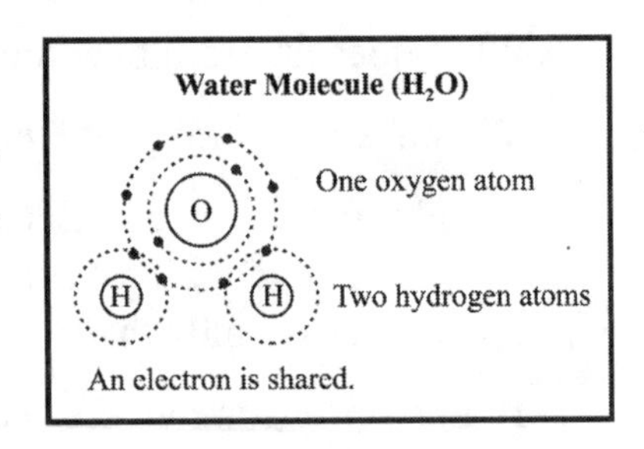

Figure 5.11 Water Molecule

A good example of covalent bonding is the organic (carbon-based) molecule, methane (CH_4). Each of the four participating hydrogen atoms shares 1 electron with a single atom of carbon. Carbon, in turn, shares its four valence electrons — one is shared with each of the four hydrogen atoms.

Water is another example of covalent bonding. Two hydrogen atoms and one oxygen atom combine to form one molecule of water. Figure 5.11 shows how the atoms in water share electrons. However, oxygen has a greater **electronegativity** than hydrogen; this means that it draws electrons away from hydrogen and toward itself. This lends an "ionic character" to the bond that will be described in the next section.

POLAR COVALENT AND HYDROGEN BONDS

The covalent bond will sometimes have an "ionic character," depending on the identity of the atoms involved in the bonding. This means that one of the two atoms participating in the bond "wants" electrons more than the other, and thus pulls them closer; this atom has a partially negative charge symbolically shown as δ^-.The other atom, which has allowed its electrons to be pulled away a bit, has a partially positive charge symbolically shown as δ^+. These molecules are called polar molecules, and water is an excellent example. The oxygen atom in water pulls electrons toward it; the hydrogen atom is left with a partially positive charge.

The presence of these partial charges creates an electrical attraction between polar molecules: the partially positive end (that is, the hydrogen) lines up in such a way that it is close to the partially negative end (the oxygen) of another polar molecule. The resulting orientation is highly stabilizing, and powerful enough to be called a form of bonding: hydrogen bonding. Hydrogen bonds are not as powerful as covalent bonds, but are highly stabilizing and represent a significant organizing force.

Section Review 4: Bonding of Atoms

A. Define the following terms.

ion	ionic bond	molecule
compound	covalent bond	hydrogen bond
valence electrons	polar molecule	electronegativity

B. Choose the best answer.

1. A covalent compound has which of the following characteristics?
   A. high melting and high boiling points
   B. atoms share electrons to bond
   C. conducts electricity
   D. all of the above

2. Hydrogen bonding takes place between
   A. polar molecules.
   B. protons.
   C. ionic compounds.
   D. valence electrons

3. What type of bond is formed when atoms transfer electrons?
   A. covalent
   B. hydrogen
   C. ionic
   D. polar

C. Answer the following questions.

1. Compare and contrast ionic, covalent and hydrogen bonds. Which is strongest? Which is the most flexible? What kind of bonding do you think is found in a crystal of table sale (NaC1)? How about octane (C_8H_{18}, a primary component of gasoline)? How about in ethanol (C_2H_5OH)?

2. Look at Figure 5.9. How many valence electrons do A1 and O each have? How many total electrons?

Chapter 5 Review

Choose the best answer.

1. Look at the element of fluorine as it appears in the periodic table shown below. How many neutrons are in the nucleus of most fluorine isotopes?

9 F Fluorine 18.998403 2,7

A. 9
B. 10
C. 18
D. 19

2. Why are Group 16 atoms extremely reactive?

   A. They want to gain one electron to become stable.
   B. They want to gain two electrons to become stable.
   C. They want to lose one electron to become stable.
   D. Their outer shell is full of electrons.

3. Look at the following blocks of atoms as found in the periodic table. Which element shown below would be most reactive?

11 Na Sodium 22.9898 2,8,1	12 Mg Magnesium 24.305 2,8,2
19 K Potassium 30.0983 2,8,8,1	20 Ca Calcium 40.08 2,8,8,2

A. sodium
B. potassium
C. magnesium
D. calcium

4. Look at the following blocks of elements as they appear in the periodic table. Which two elements would have the most similar chemical properties?

9 F Fluorine 18.998403 2,7	10 Ne Neon 20.179 2,8
17 Cl Chlorine 35.453 2,8,7	18 Ar Argon 39.948 2,8,8

A. fluorine and chlorine

B. fluorine and neon

C. fluorine and argon

D. chlorine and neon

5. Which of the following is a characteristic of an ionic bond?

A. low melting point

B. shares electrons

C. good conductor of electricity in aqueous solution

D. insoluble

6. Which of the following molecules is most likely to have a covalent bond?

A. O_2 B. NaC1 C. MgO D. Fe_2O_3

7. Look at the following block of atoms as they appear in the periodic table. Which of the elements is most reactive?

A	B	C	D
15 P Phosphorus 30.97376 2,8,5	16 S Sulfur 32.06 2,8,6	17 Cl Chlorine 35.453 2,8,7	18 Ar Argon 39.948 2,8,8

8. Elements in the same group/family have similar chemical properties because

A. their electrons are inside the nucleus.

B. they have the same number of electrons in their valence shell.

C. they have the same number of neutrons in their nucleus.

D. they have similar atomic radii.

9. Which group/family of elements contains 8 electrons in its valence shell?

A. noble gases

B. metals

C. non metals

D. metalloids

10. An ionic bond results from the transfer of electrons from

    A. one orbital to another within the same atom.

    B. a valence shell of one atom to a valence shell of another atom.

    C. the valence shell of one atom to the nucleus of another atom.

    D. the nucleus of one atom to the nucleus of another atom.

11. The element magnesium, Mg, has 12 electrons. In which energy level will its valence electrons be found?

    A. first

    B. second

    C. third

    D. fourth

12. Which of the following statements correctly describes compounds containing covalent bonds?

    A. Covalent compounds have high melting points.

    B. Covalent compounds conduct electricity well.

    C. Covalent compounds have high boiling points.

    D. Covalent compounds tend to be brittle solids.

13. How many bonds does carbon usually form when part of an organic compound?

    A. 1　　B. 2　　C. 3　　D. 4

Chapter 6
Nuclear Processes

PHYSICAL SCIENCE STANDARDS COVERED IN THIS CHAPTER INCLUDE:

SPS3 Students will distinguish the characteristics and components of radioactivity.

RADIOACTIVITY, FISSION, AND FUSION

ELEMENTS AND ATOMIC NUMBER

In this chapter, we are going to look at nuclear processes. Let's begin by defining how the nucleus of an atom helps determine that atom's identity as an element.

Atoms are made up of subatomic particles, including the positively-charged proton, the neutral neutron and the negatively-charged electron. The proton and neutron are located in the nucleus of the atom. The electron is located outside the nucleus. **Elements** consist of groups of atoms that have identical numbers of protons. All atoms with a given number of protons are representatives of one particular element, in one form or another. Here arc a few examples:

-All carbon (C) atoms have 6 protons.

- All silver (Ag) atoms have 47 protons.

- All uranium (U) atoms have 92 protons.

So, the number of protons is a defining quantity of an atom. As you will see, the number of electrons is another defining feature of an atom.

The number of protons in the nucleus of an atom is called the **atomic number (Z)** of the atom. The atomic number also corresponds to the number of electrons in the same atom if the atom is neutral. For example, the atomic number of carbon is 6. Thus, a neutral carbon atom contains 6 protons and 6 electrons.

The **mass number (A)** is the number of protons plus the number of neutrons found in the nucleus of the atom. Atoms of the same element do not always have the same number of neutrons. Atoms of an element that have different numbers of neutrons are called **isotopes**.

In other words, isotopes are atoms that have the same atomic number but different mass numbers. For example, carbon atoms always have 6 protons and 6 electrons, but they can have 6, 7, or 8 neutrons. In general, isotopes of an element (X) are denoted using the following form:

$$^{A}_{Z}X$$

where A is the mass number and Z is the atomic number. Therefore, the isotopes of carbon mentioned above are written as:

$^{12}_{6}C$	$^{13}_{6}C$	$^{14}_{6}C$
carbon-12	carbon-13	carbon-14

NUCLEAR FORCE

You know that opposites attract — that is, objects with opposite charges are drawn to each other. Likewise, objects with different charges repel each other. That is called **electrostatic force**. The nucleus of an atom contains positively-charged protons (as well as neutrons, which have no charge); all of these subatomic particles are packed closely together. Why, then, do the protons not repel one another and cause the nucleus to blow apart? The answer is that a force stronger than the electrostatic force holds the nucleus together: the nuclear force. This **nuclear force** is an attractive force that acts between the protons and neutrons at the very short distances between these particles. As long as this nuclear force is strong, an atom's nucleus is stable. If this nuclear force is small, an atom becomes **radioactive**. Radioactive decay occurs when an atom loses protons or neutrons. Nuclear force may be considered the strongest of nature's forces since under ordinary circumstances nothing separates the nucleus of stable atoms.

RADIOACTIVITY

As mentioned in the previous section, **isotopes** are atoms of the same element with different numbers of neutrons. The nucleus of an atom can be unstable if there are too many neutrons for the number of protons. An unstable nucleus is **radioactive**, and unstable isotopes are called **radioactive isotopes**. All elements with atomic numbers greater than 83 are radioactive. Figure 6.1 shows a model of a helium (He) atom, which has 2 protons, 2 neutrons, and 2 electrons. The largest radioactive emission is an alpha particle; the x particle is very similiar to a helium atom, except that it has no electrons. Radioactive atoms give off radiation in the form of alpha particles, beta particles, and gamma rays.

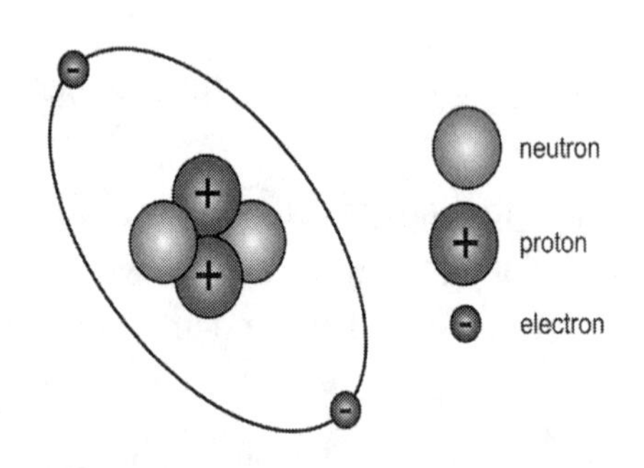

Figure 6.1 Model of a helium atom

An **alpha particle** is a helium nucleus with 2 protons and 2 neutrons, as shown in Figure 6.1. When an alpha particle is released from the nucleus of an atom, the atomic number of the parent nucleus is reduced by two. Alpha particles cannot penetrate a piece of paper or even a thin layer of cloth. However, if ingested, they will do more damage to internal tissue than other forms of radiation.

Beta particles are electrons emitted by an unstable atom. Beta particles are more penetrating than alpha particles. However, lead is capable of stopping them. **Gamma rays** are high energy X-rays, and only thick lead or concrete can stop them.

Table 6.1 Radioactive Particles

Radiation	Symbol	Particles/Waves	Electric Charge	Energy	Energy stopped by
Alpha particle	α	2 protons, 2 neutrons	positive	low	a piece of paper
Beta particle	β	1 electron	negative	medium	lead 1 cm thick
Gamma rays	γ	wave of energy	no charge	high	thick lead or concrete

A radioactive atom that emits an alpha particle, beta particle, or gamma ray is going through a process of **radioactive decay**. Radioactive decay causes an atom of one element to become a different element by reducing its atomic number.

Each isotope decays in its own characteristic way. It will emit α particles, β particles and/or γ rays in a particular order, over a particular period of time. The amount of time that it takes for ½ of the atoms of a radioactive sample to decay is called the **half life** of the isotope. For instance, radium-226 has a half-life of 1,602 years. Let's say a sample of 10 grams of ^{226}Ra is placed in a weighing dish and left in a locked vault. After 1,602 years, the vault is opened. How much ^{226}Ra is in the weighing dish now? That's right, only 5 grams remains. One half of the sample has decayed to something else. But what? That is where it becomes important to know *how* the isotope decayed.

Radium-226 decays by alpha particle emission, as shown in the following equation.

$$^{226}_{88}\text{Ra} \longrightarrow ^{222}_{86}\text{Rn} + ^{4}_{2}\alpha$$

By releasing an alpha particle, the radium-226 atom has lowered its energy and transformed itself into a radon-222 atom.

So, you have seen that unstable nuclei can emit an α particle, β particle, or γ ray to become more stable. However, there is another way for an unstable nucleus to lower its energy: the process of nuclear fission.

FISSION

Fission occurs when the nucleus of an atom that has many protons and neutrons becomes so unstable that it splits into two smaller atoms. Fission may be spontaneous or induced.

Spontaneous fission is a natural process that occurs mostly in the transactinide elements, like rutherfordium (Rf). However, some of the actinides (which are a little bit lighter than the transactinides) decay partially by spontaneous fission, including isotopes of uranium (U) and plutonium (Pu). For example, a ^{235}U atom has 92 protons and 143 neutrons. When it fissions, it may split into a krypton atom and a barium atom, plus 2 neutrons, as shown in the following equation and in Figure 6.2.

$$^{235}_{92}\text{U} \longrightarrow ^{94}_{36}\text{Kr} + ^{139}_{56}\text{Ba} + ^{1}_{0}\text{n}$$

The process of spontaneous fission wasn't well-known or understood until fairly recently. In fact, it was only discovered as a by-product of the investigation into induced fission. **Induced fission** is the process of firing neutrons at heavy atoms, to induce them to split. It was first investigated by Enrico Fermi in the 1930s. The theory was proven in 1939, with the discovery by Lise Meitner and Otto Frisch that the use of neutron projectiles had actually caused a uranium nucleus to split into two pieces, exactly as shown in Figure 6.2 (except that more neutrons were emitted). Meitner and Frisch named the process nuclear fission. Fermi proceeded to co-invent the first nuclear reactor. This design led to the invention of nuclear reactors found in nuclear power plants, as well as nuclear bombs.

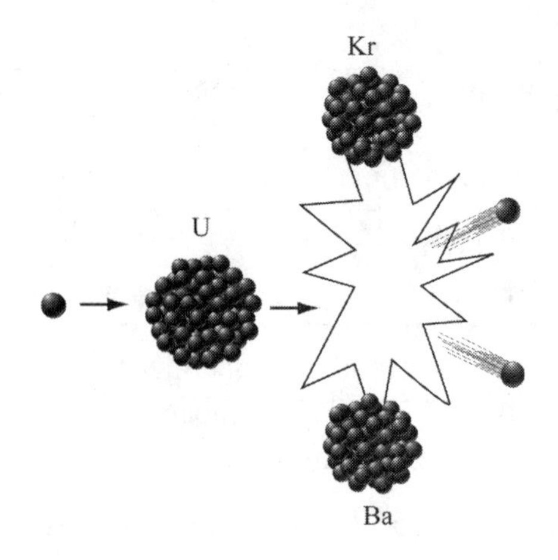

Figure 6.2 Induced Fission

FUSION

During this time another nuclear process was being investigated: nuclear fusion. **Fusion** is the exact opposite of fission, involving the joining (fusing) of two small atoms to form one larger atom. Fusion reactions occur in the sun (and other stars), where extremely high temperatures allow hydrogen isotopes to collide and fuse, releasing energy. In 1939, Hans Bethe put forth the first quantitative theory explaining fusion, for which he later won the Nobel Prize.

The most commonly cited fusion reaction involves the fusing of deuterium (^{2}H) and tritium (^{3}H) to form a helium nucleus and a neutron, as shown in Figure 6.3.

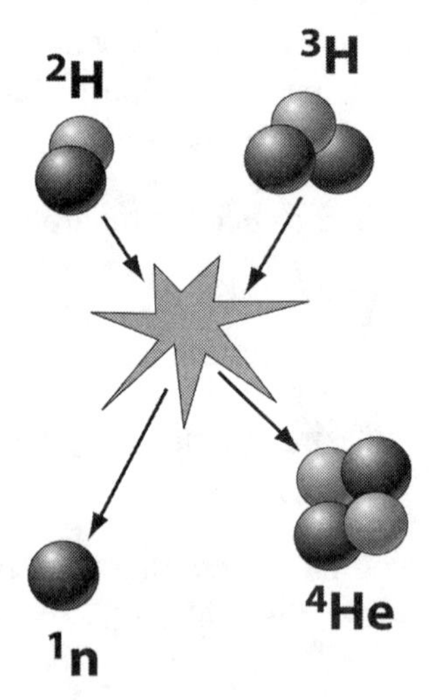

Figure 6.3 Fusion

Section Review 1: Radioactivity and Fission

A. Define the following terms.

element	radioactive isotope	radioactive decay
atomic number	gamma ray	half-life
mass number	alpha particle	fission
isotope	beta particle	

B. Choose the best answer.

1. What are isotopes?
   A. elements with the same mass number A
   B. elements with the same atomic mass
   C. elements with the same number of protons and electrons but different number of neutrons
   D. elements with the same number of protons and neutrons but a different number of electrons

Refer to the graph at right to answer questions 2 – 3.

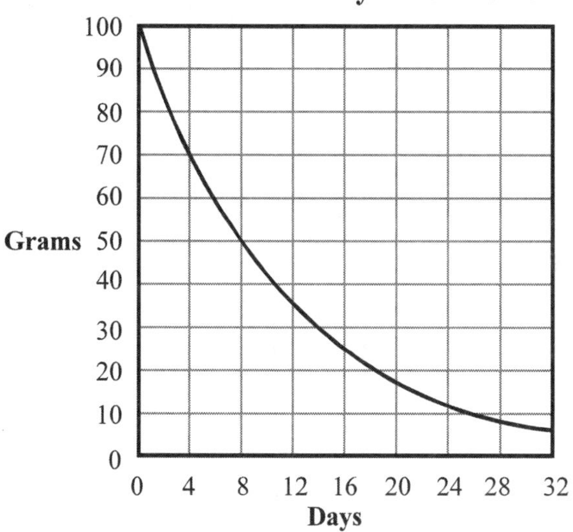

2. What is the half-life of iodine-131?
   A. 1 day
   B. 8 days
   C. 4 days
   D. 16 days
3. How long does it take for 100 grams of iodine-131 to decay to 25 grams of iodine-131?
   A. 1 day
   B. 8 days
   C. 16 days
   D. 24 days
4. Where would you find a nuclear fusion reaction occurring?
   A. in a nuclear reactor
   B. in the sun
   C. in an X-ray machine
   D. in a microwave oven
5. Palladium-100 has a half-life of 4 days. If you started with 20 grams of palladium-100, how much would remain after 12 days?
   A. 10 grams
   B. 0 grams
   C. 5 grams
   D. 2.5 grams
6. Which of the following is an example of technological design?
   A. spontaneous fission
   B. induced fission
   C. nuclear reactor
   D. fusion

7. Which of the following is an example of a scientific investigation?
   A. spontaneous fission
   B. induced fission
   C. nuclear reactor
   D. Both A & B

8. Uranium-238 has 92 protons and 146 neutrons. It undergoes radioactive decay by emitting an alpha particle. What element is the product of this decay?
   A. an isotope of uranium having 92 protons and 144 neutrons
   B. an ion of uranium having 92 protons and 91 electrons
   C. the element of neptunium, which has 93 protons and 144 neutrons
   D. the element of thorium, which has 90 protons and 144 neutrons

C. Answer the following questions.

1. The half-life of technetium-99 is 6 hours. If you began with 100 grams of Tc-99, graph the radioactive decay curve showing the amount of Tc-99 versus time. Label the *x*-axis with appropriate time intervals, and label the *y*-axis with appropriate masses.
2. Describe the contributions of the following people to the development of our nuclear understanding and technology.
   A. Lise Meitner and Otto Frisch
   B. Enrico Fermi
3. Consider the element lithium with an atomic number of 3 and an atomic mass of around 7, and compare it to the element einsteinium, which has an atomic number of 99 and an atomic mass of around 254. Which of these two elements would have the lesser nuclear force? Why?

Use the figure to answer the following questions.

4. Does carbon decay by α-particle emission?
5. An initial sample of 30 grams of ^{14}C is allowed to decay for 11,460 years. How much does the sample weigh at that time?

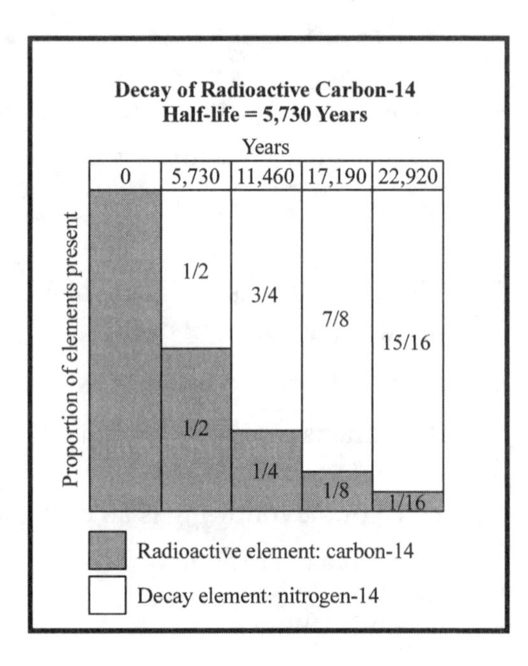

INDUCED FISSION

As you know, **induced fission** is the process of bombarding radioactive atoms with neutrons to cause them to split apart. What are the products of these processes? There are several, including fission fragments, neutrons, and energy. First, let's look at the fission fragments and neutrons.

FISSION FRAGMENTS AND NEUTRONS

Look at Figure 6.2 on Page124. It has been illustrated to simplify the fission process and depicts an atom of uranium always splitting into krypton and barium. In actuality, however, the nuclear products are much more diverse, as shown in Figure 6.4.

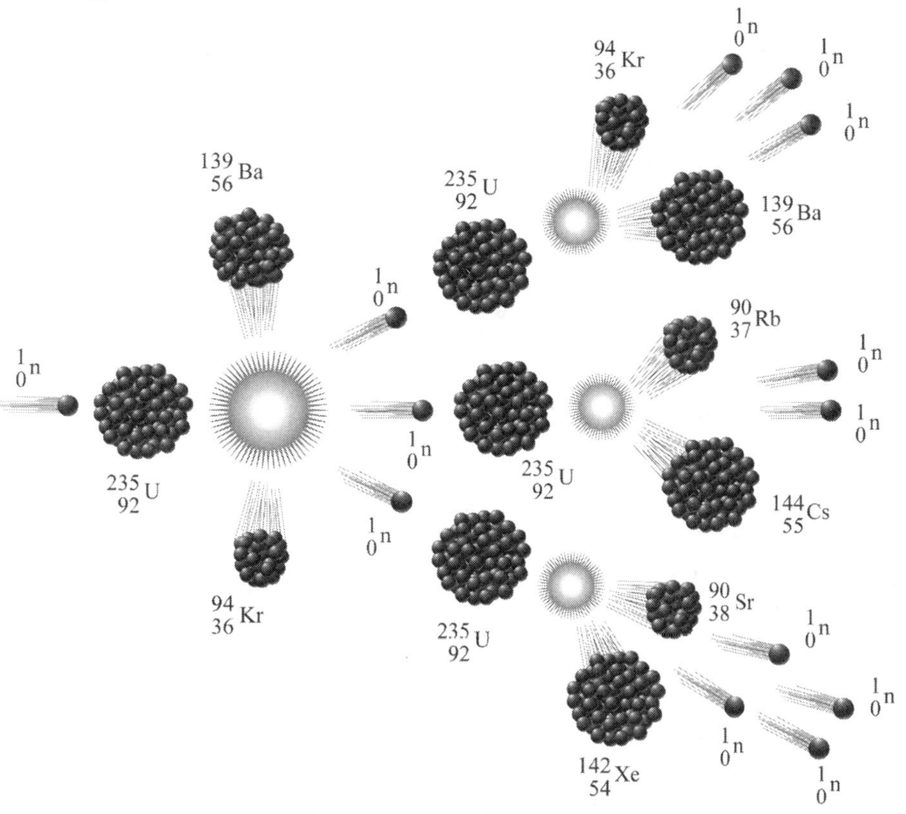

Figure 6.4 Production of Fission Fragments

Besides barium and krypton, rubidium, cesium, strontium, and xenon are also produced. These **fission fragments** are the product of an atom split by neutron bombardment. The average mass of ^{235}U fission fragments is 118, but as Figure 6.5 indicates, a fragment of mass 118 is rarely detected. Instead, ^{235}U tends to split into uneven fragment masses around 95 and 137. To see this in another way, look again at the mass numbers and atomic numbers of the fragments shown in Figure 6.4.

Each of the fission fragments is an isotope with a half-life of its own, which may range from seconds to millions of years. As the half-life of each isotope passes, the isotope decays by emitting one or more forms of radiation, like alpha and beta particles or gamma rays. The result is a new isotope called a **daughter**, which may or may not be **stable**. If the atom is **unstable** (meaning that it is still radioactive), it will decay to yet another isotope. If the atom is stable, it will remain as it is, with no further transformation. The succession of decays is called a **decay chain**.

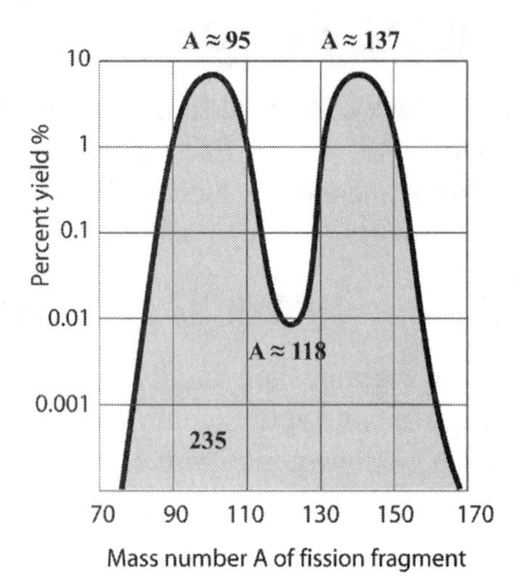

Figure 6.5 Fission fragment mass distribution

One common pair of fragments is xenon and strontium. The fission is illustrated by the following reaction.

$$^{235}U + n \rightarrow {}^{236}U^* \longrightarrow {}^{140}Xe + {}^{94}Sr + 2n$$

U-236 has an asterisk (*) because it only lasts a moment after absorbing the neutron. The forces within the nucleus redisstribute themselves allowing for the fission decay, in this case to xenon and strontium isotopes. Xenon-40 is a highly radioactive isotope with a half-life of 14 seconds. It undergoes a series of decays, finally ending with cerium-140. Strontium-94, with a half-life of 74 seconds, decays by beta emission to yttrium-94. Let's look at a partial decay chain of those isotopes, as in Figure 6.6.

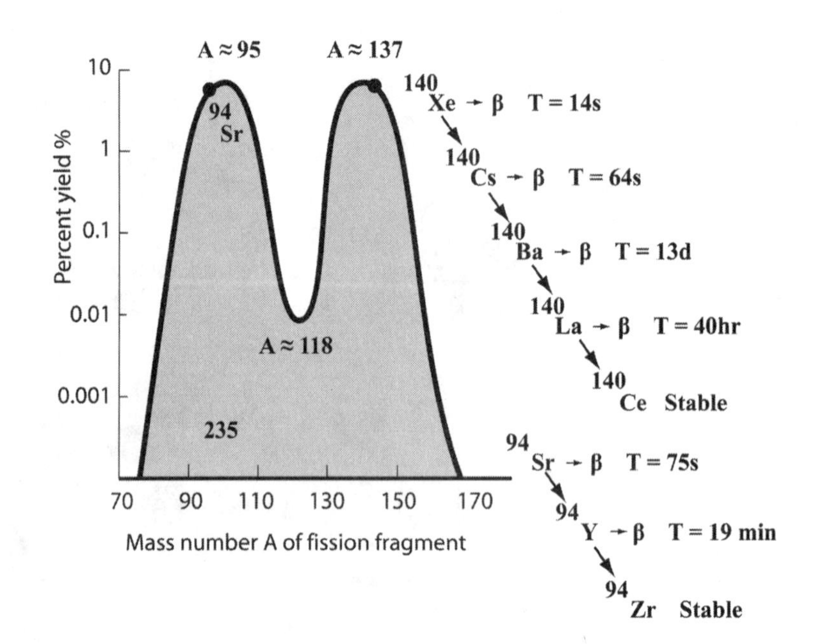

Figure 6.6 Decay Chain

These decay chains show only beta (β) emissions, but other fission fragments may have other types of radiative decay. Even different isotopes of xenon and strontium will decay differently. For instance, we have been looking at Xe-140 and Sr-94. Look back at Figure 6.4. See that the isotopes of strontium and xenon noted there are Xe-142 and Sr-90. The reaction that produces those fragments would be:

$$^{235}U + n \rightarrow {}^{236}U^* \longrightarrow {}^{142}Xe + {}^{90}Sr + 4n$$

These isotopes will have a different decay chain than that illustrated by Figure 6.6. Try going to the Internet to find their decay chains. Surf around the Brookhaven National Lab's national nuclear database at http://www.nndc.bnl.gov.

Nuclear Energy

When a nucleus splits, or fissions, a great deal of energy is also released. In fact, the scientific world was surprised by how much energy was generated. Neils Bohr, the Danish physicist who first modeled the atom, wrote to Lise Meitner to comment on how unexpectedly large the energy release was. Much larger than calculations had predicted, it turns out. Up until this point, fission research had been performed simply to understand more about the atom. Now, though, the stakes began to rise: a new energy source had been found.

How much energy are we talking about, though? The creation of **nuclear energy** in the fission and fusion reactions requires only a small amount of matter. After all, an atom is a very small amount of matter. Einstein's famous **mass-energy equation**, **$E = mc^2$**, states this fact very simply. Einstein's equation, written in a simpler form, is Energy = mass × speed of light × speed of light. Since the speed of light is 3×10^8 meters per second and this term is squared, we can still have a very small amount of matter and end up with a large amount of energy. A nuclear fission reaction, utilizing one U-235 atom, will produce 50 million times more energy than the combustion (burning) of a single carbon atom.

Today we use nuclear reactors to harness this power for the production of electricity. Figure 6.7 shows the process. Fissile material, like uranium-235, is manufactured into pellets that are bound together into long rods, called **fuel rods**. Fuel rods are bundled together with **control rods** and placed in the reactor core. Here is what happens.

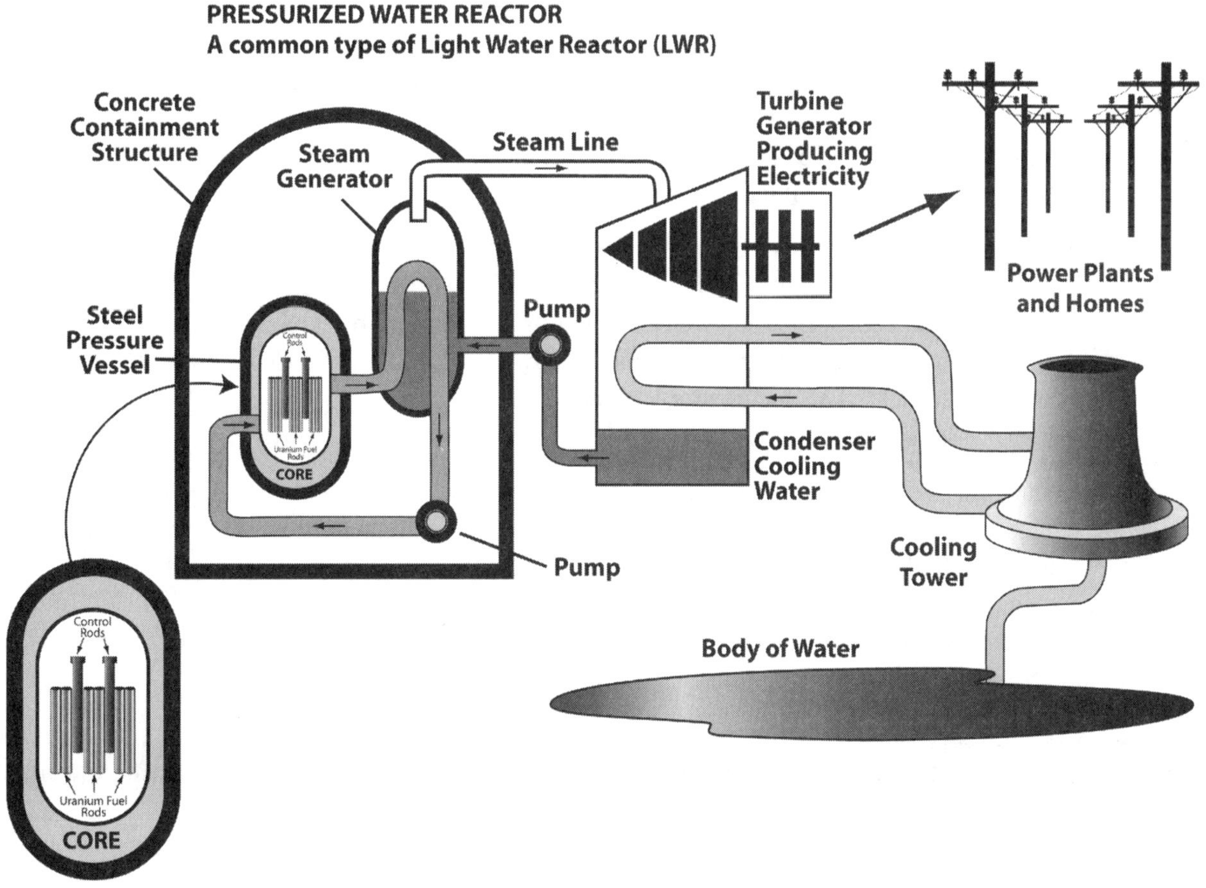

Figure 6.7 Nuclear Power Reactor

Nuclear Power Plant

Figure 6.8

Every time a uranium atom fissions, it releases more neutrons, which cause another atom to fission. This process is called a **chain reaction**, and it produces energy in the form of heat. The water surrounding the reactor core gets very hot. It is pumped through another water tank by way of a pipe. Note that the water from the core never touches any reactor component outside the core. The continuous pumping of the super-hot water from the reactor core heats the secondary tank water to produce a continuous supply of steam. The steam turns the turbines of a generator, which generates electricity. The steam is then diverted to a cooling tower. This structure, as shown in Figure 6.8, is commonly associated with nuclear power plants, though it actually has nothing to do with energy production. It is where the steam condenses and cools before its release into a body of water, like a river.

ENERGY AND ENVIRONMENTAL CONSEQUENCES

Nuclear power is a very attractive energy option because it is clean and cheap. However, as with every energy option, there are environmental consequences. The two that are the most important to prevent are **supercriticality** and **environmental contamination** from general operation.

SUPERCRITICALITY

Inside a nuclear power reactor, uranium fuel is used to create energy. Long fuel rods formed of small U-235 pellets are arranged into bundles and submerged in some coolant, usually water. In order for the reactor to work, the submerged bundles of rods must be *slightly* **supercritical**. This means that, left to its own devices, the uranium in the rods would eventually overheat and melt.

To prevent this, control rods are used. **Control rods** are made of a material, like cadmium, that absorbs neutrons. Inserted into the bundle of uranium, control rods allow operators to control the rate of the nuclear reaction. If more heat is needed, the rods are raised out of the uranium bundle. To lessen heat, the rods are lowered into the uranium bundle. The rods can also be lowered completely into the uranium bundle to shut the reactor down in the case of an accident or to change the fuel.

These control rods are the safeguard of the power plant. Without them, true supercriticality could be reached. Were this to happen, the uranium would melt the reactor core, causing a **breech** (a crack or hole) and subsequent release of radioactive isotopes, encased in superheated steam and melted metals. Depending on the scale of the breech, this could be an environmental disaster. The Three Mile Island accident in the US was not a disaster; very little radioactivity was released. It was a warning, however, for the U.S. to increase safety and maintenence precautions. The Chernobyl accident in the former Soviet Union was a disaster and one that the region has yet to recover from.

The issue of environmental impact must be studied whenever an effort to produce energy is planned. Drilling for oil, damming rivers and erecting windmills all have environmental impacts. These must be weighed against the value of the energy produced and the ultimate cost of failure.

Environmental Contamination

We have noted that Sr-90 is one product of the induced fission of U-235. This isotope of strontium has an intermediate half-life of around 30 years. This is a difficult time span for environmental contaminants. If you are asking "why?", consider this. A short half-life of minutes, days, or weeks indicates that the contaminant will be gone (decay) quickly and not have a chance to do much damage. In addition, strontium mimics the properties of calcium. Look where it is on the periodic table, in the same group as calcium. This means that strontium is taken up by living organisms that utilize calcium; those organisms incorporate Sr-90 into their bones. There the Sr-90 decays, emitting radiation that can cause cancer. While strontium is *very unlikely* to enter the environment from a nuclear power reactor, it is one of several isotopes that would have a negative environmental effect if released. In addition to the normal security and operational controls of a nuclear power plant, the area surrounding the reactor must be continually monitored to ensure that no such release has occurred.

Another, more pressing, example of environmental contamination is the issue of radioactive waste. Remember that many different kinds of radioactive isotopes, each of which decays in a different way, are the result of the fission of ^{235}U. This occurs *within the core of the reactor*; during normal operation, no radioactive components come in contact with any other part of the facility. However, a reactor core does not last forever; periodically, fuel rods and control rods must be replaced to maintain optimal function of the reactor. The spent rods still contain a great deal of radioactive material, mostly from the still-decaying daughters of the fission fragments.

The processing of this waste, to separate and neutralize the individual components, is not always possible or feasible. At present, there is no ideal storage solution for this waste. In order to avoid contamination, it must be stored in a highly absorbing material and allowed to decay in a location that will remain secure for many years. Yucca Mountain (NV) is the prospecive site for nuclear waste storage in this country. Other countries, like France, almost completely reprocess their nuclear waste; this leaves little need for waste storage.

Ongoing Research

Three kinds of research are being performed that may revolutionize the way nuclear processes are used in power production.

1. New fission reactor designs are now under construction that make nuclear power even cheaper, safer, and more efficient.
2. New waste re-processing technologies are being investigated to help us deal with dangerous and long-lived nuclear waste.
3. Fusion reactors are still being investigated. Fusion reactions, as described earlier in this section, produce a great deal of energy — potentially more than fission reactions. They have fewer reactants, fewer products, and produce little waste. Scientists are still trying to overcome the obstacle of the extremely high temperatures necessary for fusion to occur and sustain itself.

Keep an eye on these technologies, as well as other energy technologies. Remember, you will be paying the power bills one day soon.

Section Review 2: More About Nuclear Energy

A. Define the following terms.

spontaneous fission	stable	unstable	decay chain	control rod
induced fission	fission fragments	supercritical	breech	fuel rod

B. Choose the best answer.

1. In the following reaction, how many neutrons are produced?

$$^{235}U + n \rightarrow {}^{236}U^* \rightarrow {}^{90}Rb + {}^{144}Cs + \_\_\_\_\_\_$$

A. 1 B. 2 C. 3 D. 4

2. The spontaneous fission of californium-252 produces the isotopes barium and molybdenum. How many neutrons are produced in the following reaction?

$$^{252}Cf \rightarrow {}^{142}Ba + {}^{106}Mo + \_\_\_\_\_\_$$

A. 1 B. 2 C. 3 D. 4

3. Fission fragments are produced as a product of

A. fusion only.
B. fission only.
C. supercritical fission only.
D. A and B.

4. Which subatomic particle is used as a projectile to induce fission reactions?

A. the proton B. the neutron C. the electron D. the alpha particle

5. The sun is a good example of what kind of reactor?

A. a spontaneous fission reactor
B. an induced fission reactor
C. a fusion reactor
D. a supercritical fission reactor

6. Einstein's equation $E = mc^2$ gives the relationship between

A. energy and mass.
B. electron charge and mass.
C. the speed of light and mass.
D. the speed of light and electricity.

C. Answer the following questions.

1. Describe the use of control rods in a nuclear power reactor.
2. Search the terms "Three Mile Island" and "Chernobyl" on the Internet. From what you find, describe what happened and what the difference was in the two accidents.
3. Describe the environmental impact of nuclear power plants.
4. It was noted in this chapter that many different fission fragments are produced during a fission process. Does this have an impact on the handling of nuclear waste?
5. Nuclear fission reactions are used to make nuclear energy. Name one advantage and one disadvantage of using nuclear fission as an energy source.
6. A nuclear reactor uses fission to produce harnessed energy that we can use. A nuclear bomb produces a nuclear explosion of unharnessed energy. What is the difference between these two nuclear devices?
7. Why would a fusion reactor be more desirable than a fission reactor? Why are fusion reactors not used?
8. How is nuclear fission similar to nuclear fusion? How are these two types of nuclear reactions different?

Chapter 6 Review

Choose the best answer.

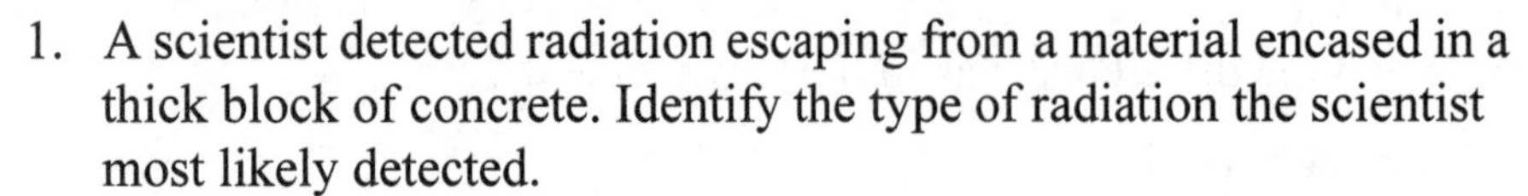

1. A scientist detected radiation escaping from a material encased in a thick block of concrete. Identify the type of radiation the scientist most likely detected.

   A. beta particles
   B. alpha particles
   C. gamma radiation
   D. high speed neutrons

2. Given 100.0 g of a radioactive isotope that has a half-life of 25 years, identify the amount of that isotope that will remain after 100 years.

   A. 50.0 g
   B. 25.0 g
   C. 12.5 g
   D. 6.3 g

3. The half-life of an isotope is the time required for half of the nuclei in the sample to undergo

   A. induced fission.
   B. spontaneous fission.
   C. fusion.
   D. radioactive decay.

4. Which of the following radioactive emissions is the most dangerous if ingested?

   A. α-particle
   B. β-particle
   C. X-ray
   D. microwave

5. Identify the element that CANNOT participate in nuclear fission reactions

   A. plutonium
   B. hydrogen
   C. uranium
   D. thorium

6. Identify the issue that has NOT been a factor in no new nuclear power plants having been built in over twenty years.

   A. construction costs
   B. political opposition
   C. availability of nuclear fuel
   D. disposal of radioactive by-products

7. Describe the reaction illustrated by:

$$^{3}H + ^{2}H \longrightarrow ^{4}He + ^{1}n$$

   A. spontaneous fission
   B. induced fission
   C. decay
   D. fusion

8. Which of the following is an appropriate material to use in making control rods?

   A. strontium
   B. cadmium
   C. calcium
   D. uranium

9. The following reaction shows the alpha decay of uranium-238 to thorium-234. The nuclear mass, in grams, is written beneath each nuclide symbol. What is the change in mass Δm for this reaction?

$$\underset{238.0003}{{}^{238}U} \longrightarrow \underset{233.9942}{{}^{234}Th} + \underset{4.00150}{{}^{4}He}$$

A. –0.0046 g   B. 0.0046 g   C. 8.0076 g   D. –8.0076 g

10. Every mass has an associated energy, and every energy has an associated mass. This is described by Einstein's equation $E=mc^2$. When the mass of a product set is different than the mass of a reactant set, what has happened to the mass?

A. It has been eliminated.

B. It has been transferred to another form.

C. It has been accelerated to the speed of light.

D. It has been accelerated to the speed of light, squared.

11. Complete the following equation. What nuclei belong in the blanks?

$${}^{1}n + {}^{235}U \longrightarrow {}^{136}I + \_\_\_\_ + \_\_\_\_ {}^{1}n$$

A. ${}^{96}Y$, 3   B. ${}^{94}Sr$, 4   C. ${}^{96}Y$, 4   D. ${}^{94}Sr$, 3

12. A decay chain ends when

A. the product nucleus decays to zero grams.

B. the product nucleus undergoes fission.

C. the product nucleus is stable.

D. the product nucleus undergoes fusion.

13. When a reaction is supercritical,

A. small amounts of neutrons are being produced.

B. large amounts of neutrons are being produced.

C. all the fission fragments in the core are unstable.

D. all of the fission fragments in the core are stable.

Chapter 7
Chemical Equations and Reactions

PHYSICAL SCIENCE STANDARDS COVERED IN THIS CHAPTER INCLUDE:

SPS2	Students will explore the nature of matter, its classifications and the system for naming types of matter.

UNDERSTANDING CHEMICAL FORMULAS

A **chemical formula** is a group of symbols that show the makeup of a compound. For example, the chemical formula for water is H_2O. The **subscript**, or small number, after the elemental symbol indicates the number of atoms of the element present in the compound. The chemical formula for water, H_2O, indicates that 2 atoms of hydrogen combine with 1 atom of oxygen. In aluminum oxide, Al_2O_3, 2 atoms of aluminum combine with 3 atoms of oxygen. In sodium chloride, NaCl, 1 atom of sodium combines with 1 atom of chlorine.

Practice Exercise 1: Chemical Formulas

For each chemical formula, give the ratio of atoms present for each element.

Formula	**Number of atoms of each element**
1. HCl	____________________
2. $MgCl_2$	____________________
3. H_2SO_4	____________________
4. C_2H_2	____________________
5. CO_2	____________________

A group of atoms can also combine in multiples. The group of atoms may be set off by parentheses in the chemical formula such as $Ca(OH)_2$ and $Al(NO_3)_3$. To find the number of atoms of each element, multiply the subscript number inside the parentheses by the subscript number outside the parentheses.

Example 1: $Ca(OH)_2$ contains 2 groups of OH atoms. Therefore, the number of atoms in a calcium hydroxide molecule equals 1 atom of calcium, 2 atoms of oxygen, and 2 atoms of hydrogen.

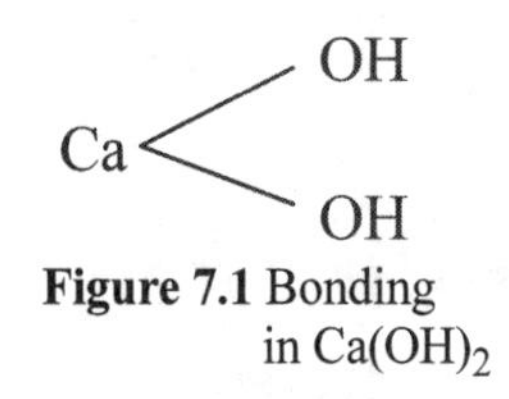

Figure 7.1 Bonding in $Ca(OH)_2$

Example 2: $Al(NO_3)_3$ has 1 atom of aluminum, 3 atoms of nitrogen, and 9 atoms of oxygen.

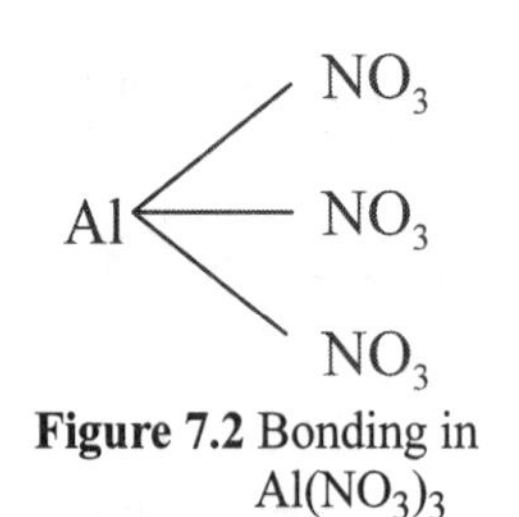

Figure 7.2 Bonding in $Al(NO_3)_3$

Practice Exercise 2: More Chemical Formulas

For each chemical formula, give the ratio of atoms present for each element.

Formula	**Number of atoms of each element**
1. $Mg(NO_3)_2$	________________
2. $Al_2(SO_4)_3$	________________
3. $Ba(OH)_2$	________________
4. $Ni(OH)_2$	________________
5. $Cr_2(C_2O_4)_3$	________________

Now that we know what elements make up each of these chemical formulas, let's try to name them. The **International Union of Pure and Applied Chemistry (IUPAC)** has systematized the naming of all chemical compounds. We will just look at diatomic and binary compounds for now, but you can always go to www.iupac.org to see more.

DIATOMIC MOLECULES

A diatomic molecule consists of two atoms of the same element, bound together. Oxygen, hydrogen and nitrogen, as well as all of the halogens form diatomic molecules as their most stable elemental form. Note these are all gases, as in $O_2(g)$, $H_2(g)$ and $C1_2(g)$.

Binary Compounds

A binary compound is one where two different elements have reacted to form a molecule. Some examples are CCl_4, HBr, NaCl and $FeCl_3$. Notice that the bonding of the binary compound may be covalent or ionic.

Binary Covalent Compounds

Covalent compounds are named using a few rules.

- Binary covalent compounds have two words in their name. Name them as their formulas are written, from left to right. The first word is simply the name of the element. For instance, in CCl_4, carbon (C) will be the first word.
- The second word is the name of the other atoms, with "ide" replacing the end of the element name. For instance, HBr would be called hydrogen bromide.
- Sometimes a prefix is needed to indicate how many of each element makes up the compound. Refer to Table 7.1 for the necessary prefixes. In CCl_4, carbon does not require the "mono" prefix; the only time that mono is used is when oxygen is being named. However, the four chlorides do require a prefix. The full, correct name is carbon tetrachloride.
- There are a few common molecules that are not named this way. Water is one of them. It is called water, rather than dihydrogen monoxide. Ammonia (NH_3) is another.

Number of Atoms	Prefix
1	mono- (use only for oxygen)
2	di-
3	tri-
4	tetra-
5	penta-
6	hexa-
7	hepta-
8	octa-

Table 7.1

Binary Ionic Compounds

- Binary ionic copies are named similarly to binary covalent compounds. The first word is the name of the first element in the formula. This will always be the cation (+ charge ion). In NaCl, for instance, the cation is Na^+ and the first word in the compound's name is sodium.
- The second word is the name of the anion (– charge ion). Since we are naming binary compounds, the anion is a single element, like chloride. So, NaCl is called sodium chloride.
- Some elements form cations that can have more than one possible charge. $FeCl_3$ and $FeCl_2$ are good examples. Iron in $FeCl_3$ has a +3 charge; iron in $FeCl_2$ has a +2 charge. At this point we would call both of them sodium chloride. (Notice that we do not distinguish a binary ionic compound by using the prefixes in Table 7.1; they are named by charge of the ions, not by the number of atoms.) If more than one charge state is possible, then the charge must be specified in the name. How do you know? Until you are more familiar with chemical formulas, you will have to use a reference like Figure 7.1 (on the next page).
- If you determine that the compound you are naming has more than one charge state, a Roman numeral is used to specify the charge state by the following formula..

$$\textit{Roman numeral} = -\frac{[(\textit{charge on anion}) \times (\textit{number of anions})]}{(\textit{number of cations})}$$

For $FeCl_3$, we get $-\frac{[(-1)\times(3)]}{(1)}=+3$, which means we use the Roman numeral (III).

The final name for $FeCl_3$ is then iron(III)chloride.

										13 Al			
21 Sc +3	22 Ti +2, +3, +4	23 V +2, +3, +4, +5	24 Cr +2, +3, +6	25 Mn +2, +3, +4, +7	26 Fe +2, +3	27 Co +2, +3	28 Ni +2, +3	29 Cu +1, +2	30 Zn +2	31 Ga +3	32 Ge +2, +4		
39 Y +3	40 Zr +4	41 Nb +3, +5	42 Mo +8	43 Tc +4, +6, +7	44 Ru +3	45 Rh +3	46 Pd +2, +4	47 Ag +1	48 Cd +2	49 In +3	50 Sn +2, +4	51 Sb +3, +5	
71 Lu +3	72 Hf +4	73 Ta +5	74 W +6	75 Re +4, +6, +7	76 Os +3, +4	77 Ir +3, +4	78 Pt +2, +4	79 Au +1, +3	80 Hg +1, +2	81 Tl +1, +3	82 Pb +2, +4	83 Bi +3, +5	84 Po +2, +4
103 Lr +3	104 Rf	105 Db	106 Sg	107 Bh	108 Hs	109 Mt							

57 La +3	58 Ce +3, +4	59 Pr +3	60 Nd +3	61 Pm +3	62 Sm +2, +3	63 Eu +2, +3	64 Gd +3	65 Tb +3	66 Dy +3	67 Ho +3	68 Er +3	69 Tm +3	70 Yb +3
89 Ac +3	90 Th +4	91 Pa +4, +5	92 U +3, +4, +5, +6	93 Np +3, +4, +5, +6	94 Pu +3, +4, +5, +6	95 Am +3, +4, +5, +6	96 Cm +3	97 Bk +3, +4	98 Cf +3	99 Es +3	100 Fm +3	101 Md +2, +3	102 No +2, +3

Figure 7.1

BASIC CHEMICAL EQUATIONS

A **chemical equation** expresses a chemical reaction. A **chemical reaction** is a process in which one or more elements or compounds (reactants) form new elements or compounds (products). The **reactants** are the starting substances. The **products** are the substances formed by the reaction. In a chemical equation, an arrow separates the reactants and the product. When reading a chemical equation aloud, you say that the reactants yield the products. The arrow represents the "yield" part of the equation. Many times, the chemical equation also contains information about the state of the reactants and products. The equation lists the physical states of the substances in parentheses. The right side of Figure 7.2 lists some of the common physical states and their abbreviations.

$NaOH(aq) + HCl(aq) \rightarrow NaCl(aq) + H_2O(l)$ **reactants** **yield** **products**	(aq) aqueous (dissolved in water) (g) gas (l) liquid (s) solid

Figure 7.2 Example of a Chemical Equation

Section Review 1: Understanding Chemical Formulas and Equations

A. Define the following terms.

symbol	chemical equation	reactants
chemical formula	chemical reaction	products

B. Choose the best answer.

1. One molecule of calcium carbonate, $CaCO_3$, has how many atoms of calcium, Ca?

   A. 0 B. 1 C. 2 D. 3

2. How many atoms of oxygen are present in one molecule of H_2SO_4?

   A. 1 B. 2 C. 4 D. 8

3. What is the ratio of atoms present in ammonia, NH_3?

   A. 1 atom of nitrogen to 3 atoms of hydrogen

   B. 3 atoms of nitrogen to 1 atom of hydrogen

   C. 1 atom of nitrogen to 1 atom of hydrogen

   D. the ratio varies

C. Answer the following questions.

1. Give the ratio of atoms present in the compound $Al_2(SO_4)_3$.
2. Determine the names of CuO_2 and AgI using the rules for binary ionic compounds.
3. What is the name of the compound $N_2(g)$?
4. Determine the names of SF_6, CO_2 and CH_4 using the rules for naming binary covalent compounds.

The following formula describes how iron and oxygen react to form iron oxide (rust). Use this equation to answer questions 2–4.

$4Fe(s) + 3O_2(g) \rightarrow 2Fe_2O_3(s)$

5. Write out the equation in words, and include the state of each substance.
6. Which substances are the reactants and which ones are the products?
7. How many atoms of iron are in one molecule of rust?

Basic Chemical Reactions

All chemical reactions involve one or more reactants that interact to form one or more products. From the huge array of elements in the periodic table, we can assume that there is an equally wide variety of possible interactions between these elements. These interactions are divided into several different categories.

Synthesis reactions: Small molecules combine to form larger ones.

$$H_2(g) + Cl_2(g) \rightarrow 2HCl(aq)$$

Decomposition reactions: The opposite of synthesis reactions. Large molecules break apart to form smaller molecules.

$$2H_2O_2(l) \rightarrow 2H_2O(l) + 2O_2(g)$$

Single displacement reaction: When a pure element switches places with one of the elements in a compound.

$$Mg(s) + 2HCl(l) \rightarrow MgCl_2(aq) + H_2(g)$$

Double displacement reaction: When the components of ionic compounds switch places.

$$MgSO_4(aq) + 2NaOH(aq) \rightarrow Mg(OH)_2(s) + NaSO_4(aq)$$

Each of these is a **balanced equation**; we will see why this is so in the next section.

Writing Balanced Chemical Equations

Conservation of Mass

Chemical equations must maintain balance. There must be the same number of atoms of each element on both sides of the equation. The **law of conservation of mass** states that matter is conserved, which means that you can neither create nor destroy it. The amount of matter remains the same before and after a chemical reaction. An atom of hydrogen remains an atom of hydrogen, but it may chemically bond to form a different compound.

Balancing Chemical Equations

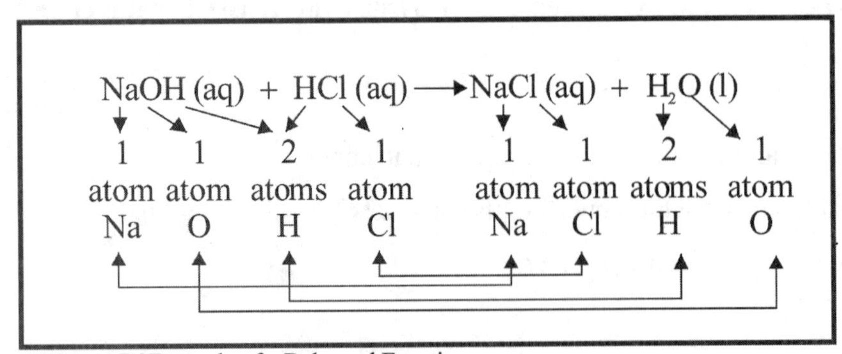

Figure 7.3 Example of a Balanced Equation

Look at the equation in Figure 7.3. On the side of the reactants, there is 1 atom of Na, 1 atom of O, 2 atoms of H, and 1 atom of Cl. On the product side of the reaction, there is 1 atom of Na, 1 atom of Cl, 2 atoms of H, and 1 atom of O. The number of atoms of each element is equal, so this is a balanced equation.

Now look at the following equation. There are 2 atoms of hydrogen reacting and 2 atoms of hydrogen as products, so the hydrogen in the equation is balanced. However, there are 2 atoms of oxygen reacting but only 1 atom of oxygen shown as product. This is not a balanced equation.

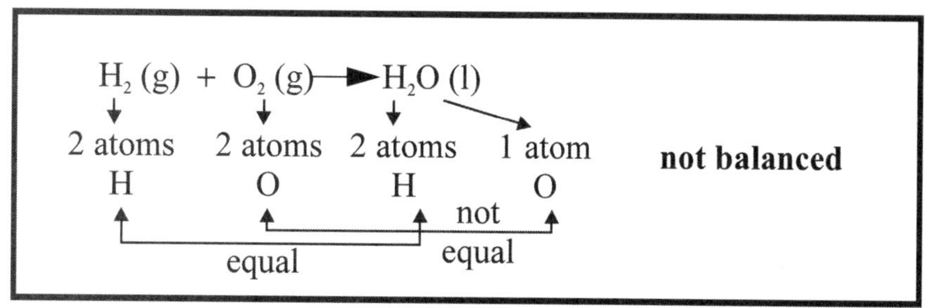

Figure 7.4 Example of an Unbalanced Equation

How can we balance the equation in Figure 7.4? We balance the equation by inspection

Step 1: Put a 2 in front of the H_2O. Now, we have 2 atoms of oxygen to balance the oxygen, but the number of hydrogen atoms increases to 4.

Step 2: Put a 2 in front of the H_2. Now, we have 4 atoms of hydrogen on each side of the equation, and the equation is completely balanced.

Balancing equations is a matter of trial and error. Change the numbers in front of each element until the number of atoms of each element is equal on both sides. Remember, you can never change the number of the subscripts. Changing the number of the subscripts changes the compound, or substance, which consequently changes the meaning of the equation.

Up until now, we have said that equations are balanced by balancing the number of atoms on the reactant side with the number of atoms on the product side. This is true. It is rare, however, that chemists work with such tiny amounts. Usually they are working with some tangible quantity, like milligrams or grams. A gram of, say, magnesium, contains a LOT of atoms of magnesium. So many, in fact, that chemists came up with the **mole concept**. One mole of an element is equal to 6×10^{23} atoms of the element. One mole of a compound equals 6×10^{23} molecules of that compound. That large number, 6×10^{23}, is called **Avogadro's number**. Do not let the size of it scare you away from using it.

Let us look at the four primary reaction types again, and use words to describe the reaction.

Synthesis reactions: In words, we can say that 1 mole of hydrogen gas reacted with 1 mole of chlorine gas to form 2 moles of aqueous (in water) hydrochloric acid.

Decomposition reactions: Here, 2 moles of hydrogen peroxide decomposed to 2 moles of water and one mole of oxygen gas.

Single displacement reaction: Here, 1 mole of elemental magnesium reacted with 2 moles of hydrochloric acid to form one mole of magnesium chloride (dissolved in water) and 1 mole of elemental hydrogen gas.

Challenge Activity

Double displacement reaction: (Now it is your turn) ______________________________

____

____

H

Section Review 2: Writing Balanced Chemical Equations

A. Define the following term.

Law of conservation of mass | balanced equation

Avogadro's number | mole concept

B. Balance the following equations.

1. $C_2H_6(g) + O_2(g) \rightarrow CO_2(g) + H_2O(l)$

2. $H_2O_{2(l)} \longrightarrow H_2O_{(l)} + O_{2(g)}$

3. $AgNO_3(aq) + NaCl(aq) \rightarrow AgCl(s) + NaNO_3(aq)$

4. $Na(s) + Cl_2(g) \rightarrow NaCl(s)$

5. $N_2(g) + H_2(g) \rightarrow NH_3(g)$

C. For each of the following, determine whether the chemical equation represents a synthesis, decomposition, single displacement or double displacement reaction.

1. $AgNO_3(aq) + KCl(aq) \longrightarrow AgCl(aq) + KNO_3(aq)$__________

2. $N_2(g) + 3H_2(g) \longrightarrow 2NH_3(g)$__________

3. $Cu(s) + AgNO_3(aq) \longrightarrow CuNO_3(aq) + Ag(s)$ __________

4. $Zn(s) + H_2CO_3 \longrightarrow H_2(g) + ZnCO_3(aq)$__________

5. $2C(s) + O_2(g) \longrightarrow 2CO(g)$__________

6. $H_2CO_3(aq) \longrightarrow H_2O(l) + CO_2(g)$__________

ORGANIC AND BIOLOGICAL MOLECULES

Organic chemistry is the chemistry of carbon atoms as they are found in living organisms. All life is built on compounds that contain carbon, called **organic compounds**. Compounds that do not contain carbon are called **inorganic compounds**. As mentioned earlier, carbon has four valence electrons that it can share with other atoms in covalent bonds. Carbon shares these electrons most commonly with hydrogen, oxygen, and/or nitrogen. Figure 7.5 is a diagram showing carbon with

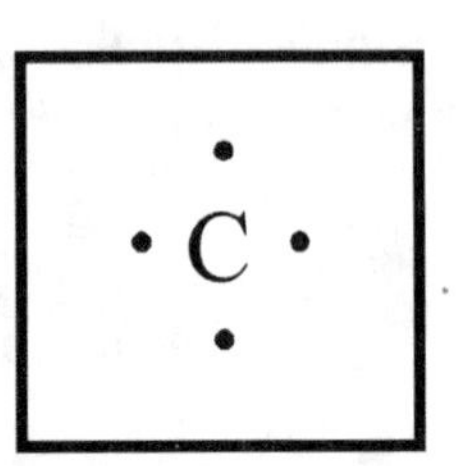

Figure 7.5 Carbon

its four valence electrons. Carbon atoms can be linked together to form chains, rings, or other branched molecules. Chains of carbon atoms are called **aliphatic** compounds. If carbon atoms form rings, they are called **cyclic** compounds.

HYDROCARBONS

One of the most common classes of organic compounds is the hydrocarbons. **Hydrocarbons** are formed when carbon combines with hydrogen. The simplest hydrocarbon is methane, which is shown in Figure 7.6. Notice how each valence electron from the carbon atom shares with one lone electron of a hydrogen atom. This kind of covalent bond is called a **single bond**. Single bonds are formed when one pair of valence electrons is shared between two elements. Carbon can also bind with itself to form long hydrocarbon chains. A chain of atoms that has only single bonds is called an **alkane**, and their names end in *-ane*. When a hydrocarbon only has single bonds, it is also called a **saturated hydrocarbon**.

The simple hydrocarbons typically exist as highly flammable liquids or gases in nature. Methane is released during the decomposition of organic matter. The simple hydrocarbons also make good fuels because they undergo combustion (burning) reactions in the presence of oxygen to produce carbon dioxide and water. Methane is the main compound in natural gas, which is used for household heating. Propane is used in gas grills to cook food, and butane is used in cigarette lighters.

```
   H            H
   ••           |
H: C :H   or  H–C–H
   ••           |
   H            H
```

Figure 7.6 Methane

Table 7.2 Naming Conventions for Single Bond Hydrocarbons

Name of Hydrocarbon	Molecular Formula	# Carbon Atoms
Methane	CH_4	1
Ethane	C_2H_6	2
Propane	C_3H_8	3
Butane	C_4H_{10}	4
Pentane	C_5H_{12}	5

```
H.      .H         H\     /H
  .C :: C.    or     C=C
H˙      ˙H         H/     \H
```

ethene (double bond)

```
H:C:::C:H    or    H–C≡C–H
```

ethyne (triple bond)

Figure 7.7 Unsaturated Hydrocarbons

A **double bond** forms when two atoms share two pairs of valence electrons. A **triple bond** forms when two atoms share three pairs of valence electrons. Figure 7.7 shows carbon forming a double and a triple bond. When a hydrocarbon forms one or more multiple bonds, it is called an **unsaturated hydrocarbon**. Unsaturated hydrocarbons containing double bonds are called **alkenes**, and their names end in *-ene*. Hydrocarbons containing triple bonds are called **alkynes**, and their names end in *-yne*.

Double bonds can also be present in cyclic hydrocarbons. An important cyclic hydrocarbon is benzene. Figure 7.8 shows a benzene ring and the different ways it can be represented.

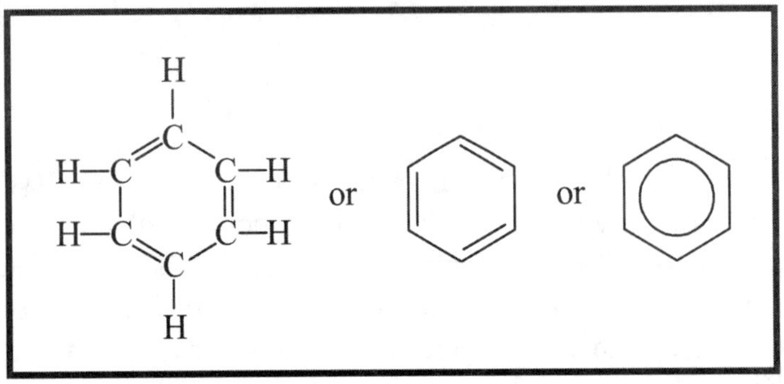

Figure7.8 Benzene

Carbon can bond with other elements besides just hydrogen. The **alcohols** are hydrocarbons with hydroxide (-OH) groups bound to carbon. Amino acids, which are the building blocks of protein, are carbon chains with nitrogen, oxygen, and hydroxide groups attached. Carbon chains with one or more chlorine (Cl) groups are commonly used to make plastics such as polyvinylchloride (PVC).

Hydrocarbons compose the major fossil fuels such as coal, petroleum, and natural gas. **Petroleum**, or crude oil, is composed of various hydrocarbons, especially alkanes, plus elements such as nitrogen, sulfur, and oxygen. Refined crude oil is used to produce kerosene, gasoline, wax, and asphalt.

BIOLOGICAL MOLECULES

Hydrocarbons are the building blocks for the complex molecules that make up living things. Four groups of important organic molecules found in cells are carbohydrates, lipids, proteins, and nucleic acids.

Carbohydrates are the most important energy source for cells. Simple carbohydrates are called **sugars**. Sugars can form long chains that serve a variety of functions in cells. These long chains are **polymers**, organic molecules made of many repeating linked units, called **monomers**. These long chains of sugars are called **complex carbohydrates**, and include starch, glycogen, and cellulose. Plants store extra sugars in the form of a carbohydrate called **starch**. Starch from wheat, corn, potatoes, and rice serves as a major food source for the world's people. Animals store excess sugars in the form of a carbohydrate called **glycogen**, which is stored in muscles and in the liver. **Cellulose** is a tough, flexible carbohydrate that gives plants much of their strength.

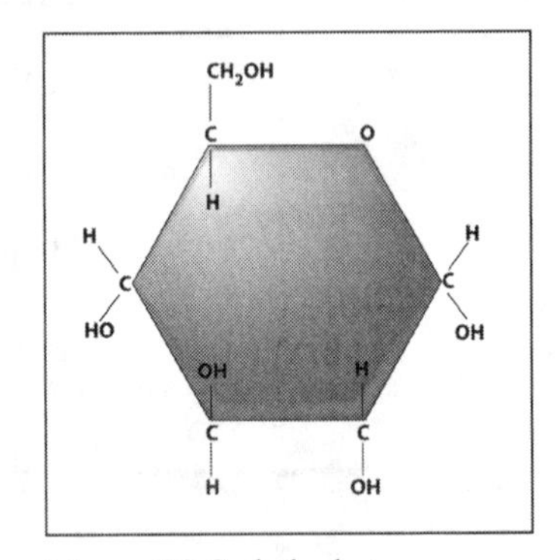

Figure 7.9 Carbohydrate

Lipids are used to store energy, and they are important parts of biological membranes. They are generally made of fatty acids, and are not soluble in water. Types of lipids include fats, waxes, and steroids.

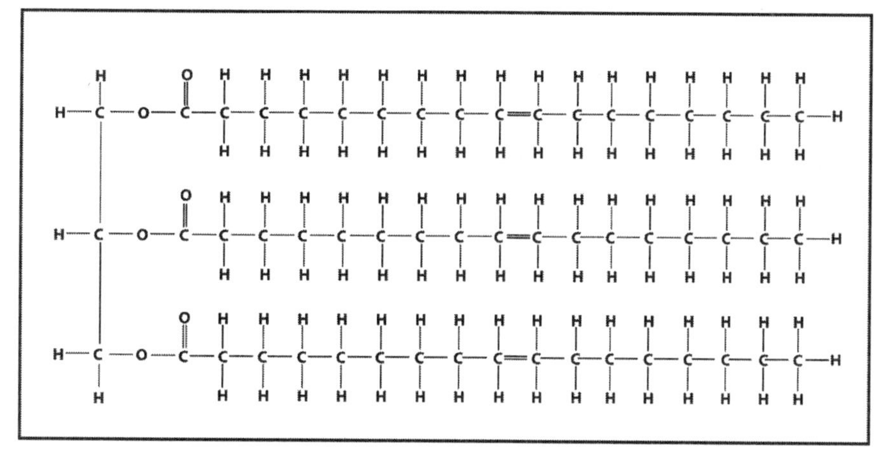

Figure 7.10 Lipid

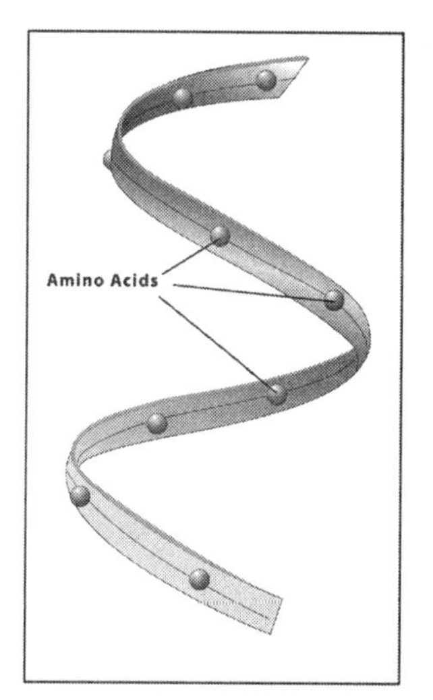

Figure 7.11 Protein

Fats are lipids made when several fatty acids combine with a molecule called glycerol. Fats store twice as much energy for their weight than carbohydrates or proteins do. Fats are also used for insulation, especially among organisms living in cold climates.

Waxes and steroids also perform important functions in organisms. Waxes are made of fatty acids and some alcohols. Waxes are used for waterproofing plants and the feathers of water birds. Steroids are made of linked rings of carbon atoms. Steroids can serve important functions in the structure of cells. An example of a steroid is cholesterol. Other steroids, called hormones, are used to send messages throughout organisms.

Proteins, polymers of **amino acids**, play an essential role in the structure of the cell, and proteins called **enzymes** help speed up chemical reactions in cells. Proteins are also used in movement, in transport of materials, and in fighting disease.

Nucleic acids store and transmit hereditary, or genetic, information. Nucleic acids are made up of monomers called **nucleotides**. The two types of nucleic acids are **deoxyribonucleic acid (DNA)** and **ribonucleic acid (RNA)**.

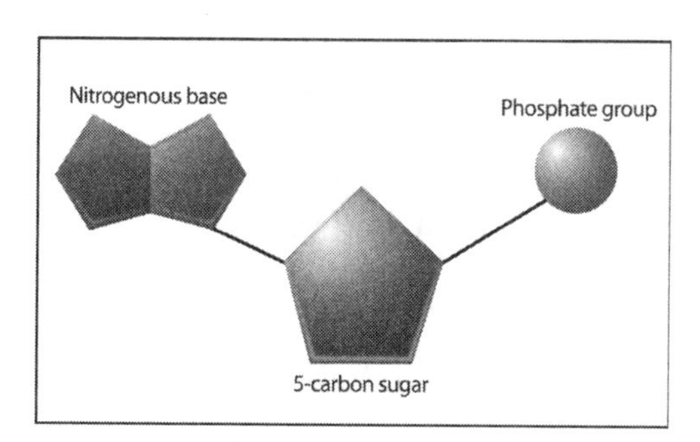

Figure 7.12 Nucleic Acid

Other Polymers

In addition to polymers that are found in nature, many common polymers today are **synthetic** (man-made). For example, ethene (C_2H_4) has two carbons connected by a double bond. If that double bond is broken, the resulting molecule is a chain called polyethylene, a plastic used for sandwich bags. The process of joining monomers together is called **polymerization**. Synthetic polymers are used in medicine, consumer products, industry, and sports. The IV bags used in hospitals are composed of polymers. Polymers can be found in such consumer products as sandwich bags, plastic utensils and plates, and storage containers. Polymers are also used in protective sporting equipment such as the shin guards you wear when playing soccer or the helmets you wear when playing football. We use polymers to form plastics, synthetic rubber, and synthetic fibers such as polyester and nylon.

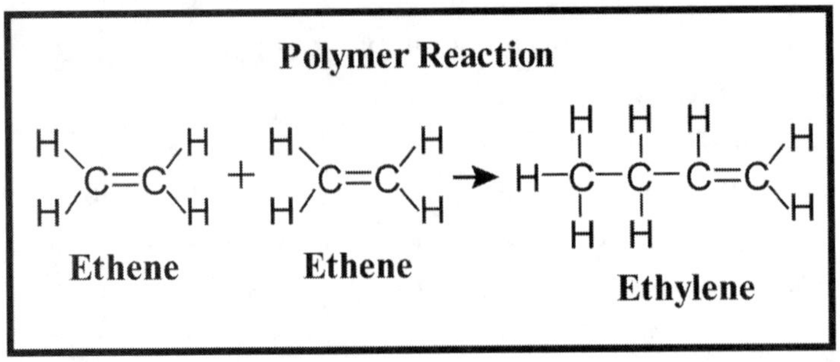

Polymer Reaction

$H_2C{=}CH_2 + H_2C{=}CH_2 \rightarrow H{-}CH_2{-}CH_2{-}CH{=}CH_2$

Ethene + Ethene → Ethylene

Figure 7.13 Polymer Reaction

Section Review 3: Organic and Biological Molecules

A. Define the following terms.

organic compound	triple bond	polymer	amino acid
inorganic compound	unsaturated hydrocarbon	monomer	enzyme
aliphatic	alkene	starch	nucleic acid
cyclic	alkyne	glycogen	nucleotide
hydrocarbon	alcohol	cellulose	deoxyribonucleic acid (DNA)
single bond	petroleum	lipid	ribonucleic acid (RNA)
alkane	carbohydrate	fat	synthetic polymer
saturated hydrocarbon	sugar	protein	polymerization
double bond			

B. Choose the best answer.

1. What kind of bonds are possible between carbon atoms?
   A. single bonds
   B. double bonds
   C. triple bonds
   D. all of the above

2. Which term best describes the molecule shown to the right?
   A. unsaturated hydrocarbon
   B. saturated hydrocarbon
   C. cyclic hydrocarbon
   D. amino acid

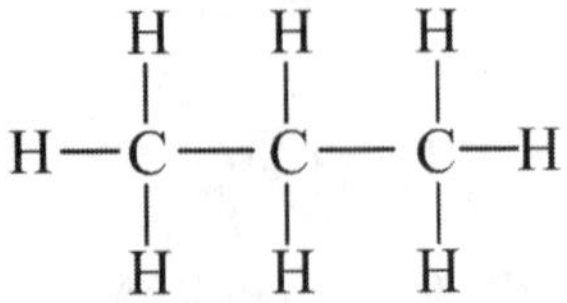

   $H{-}CH_2{-}CH_2{-}CH_2{-}H$

3. How many valence electrons does carbon have?
   A. 1
   B. 2
   C. 3
   D. 4

4. What kind of molecule is benzene?
   A. saturated cyclic hydrocarbon
   B. unsaturated cyclic hydrocarbon
   C. saturated aliphatic hydrocarbon
   D. unsaturated aliphatic hydrocarbon

5. Cellulose, which is made of repeating units of glucose, is an example of
   A. a protein.
   B. a monomer.
   C. an inorganic compound.
   D. a polymer.

6. Biological molecules that store energy and play an important role in the structure of cell membranes are
   A. carbohydrates.
   B. lipids.
   C. proteins.
   D. nucleic acids.

7. An amino acid is an example of
   A. a monomer.
   B. a synthetic fiber.
   C. a polymer.
   D. a nucleic acid.

C. Answer the following questions.

1. What characteristic of carbon allows it to be the building block of living things?
2. What are some examples of polymers that are used in everyday life?
3. What kind of bond must be present between carbon atoms for the molecule to be unsaturated?
4. Draw a diagram of the following molecules, and indicate the type of bond between the carbon atoms.
   A. C_2H_6 B. C_2H_2 C. C_2H_4
5. Label the following molecules as aliphatic or cyclic. Also indicate whether each is an alkane, an alkene, or an alkyne.

```
H       H
 \     /
  C = C
 /     \
H       H
```

A.

```
H−C−C−H
```

B.

```
      H   H
       \ /
  H     C     H
   \  /   \  /
    C       C
   /  \   /  \
  H     C     H
       / \
      H   H
```

C.

CHEMICAL PROPERTIES OF MATTER

In the first three sections of this chapter, we have seen that compounds have a distinct identity and name. These compounds may react chemically with one another to produce a chemical change. In a **chemical change**, matter changes physical and chemical properties. For example, hydrogen and oxygen are both gases at room temperature, which is a physical property. Hydrogen and oxygen are both combustible gases, meaning they will burn. When hydrogen and oxygen combine to form water, they undergo a chemical change. Water has different physical and chemical properties than either hydrogen or oxygen. So, any time a substance undergoes a change in which the end product has different physical and chemical properties than the original substance, it has been chemically changed. If a change in color, odor, or temperature occurs in a process, a chemical change may have occurred.

As a rule of thumb, if a substance undergoes a change and can then be changed back to its original state, it has undergone a physical change. If the substance cannot be changed back into its original state, it is a chemical change. This rule of thumb will not always apply when physically changing the size of a substance. For example, changing a piece of wood into sawdust is a physical change, but it would be very difficult to convert the sawdust back into its original form.

How a substance behaves in the presence of other substances determines its **chemical properties**. A chemical change must take place in order to observe the chemical properties of a substance. A **chemical change** occurs any time a substance undergoes a change in which the end product has different physical and chemical properties than the original substance. Examples of chemical properties are combustibility, reactivity, and toxicity. The following are examples of chemical changes:

Combustion (burning) is a chemical process that produces light and heat. In the process of combustion, a substance chemically combines with oxygen to form carbon dioxide, water, and possibly other oxides or by-products. Oxygen must be present for combustion to occur. **Organic** substances, which are substances made up of carbon atoms, are often components of living organisms and will usually burn. Organic substances include oil, gasoline, and wood, as well as synthetic plastics and fabrics. We call substances that undergo the combustion process **combustible** or **flammable**.

Fermentation is a chemical change in which a partial breakdown of substances occurs without the presence of oxygen. Microorganisms, like yeast, aid in the process of fermentation by using enzymes to break down different kinds of sugar. For example, adding yeast to bread makes it rise. The yeast converts the sugar in the bread to carbon dioxide and ethyl alcohol gases, which are trapped in the bread and cause it to rise.The following equation expresses the chemical reaction of fermentation;

$$C_6H_{12}O_6 \longrightarrow 2C_2H_5OH + 2CO_2 + \text{energy}$$ **Equation 7.1**

Oxidation occurs when oxygen reacts with another substance. Metals tend to oxidize. For example, when iron reacts with oxygen, it forms iron oxide, which we commonly call **rust**. Oxidation also causes fruit to turn brown. Biting into an apple exposes the inside pulp of the apple to oxygen in the air. After just a few minutes, the exposed pulp of the apple will begin to turn brown due to oxidation. The following equation gives the reaction of iron and oxygen to form rust:

$$\text{iron} + \text{oxygen from air} \longrightarrow \text{rust}$$
$$4Fe\ (0) + 3O_2\ (0) \longrightarrow 2Fe_2\ (+3)O_3\ (-2)$$ **Equation 7.2**

Corrosion is the process whereby metals deteriorate due to oxidation or chemical action. Exposing iron to oxygen forms rust, which is one example of corrosion. The other type of corrosion occurs when a metal is in contact with a material that causes it to break down, such as an acid or base. We will discuss acids and bases in Chapter 9.

Table 7.3 Examples of Common Physical and Chemical Changes

Physical Changes	Chemical Changes
Melting ice	Baking a cake
Beating an egg	Mixing baking soda and vinegar
Freezing water	Burning a candle
Stirring chocolate in milk	Rusting iron
Dissolving salt in water	Making wine
Bending wire	Souring milk

Section Review 4: Chemical Properties of Matter

A. Define the following terms.

chemical change, organic, rust
chemical properties, fermentation, corrosion
combustion, oxidation

B. Choose the best answer.

1. Which of the following processes would result in a chemical change?
   A. boiling
   B. cutting
   C. dissolving
   D. digesting

Read the following paragraph about making bread, then answer the question:

Myra puts kernels of wheat in a grinder and grinds them into flour. She then adds salt, baking soda, sugar, and dry yeast and mixes them together. Next, she adds water, milk, and oil. She kneads the mixture to form dough. She lets the dough rise until it is double in size. Then, she punches it down, forms a loaf, and places it in a bread pan. She lets it rise again, and then bakes it in the oven.

2. Which of the following steps involves a chemical change?
   A. Myra puts kernels of wheat in a grinder and grinds them into flour.
   B. She adds salt, baking soda, sugar, and dry yeast to the flour and mixes them together.
   C. She lets the dough rise until it is double in size.
   D. Then, she punches it down, forms a loaf, and places it in a bread pan.

3. Which of the following involves a chemical change?
   A. crushing a cube of sugar to form granules of sugar
   B. dissolving sugar granules in water
   C. boiling the water and sugar mixture
   D. burning a cube of sugar until it turns black and eventually disappears

4. Rusting is an example of
   A. oxidation.
   B. fermentation.
   C. a physical change.
   D. electrolysis.

5. Which of the following is ***not*** a chemical property?
   A. boiling point
   B. reactivity
   C. toxicity
   D. oxidation state

C. Answer the following questions.

1. How is a chemical change different from a physical change?
2. Give one example of how a piece of wood could undergo a physical change. Give one example of how a piece of wood could undergo a chemical change.
3. Carmen dissolves sugar into hot tea. Is this a physical change or a chemical change? Why do you think so?
4. When baking soda mixes with vinegar, the mixture produces bubbles. Is this a physical or chemical change? Why do you think so?
5. Egg white dropped into boiling water turned from a clear, almost colorless liquid to a white, rubbery solid. Did a chemical change occur? How do you know?

CHAPTER 7 REVIEW

A. Choose the best answer.

1. The chemical formula for glucose is $C_6H_{12}O_6$. How many atoms of carbon make up one molecule of glucose?

   A. 1 B. 3 C. 6 D. 12

2. The reaction for photosynthesis is $6CO_2 + 6H_2O \rightarrow C_6H_{12}O_6 + 6O_2$. How many oxygen atoms, O, are released for every 1 molecule of glucose, $C_6H_{12}O_6$, produced?

   A. 1 B. 2 C. 6 D. 12

3. Which of the following represents a balanced chemical reaction?

   A. $H_2 + O_2 \rightarrow H_2O$

   B. $CO_2 + H_2O \rightarrow 2H_2CO_3$

   C. $CH_4 + O_2 \rightarrow CO_2 + 2H_2O$

   D. $N_2 + 3H_2 \rightarrow 2NH_3$

4. Graphite, which is pure carbon, burns in an open flame until the graphite completely disappears. What happens to the carbon?

   A. All of the carbon converts to energy.

   B. The carbon combines with oxygen to form carbon dioxide gas.

   C. All of the carbon is completely destroyed by the heat.

   D. Part of the carbon converts to energy, and the rest converts to carbon dioxide gas.

5. Which of the following is a true statement?

   A. Matter is conserved in a chemical reaction.

   B. Matter is created in a chemical reaction.

   C. Matter is destroyed in a chemical reaction.

   D. All of the above statements can be true depending on the reaction.

6. The chemical formula for sugar is $C_6H_{12}O_6$. Sugar placed in a test tube and then put over a Bunsen burner turns black and eventually disappears completely. Which of the following statements explains what happens to the sugar that causes it to disappear?

   A. The heat destroys the elements that make up the sugar.

   B. As the sugar burns, it combines with oxygen in the air to form water and carbon dioxide. Water vapor and carbon dioxide escape into the atmosphere.

   C. The sugar is converted to nitrogen gas and is released into the atmosphere.

   D. All of the sugar is converted to energy that cannot be seen.

7. In chemistry class, Mr. Smoak adds a small piece of sodium metal to a glass of water. The sodium reacts violently with the water, producing a small flame. The end products are hydrogen gas and sodium hydroxide. Which of the following is the correct balanced chemical equation for the reaction described above?

   A. $2Na + 2H_2O \rightarrow 2NaOH + H_2$

   B. $Na + H_2O + O_2 \rightarrow 3NaOH + H_2$

   C. $H_2O_2 + 2Na \rightarrow 2NaOH + H_2$

   D. $2NaOH + H_2 \rightarrow 2Na + 2H_2O$

8. Consider the following chemical equation: $3HfCl_3 + Al \rightarrow 3HfCl_2 + AlCl_3$

   What are the reactants in this equation?

   A. $HfCl_3$ and $AlCl_3$

   B. $HfCl_3$ and Al

   C. $HfCl_2$ and Al

   D. Al and $AlCl_3$

9. Which of the following equations is balanced?

   A. $Ga + 3H_2SO_4 \rightarrow 2Ga_2(SO_4)_3 + 2H_2$

   B. $PdCl_2 + HNO_3 \rightarrow Pd(NO_3)_2 + HCl$

   C. $O_2 + Sb_2S_3 \rightarrow Sb_2O_4 + SO_2$

   D. $RbBr + AgCl \rightarrow AgBr + RbCl$

10. How many bonds does carbon usually form when part of an organic compound?

    A. 1　　B. 2　　C. 3　　D. 4

11. The molecules that serve as the primary energy source for living things are

    A. carbohydrates.　　B. lipids.　　C. proteins.　　D. nucleic acids.

12. Plastics, synthetic rubber, polyester, and nylon are examples of

    A. ionic crystal lattices.

    B. diatomic molecules.

    C. polymers.

    D. proteins.

13. Which of the following steps below includes a chemical change?

    A. Carlos cuts down a large oak tree.

    B. He strips the tree of its limbs and leaves and chips them into smaller pieces.

    C. He then burns all of the small twigs and leaves.

    D. Finally, he cuts the trunk into fire logs.

14. Which of the following combinations would result in a substance that is chemically different than its components?

    A. Carbon and oxygen form carbon dioxide.

    B. a bowl of Cheerios, blueberries, and milk

    C. a pitcher of lemonade made from a powdered mix and water

    D. Oxygen and nitrogen form the air that we breathe.

15. Which of the following involves a chemical change?

    A. snow melting into water

    B. an apple rotting on a tree

    C. beating egg whites until they thicken

    D. making sweet iced tea

Chapter 8 Matter and Energy

PHYSICAL SCIENCE STANDARDS COVERED IN THIS CHAPTER INCLUDE:

SPS5	Students will compare and contrast the phases of matter as they relate to atomic and molecular motion.
SPS7	Students will relate transformations and flow of energy within a system.

CLASSIFICATION OF MATTER

We have spent the past few chapters looking at the nuclear and chemical properties of matter…but we haven't discussed that yet, have we? What is matter? Matter is anything that has mass and takes up space. On a large scale, matter is easy to define as anything that you can see or touch.

On a small scale, the definition of matter becomes a little trickier. The electron is a good example. The mass of an electron is 9.11×10^{-31} kg, so it is very small. In addition, the electron moves so fast that we usually only measure its location in terms of **probability**, (the likelyhood that the electron will be found in a certain place, like an orbital) so the space that it takes up is sometimes hard to find. Nevertheless, it *does* have a mass and it *does* take up space, so theoretically it *is* matter.

In this chapter, we will look at the physical properties of matter. Matter can be divided into two main categories: **mixtures** and **pure substances**. These two categories can be further broken down into a variety of classifications, as shown in Figure 8.1.

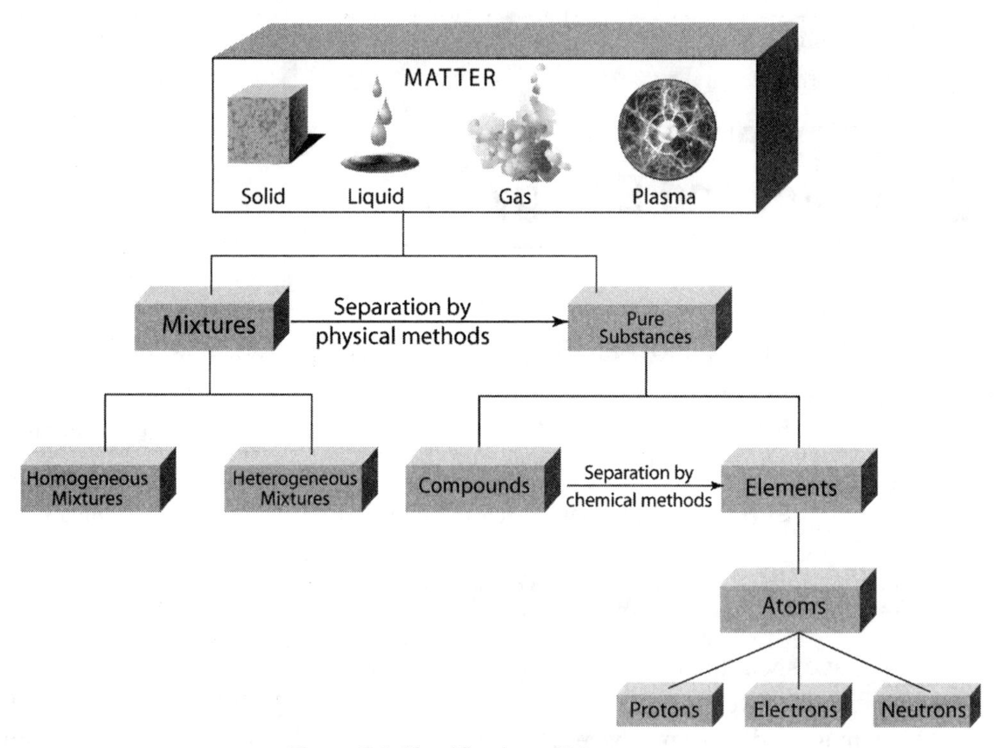

Figure 8.1 Classification of Matter

A **substance** is matter that has constant composition and distinct properties. Some examples of pure substances are oxygen, carbon, iron, sugar, and water. Notice that not all pure substances are elements. The best way to determine if something is a pure substance is to ask the question: Will we get the same sample anywhere in the substance?

Elements are substances that cannot be further broken down into simpler substances. Examples of common elements are oxygen, carbon, and iron.

As we discussed in the previous chapter, when two or more elements combine chemically, they form a **compound**. A compound has completely different properties than the individual elements that make up the compound. For example, water is a compound made up of hydrogen and oxygen. Hydrogen and oxygen, as stand-alone elements, are gases at room temperature. However, water is a liquid. It is not possible to separate a compound physically into its individual components, but it can be chemically separated.

When two or more substances (either elements or compounds) combine physically, they form a **mixture**. A mixture keeps the individual properties of the substances that make it up because the substances do not chemically combine. You can separate a mixture into its individual substances. For example, salt dissolved into water is a mixture. If you drink the salt water, you can taste the salt in the water. The salt is still "salty," and the water is still a liquid, so these substances have not changed chemically. Evaporation separates the salt and water. The liquid water turns into water vapor, so only the salt remains.

Table 8.1 Examples of Common Types of Matter

Common Elements	Common Compounds	Common Mixtures
oxygen	table salt	vinegar
carbon	water	salad dressing
helium	sugar	brass
nitrogen	baking soda	blood
aluminum	Epsom salts	gasoline
gold	carbon dioxide	soda
neon	ammonia	orange juice

We can further classify mixtures as **homogeneous** or **heterogeneous** depending on the distribution of substances in the mixture.

A **homogeneous mixture** occurs when substances are evenly distributed, and one part of the mixture is indistinguishable from the other. The ratio of "ingredients" in a mixture does not have to be in definite proportions in order to be homogeneous. Mixtures of gases are homogeneous. All **solutions** are also homogeneous. A solution consists of a substance (**solute**) dissolved in another substance (**solvent**). Solutions can be mixtures of solids, liquids, or gases. For example, brass is a solid solution of copper, tin, and other elements. Soda is a solution of carbon dioxide gas dissolved in water, a liquid. Salt water is a solution of salt, a solid, dissolved in water, a liquid. Filtering a solution cannot separate its individual parts.

Figure 8.2 Example of a Homogeneous Solution

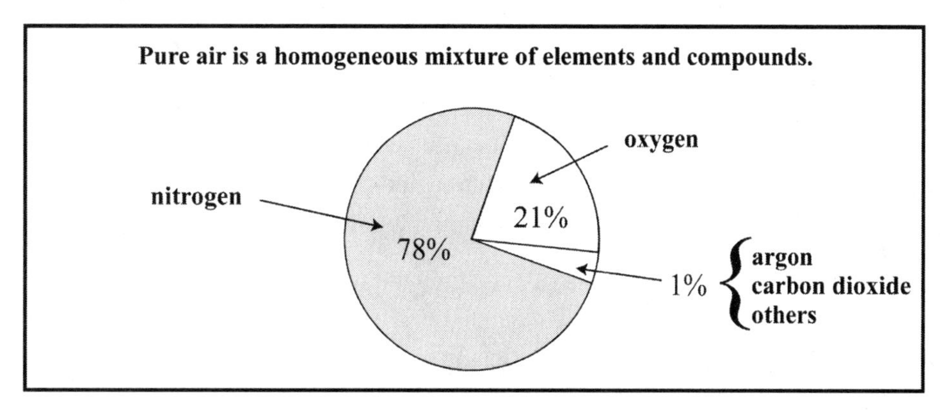

Figure 8.3 Example of a Homogeneous Mixture

A **heterogeneous mixture** occurs when one part of the mixture is distinguishable from the other. Examples of heterogeneous mixtures include granite, dirty air, oil and vinegar salad dressing, paint, blood, and soil. Filtering separates many heterogeneous mixtures.

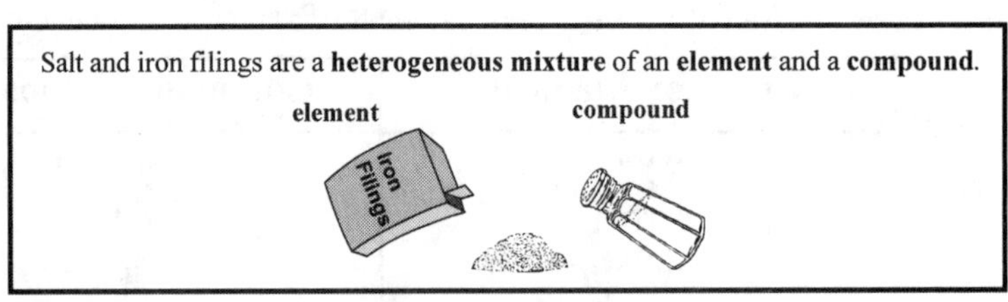

Figure 8.4 Example of a Heterogeneous Mixture

Section Review 1: Classification of Matter

A. Define the following terms.

element	mixture	heterogeneous mixture
compound	homogeneous mixture	solution

B. Choose the best answer.

1. Which of the following is a heterogeneous mixture?
   A. salt water B. carbon dioxide C. bronze D. vegetable soup

2. Which of the following can be separated by filtering?
   A. a solution
   B. a compound
   C. an element
   D. a heterogeneous mixture

3. What is a physical combination of two or more substances called?
   A. an element B. a compound C. a mixture D. an isotope

4. Which of the following could be physically separated?
   A. oxygen
   B. carbon dioxide
   C. salt dissolved in water
   D. pure water

5. Which of the following must be chemically separated to isolate individual elements?
   A. ammonia
   B. brass
   C. oil and vinegar salad dressing
   D. air

6. Which of the following combinations would result in a substance that is chemically different than its components?
   A. Carbon and oxygen form carbon dioxide.
   B. Sugar and water make a sugar-water solution.
   C. Copper and tin form bronze.
   D. Oxygen and nitrogen form the air that we breathe.

C. Answer the following questions.

1. What is the difference between an element and a compound?
2. What is the difference between a compound and a mixture?
3. Identify the following substances as element (E), compound (C), or mixture (M).

A. carbon ____
B. carbon dioxide ____
C. milk ____
D. calcium ____
E. calcium carbonate ____
F. blood ____
G. sand and sugar ____
H. chicken noodle soup ____

4. Identify the following mixtures as homogeneous (HO) or heterogeneous (HE).

A. Gasoline ____
B. chunky peanut butter ____
C. filtered apple juice ____
D. oil and vinegar ____

5. Describe two ways in which a physical combination of substances is different from a chemical combination of substances.

KINETIC THEORY AND STATES OF MATTER

Energy is the ability to do work, and **work** is the process of moving matter. Energy falls into two broad categories: potential energy and kinetic energy. **Potential energy** is stored energy due to the object's position or state, whereas **kinetic energy** is energy of motion as an object moves from one position to another. **Kinetic theory** explains how temperature and pressure affect matter. The theory assumes several things. The major assumptions are as follows:

1. All matter is composed of small particles such as atoms, ions, or molecules.
2. These particles are in constant motion, and this motion is felt as temperature and pressure.
3. Collisions between particles are perfectly elastic, meaning the average kinetic energy of a group of particles does not change when they collide.

As mentioned earlier, matter can exist as a solid, a liquid, or a gas. Particles making up the matter are in constant motion. Substances exist in certain states of matter depending on the amount of motion between the particles. In general, the particles of a substance in the gaseous state have the highest average kinetic energy, and particles in the solid state have the lowest average kinetic energy. According to **kinetic theory**, as the temperature increases, the motion of the particles also increases. Adding or subtracting heat (heating or cooling) changes matter from one state to another.

Matter exists in different states, called **phases**. The four states of matter are **solid**, **liquid**, **gas** and **plasma**.

- **Solid**- The atoms or molecules that comprise a solid are packed closely together, in fixed positions relative to each other. Therefore, the solid phase of matter is characterized by its rigidity and resistance to changes in volume. A solid does not conform to the container that it is place in.

- **Liquid**- The molecules that comprise a liquid can move relative to one another, but are fixed within the volume of the liquid by temperature and pressure. A liquid does conform to the container that it is placed in, but may not fill that container.
- **Gas**- The atoms and molecules that comprise a gas move independently of one another. The space between them is determined by the temperature and pressure of the gas, as well as the volume of the container in which it is placed. A gas placed in a container will spread out to uniformly fill that container.
- **Plasma**- A plasma is an ionized gas. This means that atoms and molecules that make up a plasma are charged. As a result of this charge, the atoms and molecules of a plasma "communicate" with each other; they move together, because each particle interacts simultaneously with many others. A plasma is characterized by its temperature, density and electrical conductivity.

You are familiar with all four phases, though you may not realize it. Figure 8.5 shows a plasma lamp, which many stores sell as a decorative item. If you have ever touched one of these lamps, you know that the filaments of ionic gas reach out toward the conducting surface- that is, your hand. This is a good visual example of how plasmas ions move together; if the lamp was just filled with unionized gas, there would be no collective movement of the state in reaction to a stimulus (your hand). For more common examples of plasma and the other states of matter, see Table 8.2.

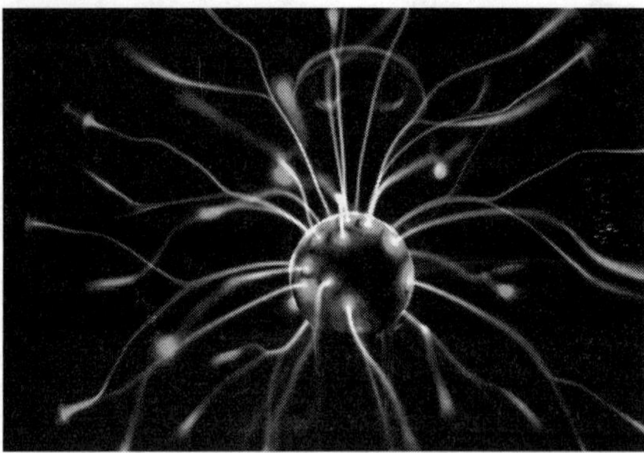

Figure 8.5

Table 8.2 Common Substances for Each State of Matter

Solids	Liquids	Gases	Plasma
silver	water	oxygen	fire
diamond	milk	helium	lightening
copper	alcohol	carbon dioxide	the sun and stars
rocks	syrup	hydrogen	the ionosphere
wood	oil	nitrogen	neon signs

The particles making up matter are in constant motion. The phase of the matter depends on the amount and type of motion of those particles. In general, the particles of the gas and plasma states have the highest kinetic energy, while solids have the least. According to kinetic theory, particle motion increases as temperature increases. Adding or subtracting energy in the form of heat changes matter from one state to another. These are called **phase changes**.

PHASE CHANGES

The phase of matter is determined by the physical condition of that matter. When the physical conditions change, a phase change may occur. Two physical conditions of primary importance are temperature and pressure. To determine how temperature and pressure changes affect phase, we must define **phase barriers**-that is, the point at which matter changes phase.

The **freezing point** of a substance is the temperature at which a liquid becomes a solid or freezes. The **melting point** of a substance is the temperature at which a solid becomes a liquid or melts. The freezing point and the melting point for a given substance are the same temperature. For example, liquid water begins to freeze at 0°C. Likewise, a cube of ice begins to melt at 0°C.

The **boiling point** of a substance is the temperature at which a liquid becomes a gas. The **condensation point** is the temperature at which a gas becomes a liquid. The boiling point and the condensation point for a given substance are the same temperature. For example, water boils at 100°C, and water vapor (steam) cooled to 100°C condenses.

Sublimation is the evaporation of a substance directly from a solid to a gas without melting (or going through the liquid phase). For example, mothballs and air fresheners sublime from a solid to a gas. Dry ice, which is frozen carbon dioxide, is also a common example of sublimation because the solid dry ice immediately sublimes into carbon dioxide gas (looking like "fog").

Deposition is the condensation of a substance directly from a vapor to a solid without going through the liquid phase. This term is mostly used in **meteorology** (the study of weather) when discussing the formation of ice from water vapor. The phase changes between solid, liquid, and gas are summarized in Figure 8.6.

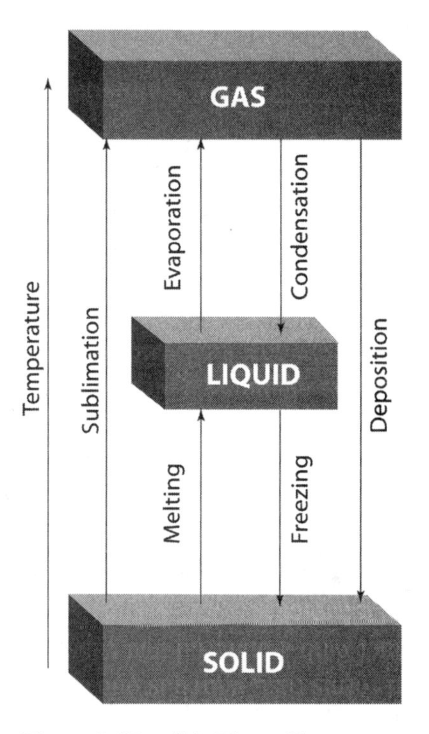

Figure 8.6 Possible Phase Changes

Depending on the temperature, water exists in all three natural states of matter; and therefore, it is a good example of how matter changes states. Figure 8.7 shows a common way to illustrate these transitions, called a **phase diagram**.

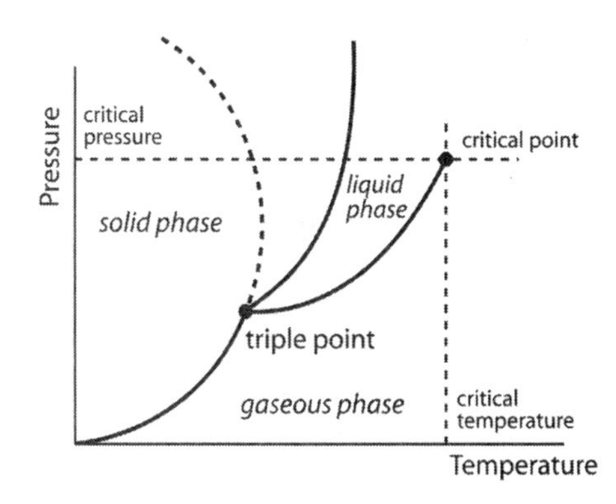

Figure 8.7

Phase changes can also be illustrated in terms of the amount of heat added. Figure 8.8 below shows this perspective. Ice remains solid at temperatures below 0ºC, but once ice reaches 0ºC, it starts to melt.

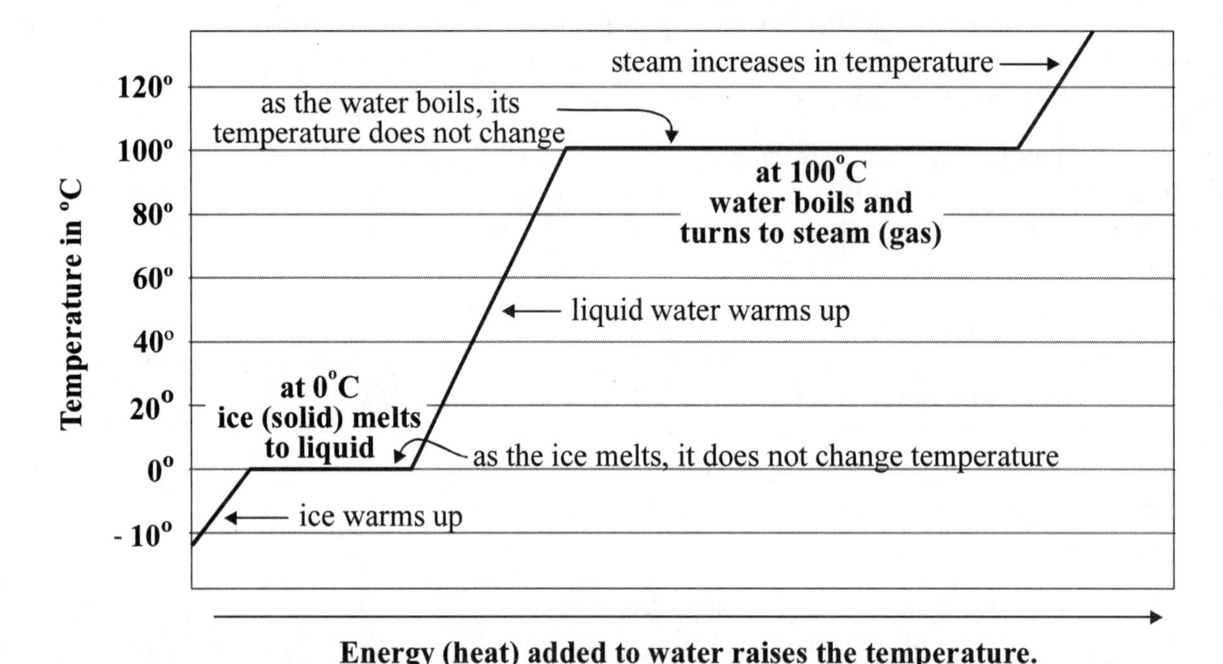

Figure 8.8 The Changing States of Water

Notice from Figure 8.8 that as ice melts, it continues to absorb energy, but the temperature of the ice-water mixture does not change. As we apply heat to the ice cube, the heat energy breaks up the molecular bonds of the ice, rather than raising the temperature of the surrounding water. The temperature does not change again until all of the ice melts. Once in a liquid state, the temperature of the water increases until it reaches 100°C. At 100°C, the water boils and turns to steam. While the liquid changes to vapor, the liquid absorbs energy, but the temperature does not increase. Once all of the liquid turns to steam, the temperature of the steam increases. In summary, the temperature remains constant through any phase change whether it be melting, freezing, or boiling. The temperature does not increase during a phase change because the energy is being used to break and/or form molecular bonds rather than to heat the substance.

Section Review 2: Kinetic Theory and States of Matter

A. Define the following terms.

kinetic theory	liquid	boiling point
solid	gas	condensation point
melting point	freezing point	sublimation
	plasma	deposition

B. Choose the best answer.

1. The kinetic energy of a substance is greatest in which state?
   A. solid
   B. liquid
   C. gas
   D. each state has equal energy

2. When might a substance absorb heat but not change temperature?
   A. when it is in its solid state
   B. when it is changing from one physical state to another
   C. when it is in its gaseous state
   D. Under no circumstances will a substance absorb heat but not change temperature.

3. What state of matter has a definite volume but no definite shape?
   A. solid
   B. liquid
   C. gas
   D. all states of matter

4. What state of matter can expand or contract depending on the volume of its container?
   A. solid
   B. liquid
   C. gas
   D. none of the states of matter

C. Answer the following question.

1. List the states of matter in order of most kinetic energy to least kinetic energy.
2. Describe the movement of molecules in each of the three states of matter.
   A. solid
   B. liquid
   C. gas
3. Can you think of a way that liquids are more similar to plasmas than they are to gases?

PHYSICAL PROPERTIES OF MATTER

Physical properties help to describe matter. The state (or phase) of matter is of primary interest when observing and recording the physical properties of a sample. There are many other physical properties that can be observed and measured. These are divided into two categories: **extrinsic** (or extensive) properties and **intrinsic** (or intensive) properties. We can measure and observe physical properties without changing the composition or identity of a substance. Physical properties are broken up into intrinsic and extrinsic properties.

Extrinsic properties depend on the amount of matter present. Mass, volume and energy are all extrinsic properties of matter. **Intrinsic properties** of a substance do not depend on the amount of matter present in the sample. Color, melting point, boiling point, hardness, and electrical conductivity are all intrinsic properties. Another important intrinsic property is **density**.

DENSITY

Each pure substance has particular properties unique to that substance. Density is one of these properties. **Density** (D) is the mass (m) per unit volume (V) of a substance. We express density in units of kg/m^3 or g/cm^3. At the atomic level, the atomic mass of the element and the amount of space between particles determines the density of a substance. Use the following formula to calculate density:

$$D = \frac{m}{V} \qquad \textbf{Equation 8.1}$$

When comparing objects of the same volume, the denser something is the more mass it has and, therefore, the greater its weight. This explains why even a small amount of pure gold is very heavy.

The following are general rules regarding density:

Rule 1. **The amount of a substance does not affect its density. The density of iron at 0°C will always be 7.8 g/cm^3. It does not matter if we have 100 g or 2 g of iron.**

Rule 2. **Temperature affects density. In general, density decreases as temperature increases. Water is an exception to this rule. The density of ice is less than the density of liquid water; therefore, ice floats.**

Rule 3. **Pressure affects the density of gases, but it does not affect solids or liquids since those two states are not compressible. As the pressure on a gas increases, density also increases.**

Mixing substances of different densities changes the density of the mixture. For example, the density of fresh water is less than the density of salt water. We will explore density in more detail as we look at other physical properties of different phases of matter.

Practice Exercise 1: Calculating Density

Find the densities of the following objects at 0°C and 1 atmosphere.

1. A block of stone material with a mass of 30 grams and a volume of 12 cm^3.
2. A volume of liquid water with a mass of 12 kilograms and a volume of 12 m^3.* (temperature is at 4°C)
3. A mineral of unknown origin with a mass of 42 grams and a volume of 13 cm^3.
4. A metallic gold nugget with a mass of 99 grams and a volume of 5.14 cm^3.
5. A metallic iron box with a mass of 64 kilograms and a volume of 8.2 m^3.
6. A metallic copper tube with a mass of 46 kilograms and a volume of 5.14 m^3.
7. A large ice block with a mass of 100 grams and a volume of 109 cm^3.
8. Helium gas in a balloon with a mass of 4 grams and a volume of 22.4 cm^3.
9. A volume of salt water with a mass of 16 grams and a volume of 15.61 cm^3.
10. A volume of gasoline with a mass of 5 kilograms and a volume of 7.4 m^3.

Lab Activity 1: Density Column

In the lab, a density column can be used to determine the relative density of various liquids. To demonstrate this, measure equal amounts of cooking oil, liquid laundry detergent, honey, and vinegar. Weigh each measured liquid to determine its mass. Take a large graduated cylinder and carefully pour in each liquid, letting it run down the inside of the cylinder. The different liquids should separate into layers.

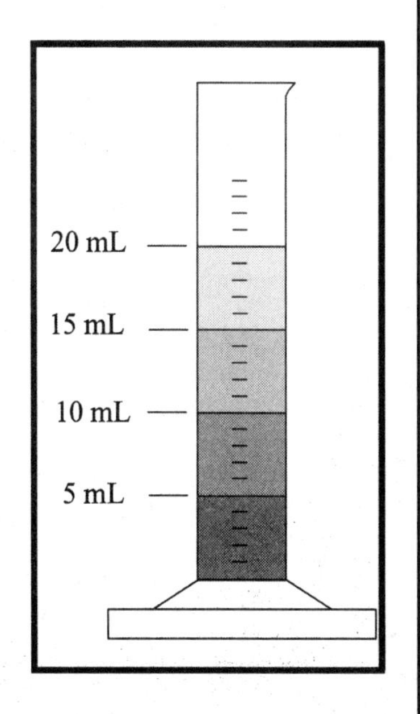

To verify the experiment, use your previous measurements of each liquid and the formula for calculating density (Equation 8.1). Compare your density calculation for each liquid against the position of that layer in the density column. The one on the bottom will be the highest density and moving up the column, each layer will be progressively less dense.

INTERACTIONS IN SOLIDS

Ionic solids. Ionic compounds have strong bonds because of the electrostatic attractions between positive and negative ions. These compounds are usually hard and have high melting points. For example, table salt is an ionic compound, and its melting point is around 800 °C (over 1400 °F). Ionic solids form geometric crystals (or **crystal lattices**) based on the arrangement of positive and negative ions. These solids are usually soluble in water, which means they will dissolve. When in a dissolved state, the ionic compound separates into ions. In a solid state, ionic compounds are not good conductors of electricity, but they will conduct electricity in their **molten** (melted) state or when dissolved in water.

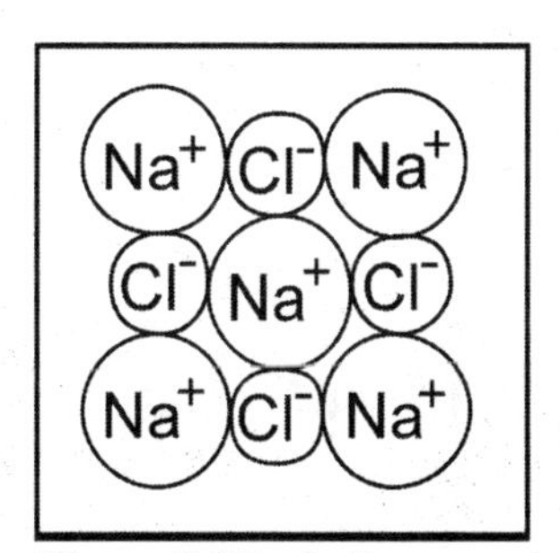

Figure 8.9 Ionic Compound Salt (NaCl)

Covalent solids. When nonmetal atoms share electrons to form a geometric crystalline structure, they form a **covalent solid**. This kind of solid does not form individual molecules. Instead, the atoms form a network of covalent bonds. For example, graphite sheets are formed from covalent bonding between carbon atoms as seen in Figure 8.10. Covalent solids are not good conductors of electricity. The bonds formed by the sharing of electrons are stronger than their attraction to any other substance that could act as a solvent. Therefore, these solids are not soluble. Some other examples are wax and diamond.

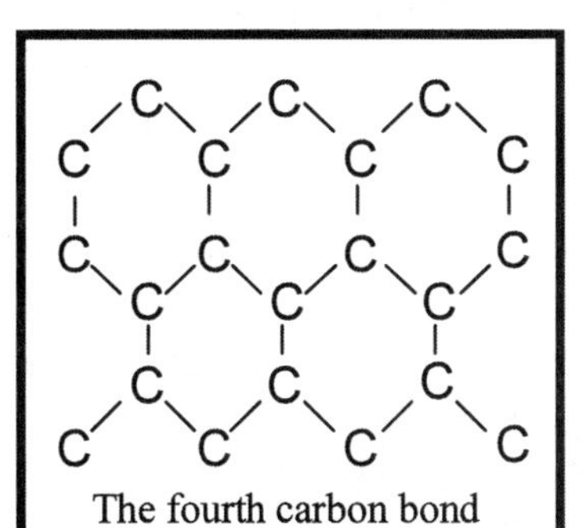

Figure 8.10 Molecular Structure of a Diamond

Molecular solids. Molecular solids are also solids formed by covalent bonding between atoms, but the atoms form individual molecules instead of a network of bonds. For example, a sugar molecule, which has the chemical formula of $C_6H_{12}O_6$, forms a molecule with 6 atoms of carbon, 12 atoms of hydrogen, and 6 atoms of oxygen. Molecular solids may still form some type of crystalline structure, but there is no sharing of electrons between molecules. Molecular solids are not good conductors of electricity. Molecular solids fall into two categories: polar molecular solids and nonpolar molecular solids. **Polar molecular solids** are usually soluble and have moderate melting points. **Nonpolar molecular solids** are usually insoluble, and their melting points are moderate or low. Sugar is an example of a polar molecular solid, and benzene is an example of a nonpolar molecular solid. Often they are liquids at room temperature and only become solids when cooled.

H O
C1
H–C2–OH
HO–C3–H
H–C4–OH
H–C5–OH
C6
H H OH

Figure 8.11 Molecular Structure of Sugar

Metallic solids. Metallic solids have a special type of bonding in which the electrons are free to move. This specific interaction of electrons in relation to the metallic nuclei gives metals some special characteristics. They have **luster** which means they reflect light. They are **ductile** which means they can be drawn out into a thin wire. They are **malleable**, which means they can be pounded into sheets. Because of the arrangement of electrons, they are also good conductors of electricity. Many metals exhibit magnetism. Melting points vary, but most metals are solids at room temperature.

PROPERTIES OF FLUIDS

Fluids are liquids, gases, and plasmas. The physical properties of a fluid are determined by the interactions among its particles. Particle interactions include the interactions between atoms that form molecules as well as the interactions between molecules of a substance.

INTERACTIONS IN LIQUIDS

The interactions of molecules in the liquid state affect its physical properties. Substances that are liquids at room temperature are most often molecular compounds. Molecular compounds can be polar or nonpolar.

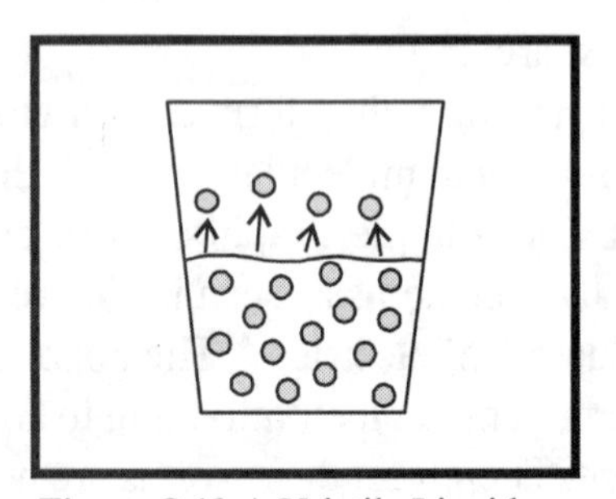

Figure 8.12 A Volatile Liquid

Volatility of Liquids. Volatility is the tendency of a liquid to vaporize (or evaporate) at a low temperature. Volatility of a liquid depends on molecular interactions. The weaker the interactions between molecules, the more volatile the liquid. In general, smaller molecules have a higher volatility than larger molecules. For example, nonpolar liquids with small molecules tend to be very volatile. The interactions between small nonpolar molecules are very weak, so it takes very little energy for the molecules to escape from their liquid state. Polar molecules have more of an attraction between particles because of the partial positive and negative charges, so they tend to be a little less volatile.

Boiling Points and Freezing Points of Liquids. Molecular interactions also affect boiling points and freezing points of liquids.

- A molecule that has little attraction for other molecules of its kind, such as nonpolar molecules, will have a low boiling point and a low freezing point.
- A molecule that has a strong attraction for other molecules of its kind, such as polar molecules, will have a high boiling point and a high freezing point.

For example, water is a molecule that exhibits polar covalent bonding. The polarity of the O-H bond induces a second type of bonding, called **hydrogen bonding**. As shown in Figure 8.11, hydrogen bonding occurs between hydrogen and oxygen molecules of neighboring molecules. Hydrogen bonds are weaker than covalent bonds, but they do give water a long-range stability. This increased stability results in a higher boiling point and a higher freezing point than might be expected for this relatively small molecule.

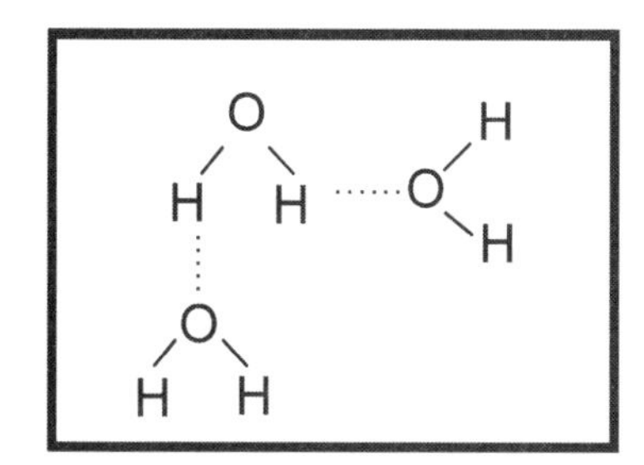

Figure 8.11 Hydrogen Bonding

Interactions in Gases

Size and **polarity** of molecules also affect the interaction of gases. Substances that are gases at room temperature are most often stand-alone atoms, diatomic molecules, or small, nonpolar molecular compounds. A small, stable atom such as helium has little attraction for other helium atoms; therefore, it exists as a gas at room temperature. The condensation point for helium or other atmospheric gases is very low. Hydrogen, oxygen, and nitrogen exist as diatomic molecules. Carbon dioxide is a small, nonpolar molecular compound. *In general, the smaller the gas molecule, the lower the condensation and freezing point it has.*

Interactions in Plasmas

Plasma is an electrically conductive collection of charged ions. Plasmas are generated by varying degrees of heat which produce varying degrees of ionization. The primary feature of the plasma is that enough energy must be added to keep the ions ionized. Lowering the energy added results in a plasma that reverts to a neutral gas. Increasing the pressure on a plasma forces ions and electrons into closer proximity and will also force them to recombine into a neutral gas. Decreasing the pressure on a plasma generally allows for a greater separation of ions and electrons and allows less energy to be added to maintain the plasma state.

On a separate note, the charged nature of the plasma means that it is strongly affected by both electric and magnetic fields. Plasmas can be shaped into sheets and filaments by the application of an electromagnetic field or any electrical stimulus.

Gas Laws

The collisions of the gas particles against the surface of the container cause the gas to exert pressure upon the container. **Pressure** is a force (push or pull) applied uniformly over an area. The SI unit of pressure is called a **pascal (Pa)**, and the English unit is called an **atmosphere (atm)**. The velocity of the gas particles relates to the temperature of the gas. The **gas laws** describe the relationship between the temperature, pressure, and volume of gases. The gas laws are summarized on the next page.

Gas Law	Type of Relationship	Relationship
Boyle's law	**Pressure-Volume (P-V) Relationship:** Increasing the pressure at a constant temperature decreases the volume of the gas. Conversely, decreasing the pressure at a constant temperature will increase the volume of the gas.	$V \propto \frac{1}{P}$
Charles' law	**Temperature-Volume (T-V) Relationship:** Heating a fixed amount of gas at constant pressure causes the volume of the gas to increase, and vice versa: Cooling a fixed amount of gas at constant pressure causes the volume of the gas to decrease.	$V \propto T$
Avogadro's law	**Volume-Amount (V-n) Relationship:** At constant pressure and temperature, the volume of a gas increases as the number of molecules in the gas increases.	$V \propto n$

The symbol $\propto$ means "is proportional to." When we say two quantities are **directly proportional**, we mean that when one quantity increases, so does the related quantity. The opposite is also true: when one quantity decreases, so does the related quantity. The quantities in Charles' and Avogadro's laws are directly proportional. Notice that Boyle's law states the volume is proportional to one divided by pressure. So, if volume increases, then the fraction 1/P must also increase. The value of the fraction 1/P is greater when the value for P is smaller. For instance, 1/2 is greater than 1/3, which is greater than 1/4. Thus, in Boyle's law, when the volume increases, the pressure must decrease. The converse is also true: when the volume decreases, the pressure must increase. In order for the fraction 1/P to decrease, the value of P must increase. The relationship in Boyle's law is called an **inversely proportional** relationship.

For the purpose of solving problems, Boyle's law and Charles' law are written in a different form. Another way of stating Boyle's law is P×V is always equal to the same constant as long as the temperature and amount of gas does not change. Boyle's law is expressed as

$$P_1V_1 = P_2V_2 \quad \textbf{Equation 8.2}$$

For a sample of gas under two different sets of conditions at constant temperature. V_1 and V_2 are the volumes at pressures P_1 and P_2, respectively.

Example: A sample of sulfur hexafluoride gas exerts 10 atm of pressure in a steel container of volume 5.5 L. How much pressure would the gas exert if the volume of the container was reduced to 2.0 L at constant temperature?

Step 1. Set up the equation: $P_1V_1 = P_2V_2$

Step 2. Insert the known information. In this problem, we know that the initial conditions were a pressure of 10 atm and a volume of 5.5 L. The final volume is 2.0 L. Therefore, the equation becomes:

$$(10\text{ atm})(5.5\text{ L}) = P_2(2.0\text{ L})$$

Step 3. Solve for P_2. $$P_2 = \frac{(10\text{ atm}) \cdot (5.5\text{ L})}{(2.0\text{ L})} = 27.5\text{ atm}$$

Charles' law can be rewritten as

$$\frac{V_1}{T_1} = \frac{V_2}{T_2} \qquad \textbf{Equation 8.3}$$

for a sample of gas under two different sets of conditions at constant pressure. V_1 and V_2 are the volumes at temperatures T_1 and T_2, respectively. Both temperatures are in Kelvin.

Example: Under constant-pressure conditions a sample of methane gas initially at 400 K and 12.6 L is cooled until its final volume is 9.3 L. What is its final temperature?

Step 1. Set up the equation: $$\frac{V_1}{T_1} = \frac{V_2}{T_2}$$

Step 2. Insert the known information. In this problem, we know that the initial conditions were a volume of 12.6 L and a temperature of 400 K. The final volume is 9.3 L.

Therefore, the equation becomes: $$\frac{12.6\text{ L}}{400\text{ K}} = \frac{9.3\text{ L}}{T_2}$$

Step 3. Solve for T_2. $$T_2 = \frac{(9.3\text{ L}) \cdot (400\text{ K})}{(12.6\text{ L})} = 295\text{ K}$$

Ideal Gas Equation (pressure-temperature-volume relationship): Heating a gas in a fixed volume container causes the pressure to increase. Conversely, cooling a gas in a fixed volume container causes the pressure to decrease. The ideal gas equation combines the three gas laws into one master equation to describe the behavior of gases,

$$PV = nRT \qquad \textbf{Equation 8.4}$$

where R is the gas constant.

Try to remember this equation using the pnemonic **"Piv Nert,"** which is how you would pronounce Equation 8.4 if you were to say it aloud as a word instead of as an equation. Each consonant in Piv Nert corresponds with a term in the Ideal gas equation as shown in Figure 8.12 to the right.

Results from these properties can be observed around us. Pressurized gases pose hazards during handling and storage. Pressurized gases should not be stored in hot locations or be handled near flames.

Balloons, whose volumes are not fixed, can also illustrate the behavior of gases. For example, if you put a balloon in a freezer, it will shrink because of the decreased pressure inside the balloon resulting from the lower temperature. If you take it to the top of a very high mountain, it will expand because of the decreased atmospheric pressure

Pressure

I

Volume

Number of molecules

E

R ⟶ gas constant

Temperature

Figure 8.12 Piv Nert

Practice Exercise 2: Gas Laws

1. A gas occupying a volume of 675 mL at a pressure of 1.15 atm is allowed to expand at constant temperature until its pressure becomes 0.725 atm. What is its final volume?
2. A 25 L volume of gas is cooled from 373 K to 300 K at constant pressure. What is the final volume of gas?
3. A sample of gas exerts a pressure of 3.25 atm. What is the pressure when the volume of the gas is reduced to one-quarter of the original value at the same temperature?
4. A sample of gas is heated under constant-pressure conditions from an initial temperature and volume of 350 K and 5.0 L, respectively. The final volume is 11.6 L. What is the final temperature of the gas?

Section Review 3: Physical Properties of Matter

A. Define the following terms.

physical properties	volatility	gas laws
extensive properties	ideal gas equation	directly proportional
intensive properties	pressure	inversely proportional
density	pascal	Boyle's law
fluids	atmosphere	Charles' law
		Avogadro's law

B. Choose the best answer.

1. If a physical property depends on the amount of matter present, it is a(n) __________ property.
   A. intrinsic B. extrinsic C. chemical D. dense

2. Density __________ as temperature increases.
   A. increases B. decreases C. stays the same D. is less noticeable

3. Pressure can affect the density of
   A. solids. B. liquids. C. gases. D. none of the above

4. Based on particle interactions, which of the following types of substances is most volatile?
   A. small, nonpolar molecules
   B. ionic compounds dissolved in a polar liquid
   C. large, polar molecules
   D. small, polar molecules which also exhibit hydrogen bonding

5. Cartridges used to fire paint balls are filled with carbon dioxide gas. Each time a paint ball is fired, some carbon dioxide gas escapes. The volume of the cartridge is rigid and does not change. Hampton buys a new carbon dioxide cartridge. Lisa has the same cartridge, but hers has been used to fire several paint balls. Which of the following is true of the cartridges assuming both cartridges are at the same temperature?
   A. The pressure in Hampton's cartridge is greater than the pressure in Lisa's cartridge.
   B. The pressure in Hampton's cartridge is less than the pressure in Lisa's cartridge.
   C. The pressure in Hampton's cartridge is equal to the pressure in Lisa's cartridge.
   D. No relationship can be determined from the given information.

6. What would be the best way to convert a sample from a plasma to a gaseous state?
   A. Increase the pressure
   B. Decrease the pressure
   C. Increase the temperature
   D. Apply an electric field

7. Which type of solid is the best conductor of electricity?
   A. ionic solid
   B. covalent solid
   C. metallic solid
   D. molecular solid

8. Which of the following states of matter would be least likely to respond to a change in pressure?
   A. liquid
   B. gas
   C. plasma
   D. All of these will respond to a change in pressure.

9. Which of the following types of solid is the most soluble in water?
   A. ionic solid
   B. nonpolar molecular solid
   C. polar molecular solid
   D. A or C are both water soluble

10. Which Gas Law gives a proportional relationship between temperature and volume?
   A. Boyle's Law
   B. Charles' Law
   C. The Ideal Gas Law
   D. B and C

C. Answer the following questions.

1. Using your understanding of particle interactions, explain why an oil and vinegar salad dressing separates. If placed in a density column, which one would settle to the bottom?
2. One molecule of carbon dioxide, CO_2, and one molecule of water, H_2O, are each made up of 3 atoms. Use your knowledge of particle interactions to explain why carbon dioxide is a gas at room temperature, but water is a liquid at room temperature.
3. Why are diatomic molecules like O_2 nonpolar?
4. Look at the two containers below. Assuming they have identical contents, which container is most likely to have the highest pressure? Why?

A. 

B. 

5. Look at the two pictures right. Which position of the piston creates the least pressure in the container? Why?

A. 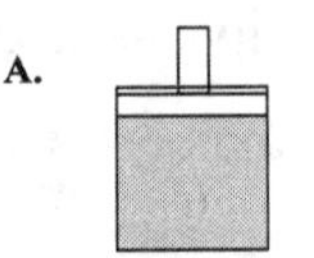

B. 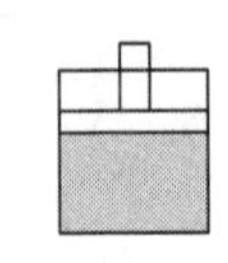

ENERGY CHANGES IN MATTER

We now know that matter can undergo physical as well as chemical changes. Both physical and chemical changes involve a change in energy. This section will deal primarily with energy changes in chemical reactions.

As we previously discussed in the chapter, energy is the ability to do work, and work is the process of moving matter. Energy falls into two broad categories: potential energy and kinetic energy. Potential energy is stored energy due to the object's position or state. Kinetic energy is energy in motion; it changes as an object moves from one position to another. There are many forms and sources of these types of energy, several of which we will discuss throughout the book. Right now, we are only concerned with the chemical and thermal energy associated with chemical reactions.

When a chemical reaction takes place, atoms exchange, transfer, or lend electrons to another atom. During these processes, chemical energy is released or absorbed in the form of light (electromagnetic energy) and heat (thermal energy). **Thermal energy** is the energy associated with the random movements of atoms and molecules; it is experienced as heat. You have learned that all matter is made up of atoms that are in constant motion. The vibrations, rotations, and velocity of an atom make up its kinetic energy. The total of these energies in a substance is called its **enthalpy** or **heat content**.

If you have a container of gas in which the atoms are moving slowly, and you combine that container of gas with a container of gas with faster moving atoms, the atoms will bump into each other. As they do, the faster moving atoms will transfer energy to the slower moving gas atoms. When the gas atoms are all colliding at the same speed, they have reached **equilibrium**. At that point, they will have the same internal energy.

The amount of energy, like matter, is always conserved. The **law of conservation of energy** states that you cannot destroy energy, but it can transfer from one form into another.

Exothermic and Endothermic Reactions

Chemical changes in matter occur because of chemical reactions. Chemical reactions are either **exothermic**, which means they give off energy or **endothermic**, which means they absorb energy. Physical changes in matter can also be exothermic or endothermic. In an exothermic reaction, the products have less energy than the reactants since energy is given off. In an endothermic reaction, the products have more energy than the reactants since energy is absorbed. Most often, the energy given off or absorbed is in the form of heat. **Heat** is the transfer of thermal energy between two bodies at different temperatures. In all cases, energy is conserved, which means it is neither created nor destroyed. The energy is just transferred from one form to another.

Exothermic reactions release heat energy. Exothermic reactions or processes are often spontaneous. This release or production of heat warms the surrounding area. One example of an exothermic reaction is combustion. The combustion of fossil fuels is the source of most of the world's energy. Condensing steam is an example of an exothermic process. The steam gives up energy to condense into a liquid form. The liquid state of a substance has less enthalpy than its gaseous state, so going from a gas to a liquid is exothermic. An example of an exothermic chemical reaction is the decomposition of food in a compost pile. Compost made up of grass clippings and leftover vegetable peels gives off heat because bacteria and other organisms break down the matter into simpler substances. Another example is rusting. Iron exposed to oxygen will react to form rust and give off heat. Rust, the product, has less energy than iron and oxygen, the reactants. Other examples of exothermic reactions are personal hand warmers and portable heating pads.

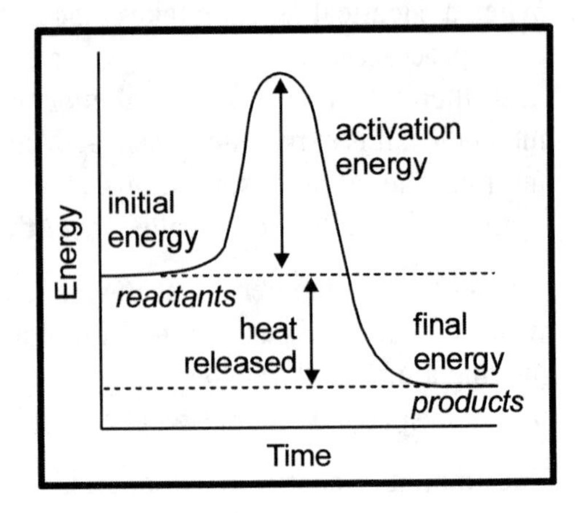

Figure 8.13 Exothermic Reaction

Often, a chemical change or reaction must be started by adding energy or heat. Once the chemical reaction begins, it gives off more energy than was added to start the reaction. The energy needed to start the reaction is called the **activation energy**. If the energy at the end of the reaction is less than the energy at the beginning of the reaction, it is still exothermic. Burning wood gives off energy in the form of light and heat, but wood does not burn spontaneously under ordinary conditions. For example, a match does not burn until friction is added to form a spark. Once the spark ignites the match, it burns and gives off more energy in the form of heat and light than the initial spark. Once the match has burned, it cannot be used again to give off energy. The total energy of the match after it has burned is less than the energy before it burned; therefore, it is an exothermic reaction.

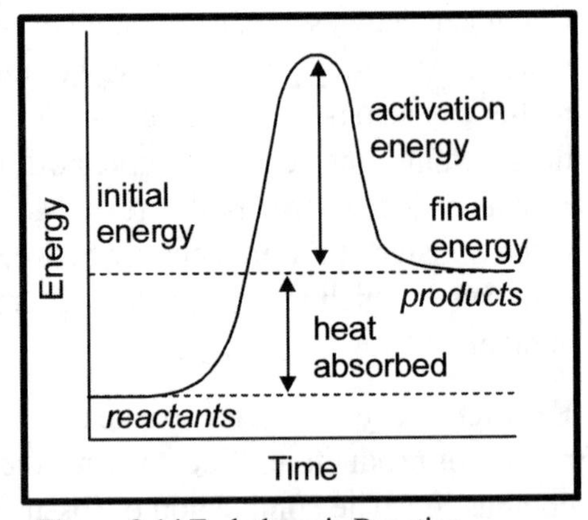

Figure 8.14 Endothermic Reaction

Some reactions absorb energy causing the products to have more energy than the original reactants. These endothermic reactions are not as common. In an **endothermic** reaction, heat energy is absorbed. This absorption of energy results in the cooling of the surrounding area. Heat transfers from the surrounding area to the point of the chemical reaction. An example of an endothermic process is melting ice. Ice must absorb energy to melt. The liquid water has more energy than the ice; therefore, it is an endothermic process. Another example of an endothermic chemical reaction is the medical cold pack included in some emergency first aid kits. A membrane separating two chemicals must be broken by bending or kneading the pack. Once the chemicals mix, they react and absorb heat causing it to feel cold.

Not all chemical reactions give off or absorb energy in the form of heat. Some generate or absorb energy in the form of light or electricity. For example, a burning candle gives off heat, but it also gives off light energy. The burning of fossil fuels also releases heat and light energy. A car battery produces electrical energy, and a recharged battery stores electrical energy.

You learned earlier in this chapter that catalysts can increase the rate of reaction. Many times, they do this by decreasing the activation energy needed for a reaction. Figure 8.15 shows what an energy graph of a reaction might look like with and without a catalyst. Notice that the beginning energy and the final energy are the same for both reactions, but the energy needed for the reaction to occur is less with a catalyst.

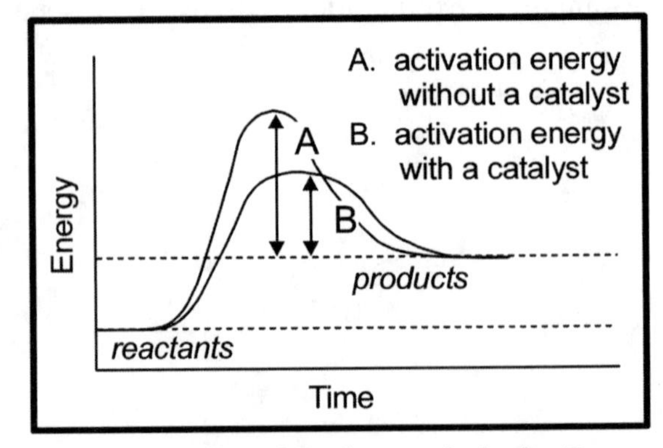

Figure 8.15 Effect of Catalysts on Activation Energy

HEAT TRANSFER

When an area of high temperature comes in contact with an area of low temperature, heat will transfer from the high temperature area to the low temperature area. The high temperature area will cool, and the movement of particles will decrease. The low temperature area will warm, and the movement of particles will increase.The transfer of heat occurs in three ways: **conduction**, **convection**, and **radiation**.

In the heat of **conduction**, kinetic energy transfers as particles hit each other directly. During this type of heat transfer, the two bodies are in direct contact with one another.

Example: The burner of the stove conducts the heat to the bottom of the teapot.

Figure 8.16 Conduction

Figure 8.17 Convection

The transfer of heat by **convection** in liquids and gases produces currents in the heated substance.

Example: The water at the bottom of the teapot becomes hot. These heated water particles move to the top of the water where they cool and fall to the bottom of the teapot to become re-heated. The heating and cooling of particles creates currents that circulate the heat throughout the water.

Radiation is the transfer of heat energy by waves. Radiant energy travels in a straight line at the speed of light. Microwave ovens use radiant energy to cook food. The sun is a source of radiant energy that heats and lights the earth.

Example: The heat from the burner and the teapot move through the air.

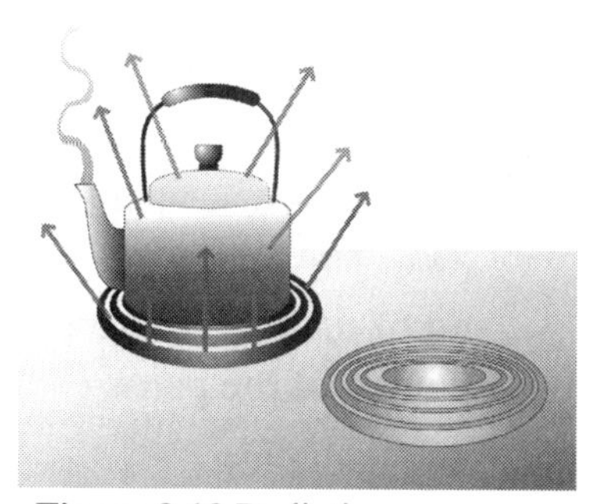

Figure 8.18 Radiation

Thermal insulators such as cork, fiberglass, wool, or wood slow the transfer of heat. **Thermal conductors** are substances that allow heat energy to transfer quickly. Many types of metal, such as copper and aluminum, are good thermal conductors.

Section Review 4: Energy Changes in Matter

A. Define the following terms.

energy	temperature	activation energy
work	enthalpy (heat content)	conduction
potential energy	law of conservation of energy	convection
kinetic energy	exothermic	radiation
chemical energy	endothermic	thermal insulators
thermal energy	heat	thermal conductors

B. Choose the best answer.

1. The energy produced by a fire is
   A. thermal energy.
   B. light energy.
   C. chemical energy.
   D. all of the above.

2. Which of the following is true concerning the law of conservation of energy?
   A. Energy is conserved when converting from potential to kinetic energy, but friction destroys it.
   B. Energy cannot be destroyed, but it can transfer from one form to another.
   C. Converting from one form of energy to another destroys a negligible amount of energy.
   D. Energy is conserved as long as mass is also conserved, but when mass is destroyed, energy is also destroyed.

3. Francisco was heating soup in a metal pan on the stove. He noticed that the soup was about to boil over. He quickly grabbed the handle of the pan to remove it from the heat. Just as quickly, he let go of the pan because he burned his hand. What kind of heat transfer occurred through the metal handle of the pan?
   A. radiation
   B. convection
   C. conduction
   D. chemical transfer

4. Which of the following processes gives off energy?
   A. melting ice
   B. burning propane in a gas heater
   C. sublimation of carbon dioxide ice to carbon dioxide gas
   D. recharging a car battery

5. Which of the following is true of a chemical change?
   - A. Chemical changes always absorb energy.
   - B. Chemical changes always release energy.
   - C. Chemical changes can either give off energy or absorb energy.
   - D. Chemical changes do not cause a change in energy.

6. The vegetable matter in Pierre's compost pile changes into black "soil." He observes that the compost pile is always warmer than the outside air temperature. Which of the following describes this change?
   - A. The vegetable matter undergoes a physical change that releases energy.
   - B. The vegetable matter undergoes a chemical change that releases energy.
   - C. The vegetable matter undergoes a physical change that absorbs energy.
   - D. The vegetable matter undergoes a chemical change that absorbs energy.

7. Look at the graph at the right. Which of the following is true about the reaction shown in the graph?
   - A. The reaction requires energy to begin, and the final energy is greater than the beginning energy.
   - B. The reaction requires energy to begin, and the final energy is less than the beginning energy.
   - C. The reaction is spontaneous without any addition of energy.
   - D. The reaction absorbs energy and stores it in its products.

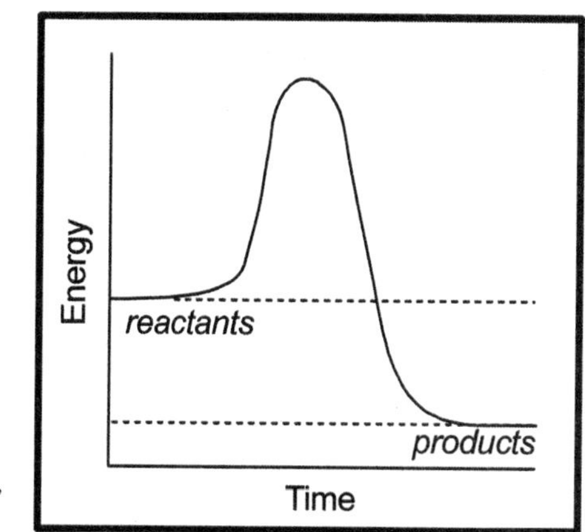

C. Answer the following questions.

1. Explain how heat is different from energy.
2. List and describe the four types of energy discussed in this section.
3. Compare and contrast endothermic and exothermic reactions.
4. Give an example of a chemical reaction that gives off energy. What kind of energy is released in your example?
5. Give an example of a physical change that absorbs energy. What kind of energy is absorbed in your example?
6. Give an example of a chemical reaction that absorbs energy. What kind of energy is absorbed in your example?
7. Give an example of a physical change that gives off energy. What kind of energy is released in your example?
8. In chemistry lab, Claudette mixes two chemicals in a test tube, and the tube gets very cold. Explain what kind of energy change is occurring in the test tube.

9. In a battery-powered flashlight, a chemical reaction in the battery produces electricity. Is this reaction endothermic or exothermic? Cite evidence for your choice.

10. Consider the following equation for photosynthesis:

$$6H_2O + 6CO_2 + \text{sunlight} \rightarrow C_6H_{12}O_6 + 6O_2$$

Is energy absorbed or released in this reaction? Justify your answer.

CHAPTER 8 REVIEW

A. Choose the best answer.

1. An object with a mass of 30 g and a volume of 6 cm^3 has a density of

   A. 5 g/cm^3. B. 15 g/m^3. C. 180 g/cm^3. D. 180 $g \cdot cm^3$.

2. When might water absorb heat but not change temperature?

   A. when it is in a liquid state

   B. when it is changing from a liquid to a gas

   C. when it is ice

   D. Under no circumstance will water absorb heat but not change temperature.

3. In which of the following situations would water molecules have the most energy?

   A. when water is frozen as ice

   B. in a mixture of ice and water

   C. when water is boiling

   D. when water is superheated steam

4. The term "fluid" applies to

   A. liquids only.

   B. gases and liquids.

   C. gases, liquids and plasmas .

   D. gases only.

5. Look at the two pictures at right. Both cylinders contain the same volume of the same gas at the same temperature. Which of the following statements is true?

   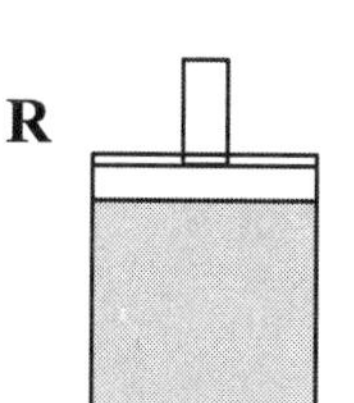

   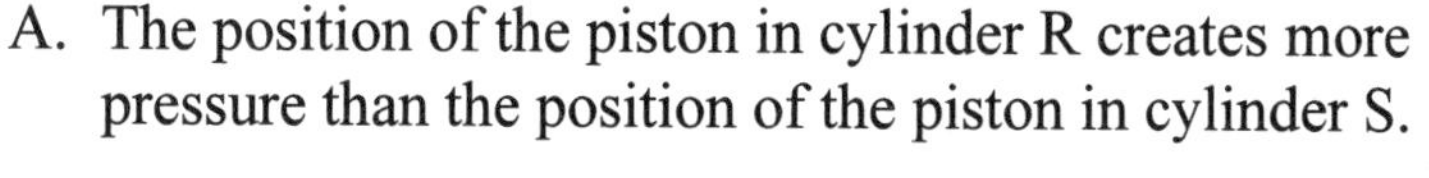

   A. The position of the piston in cylinder R creates more pressure than the position of the piston in cylinder S.

   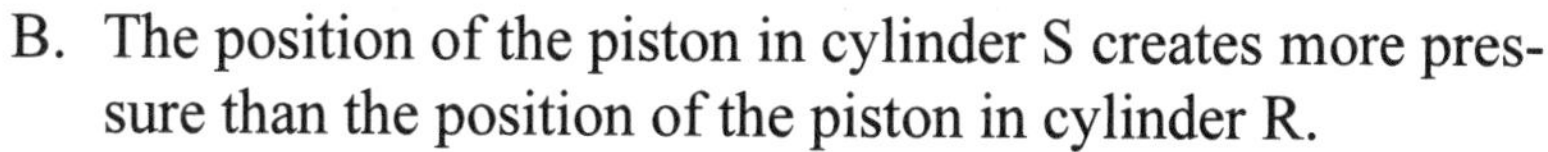

   B. The position of the piston in cylinder S creates more pressure than the position of the piston in cylinder R.

   C. There is no way to compare the pressure in the cylinders with the information given in the problem.

   D. The position of the pistons does not affect the pressure within the cylinders.

6. When the temperature of a particle of matter rises, there is an increase in the

   A. size.

   B. potential energy.

   C. mass.

   D. kinetic energy.

7. Volatility is the tendency of a liquid to
   A. disappear.
   B. vaporize.
   C. burn.
   D. explode.

8. Mixing salt and sugar is a
   A. physical change.
   B. chemical change.
   C. new substance.
   D. new compound.

9. When a substance condenses, it changes from
   A. a liquid to a solid.
   B. a liquid to a gas.
   C. a gas to a liquid.
   D. a gas to a solid.

10. Mixtures can be separated by physical means. Which is ***not*** a way to separate mixtures?
    A. evaporation
    B. filtering
    C. magnetic separation
    D. cooking

11. A substance that can be separated into its simplest parts by physical means is
    A. water.
    B. salt.
    C. salt water.
    D. hydrogen dioxide.

12. Two equivalent samples of argon gas are placed in two containers of equal and constant volume. The temperature of Sample A is increased by 10°C. The temperature of Sample B is kept constant. Which statement is true?
    A. The pressure of Sample A increases.
    B. The pressure of Sample A decreases.
    C. The pressure of Sample A is constant.
    D. The pressure of Sample B and Sample A are equal.

13. Carbon dioxide, CO_2, is an example of a(n)
    A. solution.
    B. compound.
    C. element.
    D. mixture.

14. Which of the following does ***not*** create a mixture?
    A. melting ice
    B. stirring flour in water
    C. salting rice
    D. making a salad

15. Use the graph below to answer the following question.
Which of the following statements is ***not*** true?

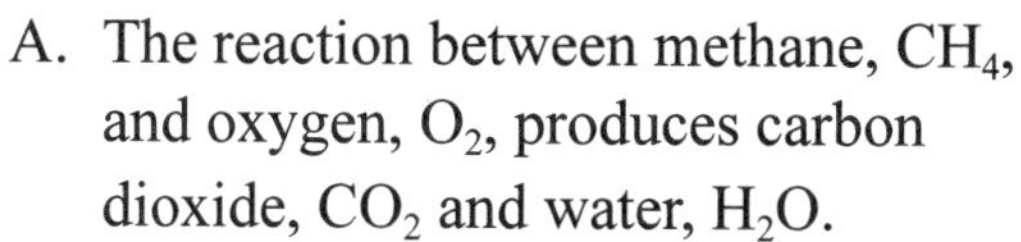

A. The reaction between methane, CH_4, and oxygen, O_2, produces carbon dioxide, CO_2 and water, H_2O.

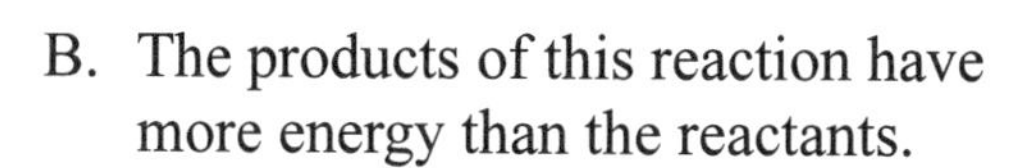

B. The products of this reaction have more energy than the reactants.

C. The reaction is exothermic and, therefore, releases energy.

D. The reaction requires energy in order to begin.

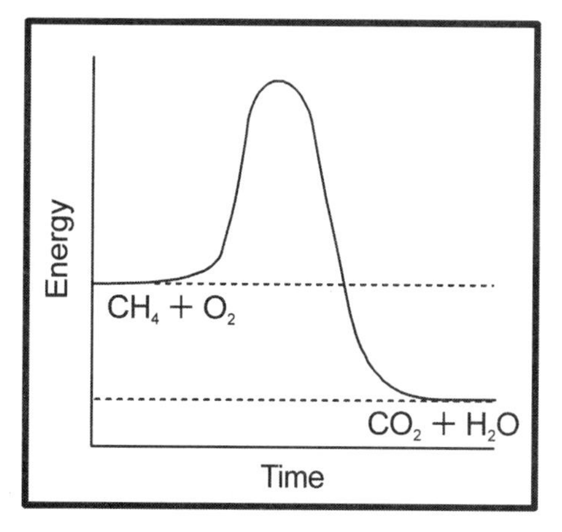

16. A package of frozen gravy is placed in a microwave. This is an example of heat transfer by

A. conduction. B. convection. C. radiation. D. insulation.

17. Water heating in a water heater is an example of heat transfer by

A. conduction. B. convection. C. radiation. D. insulation.

18. Water is heated in an Erlenmeyer flask. After the water boils, the flask is removed from the burner and placed on a wooden cooling trivet. The wood __________ the transfer of ____________ from the glass to the laboratory benchtop.

A. speeds up, thermal energy
B. slows, radiant energy
C. slows, thermal energy
D. speeds up, radiant energy

19. Which state of matter usually consists of molecules, rather than atoms?

A. solid B. liquid C. gas D. plasma

20. Which state of matter consists of ions, rather than atoms or molecules?

A. solid B. liquid C. gas D. plasma

21. A group of students boiled saltwater and condensed the vapor given off while the saltwater was boiling. They collected the condensate. Select the best description of that condensate.

A. pure water
B. deionized water
C. saltwater
D. salt

Chapter 9
Solutions

PHYSICAL SCIENCE STANDARDS COVERED IN THIS CHAPTER INCLUDE:

SPS6 Students will investigate the properties of solutions.

SOLUTION PROPERTIES

All solutions are homogeneous because their parts are evenly distributed. The parts of a solution cannot be separated by filtering. A solution is made up of a **solute** and a **solvent**. The solute is the substance that dissolves or disappears and the solvent is the part that does the dissolving. A solution can become saturated. When **saturated**, no more solute will dissolve in the solvent.

Solutions can be mixtures of solids, liquids, or gases. For example, bronze is a solid solution of copper and tin. Soda is a solution of carbon dioxide, a gas, dissolved in water, a liquid. Salt water is a solution of salt, a solid, dissolved in water, a liquid.

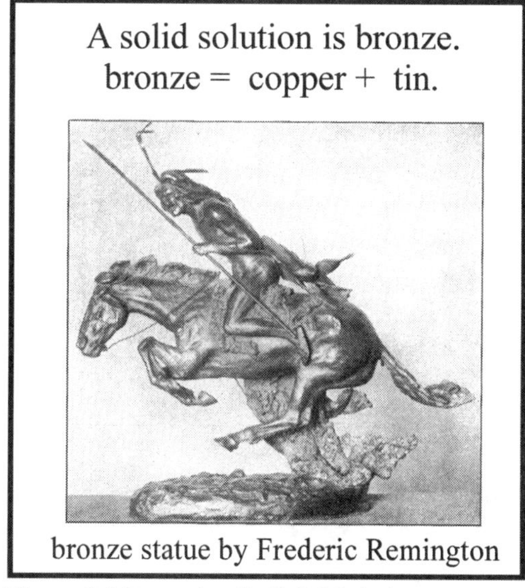

Figure 9.1 Example of a Solid Solution

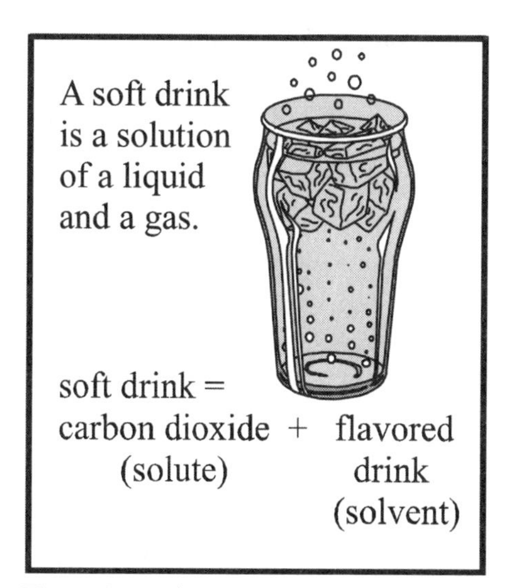

Figure 9.2 Soda as an Example of a Solution

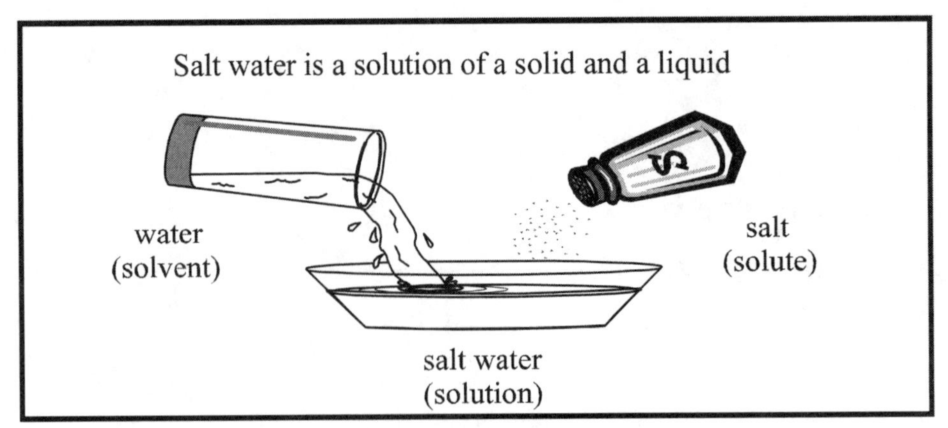

Figure 9.3 Salt Water as an Example of a Solution

SOLUBILITY OF MATTER

A solution can contain dissolved molecules or ions or a combination of the two. Some ionic solutions, such as salt water where NaCl dissociates into Na^+ ions and Cl^- ions, can conduct electricity (this explains why swimming during a thunderstorm is dangerous). The solubility of a substance is one property that is used to distinguish one substance from another. The solubility is measured by the **concentration** of the solute in the solvent. The concentration is the grams dissolved per volume of H_2O. Many factors affect the solubility of solutes in solvent, which we will look at next.

IDENTITY OF SOLUTE AND SOLVENT

There is a saying among scientists that explains why a solute will dissolve in some solvents but not in others. The saying is: **Like Dissolves Like**. It means that solutes and solvents that have similar molecular polarity will interact. Let's use a few examples.

Polar/Polar: Water is a polar solvent and easily dissolves the polar NaC1 molecule, as in Figure 9.4

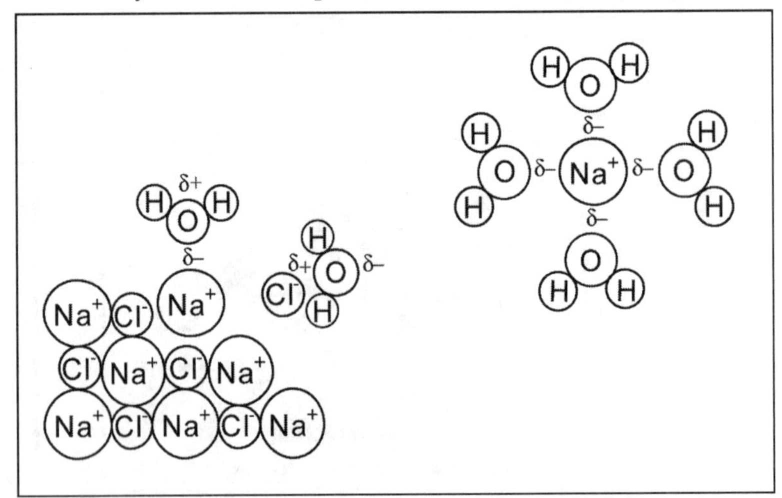

Figure 9.4 Salt Dissolving in Water

Polar/Nonpolar: Water will not dissolve the non-polar solute wax.
Nonpolar/Polar: The nonpolar solvent gasoline will not dissolve polar sugar molecules.
Nonpolar/Nonpolar: Gasoline will dissolve the nonpolar solute oil, like the oil stains on a driveway.

Keep in mind the following general rules:

- Most organic (carbon-based) compounds are nonpolar, and will not dissolve in water.
- Most ionic solids are polar, and will dissolve in water.
- Most importantly: LIKE DISSOLVES LIKE!

PRESSURE

Air pressure has no effect on solid or liquid solutes. However, an increase in pressure of a gaseous solute above the solvent pressure increases the solubility of the gas. For example, when a carbonated drink is placed in a can, pressure is added to keep the carbon dioxide in the liquid solution. However, when the tab is popped and the pressure is released, the carbon dioxide begins to escape the liquid solution.

The effect of pressure on gas solubility has important implications for scuba divers. Underwater, pressure increases rapidly with depth. The high pressure allows more nitrogen than usual to dissolve in body tissues. If divers ascend too rapidly, the lower pressure causes the nitrogen gas to come out of solution, forming gas bubbles in the blood and tissues. The gas bubbles result in a condition called "the bends," which can cause severe pain, dizziness, convulsions, blindness, and paralysis. Divers must ascend to the surface slowly in order to keep air bubbles from forming.

SURFACE AREA

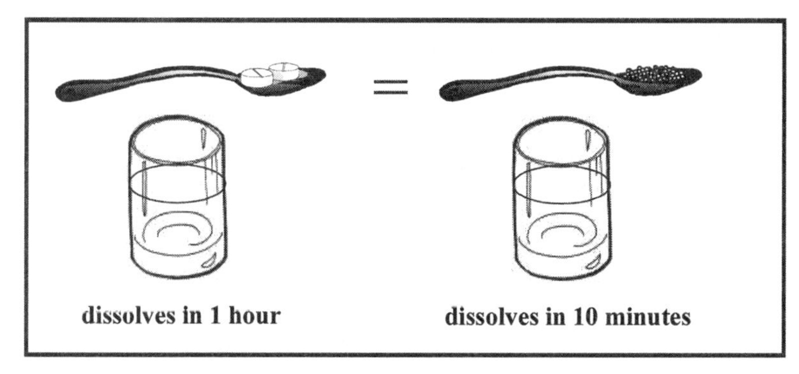

Figure 9.5 Effect of Surface Area on Solubility

The surface area of a solid solute also affects the rate of its solubility. The more surface area that is exposed to the solvent, the more readily the solute can interact with the solvent. This increased rate of reaction occurs because there is an increased chance of collisions between reactant particles. Since there are more collisions in any given time, the rate of reaction increases. For example, suppose you had a medicine which you can take in the form of a pill or a powder. Which substance would enter the body more quickly, the pill form or the powder? The answer is the powder because there is more surface area available for interaction with the solvent — in this case, stomach acid.

TEMPERATURE

Have you ever noticed that you can dissolve more sugar in hot tea than you can in cold? As you increase the temperature of a solvent, you can increase the solubility of liquids and solids. Viewing the graph in Figure 9.6, you see how the solubility of the salt and potassium nitrate increases with higher temperatures.

The solubility of gases, however, has the opposite relationship with temperature. As the temperature increases, the solubility of gases in solution decreases. For example, an open carbonated beverage will lose its fizz quickly in a hot environment, while the fizz escapes slowly in a cool environment. A decrease in temperature gives gas a greater solubility.

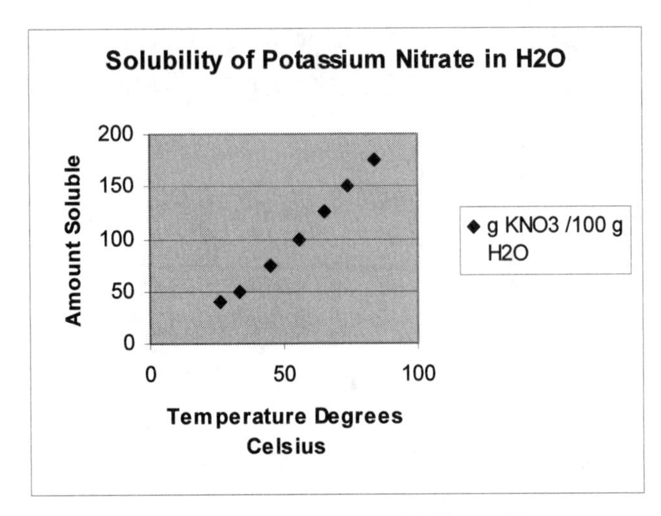

Figure 9.6 Relationship between Solubility and Temperature

DEGREE OF SOLUBILITY

NaC1 dissolves very well in water, because they are both polar. Granulated NaC1 will dissolve more quickly than a big block of the same mass. Also, the salt will always dissolve more quickly when the water is heated. But how much salt can be dissolved? At some point, the solution becomes **saturated** — that is, it cannot dissolve any more solute. The solubility of sodium chloride in water is 36.0 g/100 mL at 20°C. That means if you add 37.2 grams of NaC1 to 100 mL of water in your beaker, 1.2 grams will not dissolve. Instead they will settle to the bottom of the beaker.

However, we know that if we heat a solution, that we can dissolve more solute in it, more quickly. If we heat the NaC1 solution, we should be able to dissolve more salt, creating a **supersaturated** solution. Is this true for every solute? Look at the **solubility curves** in Figure 9.7.

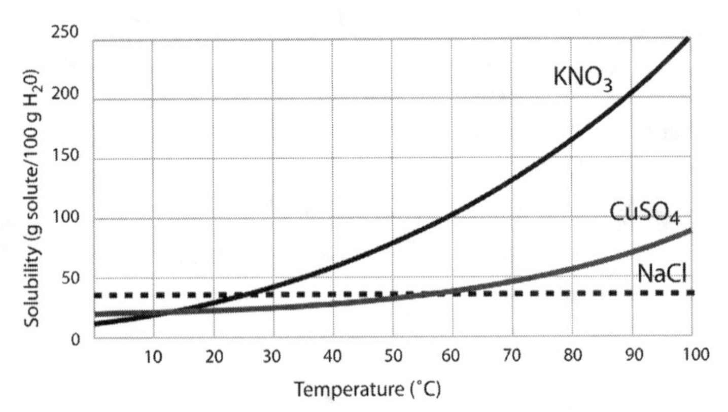

Figure 9.7

It is quite clear that all salts do not respond in the same way to heating. KNO_3 (potassium nitrate) becomes dramatically more soluble as temperature increases. Copper sulfate shows a more modest rise and NaC1 barely increases at all. You might be surprised to learn that increasing the temperature actually decreases the solubility of some ionic compounds, like $CaSO_4$ (calcium sulfate).

Section Review 1: Simple Solutions

A. Define the following terms.

solution	saturated	supersaturated
solute	Like Dissolves Like	solubility curve
solvent	polar	

B. Choose the best answer.

1. The fizzing in a soft drink is due to
   A. oxygen dissolved in liquid.
   B. alcohol in a liquid.
   C. carbon dioxide dissolved in liquid.
   D. liquid changing to a gas.

2. In which of the following will sugar be harder to dissolve?
   A. hot tea B. warm milk C. hot coffee D. iced tea

3. The substance that dissolves the solute is called the
   A. solution. B. solvent. C. solid. D. salt water.

4. Which form of matter usually increases its solubility as the temperature decreases?
   A. liquid B. gas C. solid D. ice

5. No matter how hot you make the tea, you cannot dissolve any more sugar into the solution. This is an example of a ___________ solution.
   A. solvent B. hot C. heterogeneous D. saturated

6. Which form of matter increases its solubility as pressure is increased?
   A. solid B. gas C. liquid D. powder

C. Answer the following questions.

1. Draw a water molecule. Label the hydrogen and oxygen atoms. Which part of a water molecule is partially negative δ^-? partially positive δ^+?
2. Name two ways to increase the solubility of a gas.
3. Which form of matter experiences increased solubility as a greater surface area is exposed to a solvent, and why?

IONS IN SOLUTION

ELECTROLYTES

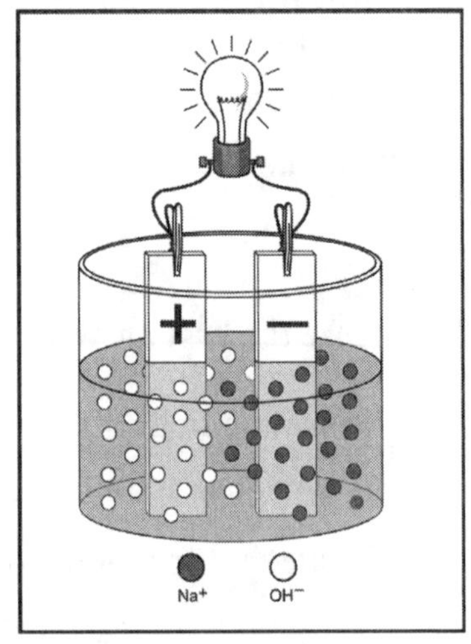

Figure 9.8 Conductivity Device Powered by Electrolytes

An **electrolytic solution** is a solution that can act as an electrical conductor in which current is carried by ions rather than by free electrons, which is how electricity is conducted through metals. The negatively and positively charged ions that carry the electrical current are called **electrolytes**. Electrolytes are also commonly called salts. Let's consider NaOH as an example. As shown in Figure 9.8, NaOH dissociates into Na^+ ions and OH^- ions in an aqueous solution. If a conductivity device is placed in the solution, the positively charged Na^+ ions will move toward a negative electrode, or **cathode**. The negatively charged OH^- will move toward a positive electrode, or **anode**. The dissolved ions in solution act as the power source conducting electricity between the electrodes and, therefore, the light bulb glows. The more ions there are present in the solutions, the stronger the electrical current, and the brighter the light bulb will glow. The presence of electrolytes in the human body is essential for the proper functioning of cells, especially the nerve, heart, and muscle cells. Nerve impulses and muscle contractions are both results of electrical currents carried by electrolytes. Electrolytes in the body are found in the blood, plasma, and interstitial fluid, which is the fluid between cells. Table 9.1 lists the major electrolytes in the body.

Table 9.1 Electrolytes in the Body

sodium Na^+	magnesium Mg^{2+}
potassium K^+	bicarbonate HCO_3^-
chloride Cl^-	phosphate PO_4^{2-}
calcium Ca^{2+}	sulfate SO_4^{2-}

Electrolytes, especially sodium, are lost through sweat during exercise. Vomiting and diarrhea can also cause rapid loss of electrolytes. Symptoms of electrolyte loss include headache, thirst, and muscle cramps. Severe loss of electrolytes can lead to disorientation, heat stroke, and even death. During exercise it is important to drink fluids to replace the water lost in sweating. Many sports drinks are formulated to help replace electrolytes lost during exercise and sports activities. Eating something salty, such as pretzels, will also help to restore sodium to the body.

ELECTROLYSIS

Electrolysis is the decomposition of a compound into simpler substances by passing an electric current through the compound. For example, water chemically separates into hydrogen and oxygen by electrolysis as seen in Figure 9.9. The negative terminal of the battery attracts positive hydrogen ions, whereas the positive terminal of the battery attracts negative oxygen ions.

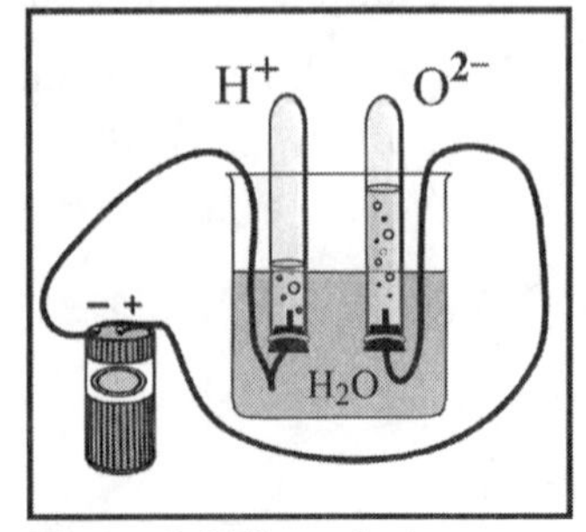

Figure 9.9 Electrolysis

Section Review 2: Ions in Solution

A. Define the following terms.

electrolytic solution electrolytes cathode anode

B. Choose the best answer.

1. In an electrolytic solution, electrical current is carried by

   A. isotopes. B. ions. C. iodine. D. indicators.

2. Amanda set up a conducting device in her chemistry lab and was using saltwater as her power source. When she placed the electrodes in the solution, the light bulb glowed very dimly. What can Amanda do to make her light bulb glow more brightly?

   A. Add more water to the solution.
   B. Cool the solution down.
   C. Use pure water instead of saltwater.
   D. Add more salt to the solution.

3. K^+, Na^+, and Ca^{2+} are ions found in the body and are necessary for proper cellular functions involving electrical impulses. These ions are

   A. vitamins. B. acids. C. organic. D. electrolytes.

C. Answer the following questions.

1. Name two functions of the human body that are dependent on the presence of electrolytes.
2. What are the symptoms of electrolyte loss? How can electrolytes be restored to the body?

SEPARATING SOLUTION COMPONENTS

We know that a solution is a physical mixture of components: a solute dissolved in a solvent. How can we separate the mixture — that is, can we get the solute out of the solvent? Since neither the solute nor the solvent has chemically changed, they can be separated. There are several methods to choose from.

DISTILLATION

Distillation is the process of changing a substance from a liquid to a gas, and then allowing the gas to become a liquid again for the purpose of separating a mixture of substances or taking out impurities in the substance. For example, Figure 9.10 shows the distillation of salt water to separate the water from the salt. Since the salt will not evaporate but the water will, the water can be boiled, and the pure water vapor condensed.

In the field of manufacturing, distillation is the most widely used method for separating substances. When separating a mixture of liquids, distillation takes advantage of different boiling points of the substances. For example, in an alcohol/water solution, alcohol boils at a lower temperature than water. Therefore, it will vaporize first, and its vapors can be collected and condensed to separate it from the water. Distillation is also used to separate and purify petroleum products.

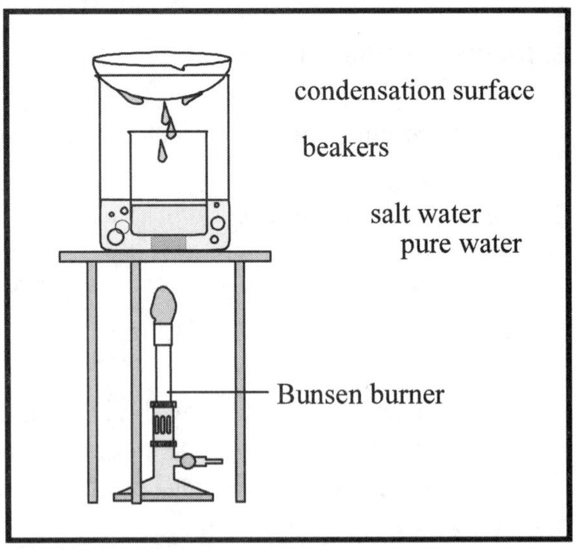

Figure 9.10 Distilling Water from a Salt Water Solution

Filtration

Filtration separates a solid from a liquid by passing the liquid through a porous filter. If the solid particles are larger than the pores in the filter, the solid particles collect on the filter surface, but the liquid passes through. Filters will not separate particles that are dissolved in the liquid, but will separate particles suspended or otherwise mixed in with the liquid. For example, a mixture of dirt and water could be separated by filtration. Filtration is usually the first step in water purification systems.

Dialysis

Dialysis is similar to filtration but at a molecular level. Dialysis uses a semipermeable membrane that allows small molecules to pass through but not larger molecules. The membrane is like a filter for molecules. Using reverse osmosis to purify drinking water is a type of dialysis. Dialysis machines are used in the medical field to purify blood.

Condensation

As distillation takes advantage of different boiling points of liquids, condensation takes advantage of different condensation points of gases. By lowering the temperature of a gas mixture, a gas with a higher condensation point will condense first and can then be collected. For example, water vapor in the air can be removed by lowering the temperature. For this reason, air conditioners decrease the amount of humidity in a room. Condensation can also be used to separate other atmospheric gases; i.e., oxygen and nitrogen.

Separating a Gas from a Liquid

Gases also dissolve in liquids. For example, soda is carbonated which means carbon dioxide is dissolved in the liquid. Gases become more soluble (more will dissolve) at lower temperatures and higher pressures. Think about a bottle of soda. Soda is canned or bottled under pressure and is kept cold. When it gets warm, it goes flat because the carbon dioxide becomes a gas and is no longer in the soda. When opened, it fizzes as the pressure is released. The fizz is caused by the carbon dioxide escaping from the soda. So, to separate a gas from a liquid, increase the temperature or decrease the pressure.

Section Review 3: Separating Physical Mixtures

A. Define the following terms.

physical change	filtration	condensation
distillation	dialysis	

B. Choose the best answer.

1. Which of the following mixtures would be most easily separated by filtering?
   A. separating clean air from dust particles
   B. separating oxygen and nitrogen from the air
   C. separating salt from water in ocean water
   D. separating water vapor from the air in a house

2. What is usually the first step in water purification systems?
   A. distillation B. condensation C. dialysis D. filtration

3. The kidneys purify blood by allowing some molecules to pass through and remain in the blood but not allowing waste materials to flow through. These waste materials are then excreted as urine. Which of the following best describes this function of the kidneys?
   A. distillation B. condensation C. dialysis D. filtration

4. In making commercial grade alcohol for cleaning, the alcohol must be separated from a water solution. Which of the following methods of separation would be best?
   A. distillation B. condensation C. dialysis D. filtration

5. How are petroleum products often purified?
   A. distillation B. condensation C. dialysis D. filtration

C. Answer the following questions.

1. A dehumidifier removes moisture from the air. Describe how a dehumidifier might work. Write down which separation method a dehumidifier might use and how it would separate water vapor from the air.
2. Ice cubes appear cloudy when the water that freezes contains a lot of dissolved gases. Suggest how you could make clear ice cubes.
3. Suggest steps to separate a mixture of salt, sand, and water.

ACIDS AND BASES

One particular area of solution chemistry deserves special attention. Acid-base reactions are very important to almost every chemical process on Earth.

Acid-Base Reactions

According to the Arrhenius theory, an **acid** is a compound that contains hydrogen and dissociates in water to produce hydronium ions (H^+ or H_3O^+). To **dissociate** means to break down into smaller parts. **Strong acids** are acids that almost completely dissociate in water. Hydrochloric acid (HCl) is a strong acid because the hydrogen ion separates from the chloride ion in water. The list of strong acids is short. See Table 8.1 to the right. **Weak acids** are acids that partially dissociate in water. Almost all acids are weak. Examples of common acids are citric acid in a lemon, tannic acid in tea, lactic acid in sour milk, and acetic acid in vinegar.

Table 9.2 Strong Acids

hydrochloric acid	HCl
nitric acid	HNO_3
sulfuric acid	H_2SO_4
hydrobromic acid	HBr
hydroiodic acid	HI
perchloric acid	$HClO_3$

Table 9.3 Examples of Bases

Hydroxide ion	OH^-
Silicate ion	SiO_3^{2-}
Phosphate ion	PO_4^{3-}
Carbonate ion	CO_3^{2-}
Ammonia	NH_3

The Arrhenius theory states that a **base** is a compound that produces hydroxide ions (OH^-) in a water solution. Solutions containing a base are **alkaline**. Examples of common bases are sodium hydroxide in lye, ammonium hydroxide in ammonia, magnesium hydroxide in milk of magnesia, aluminum hydroxide in anti-perspirant, and calcium hydroxide in limewater. Most bases do not dissolve in water. However, those that do dissolve in water are called alkalis.

A more inclusive theory of acids and bases is the Bronsted-Lowry theory. According to the **Bronsted-Lowry theory**, an acid is a proton donor, and a base is a proton acceptor. Remember that a proton is a hydrogen ion (H^+). This theory explains why substances like ammonia, NH_3, are bases even though they don't have a hydroxide (OH^-) group. The NH_3 compound becomes a proton acceptor.

pH Scale

We measure acidity and alkalinity using the **pH scale**, pH being short for "potential of hydrogen." The pH scale is logarithmic and ranges from 0 to 14. A difference of one pH unit represents a tenfold change in ion concentration. As the pH values decrease, the concentration of hydronium ions (H_3O^+) increases. For instance, a substance with a pH of 2 has 10 times the hydronium ion concentration as a substance with a pH of 3. As the pH values increase, concentration of hydroxide ions (OH^-) increases. A substance with a pH of 11 has 100 times the hydroxide ion concentration as a substance with a pH of 9.

Water is a neutral compound with a pH of 7. Acids have pHs lower than 7, and bases have pHs higher than 7. For example, Figure 9.11 on the next page shows that pure vinegar, a weak acid, has a very low pH of 2.8. Many soda drinks have acidic pHs lower than 3. These acidic drinks are known to have an erosive effect on the enamel of human teeth. Bleach has a pH of 10. Many consumer products have acidic pHs, which makes it difficult to maintain pH balance in the body. Foods containing baking soda, such as saltine crackers and pretzels, are good for restoring an acidic body back to a neutral pH.

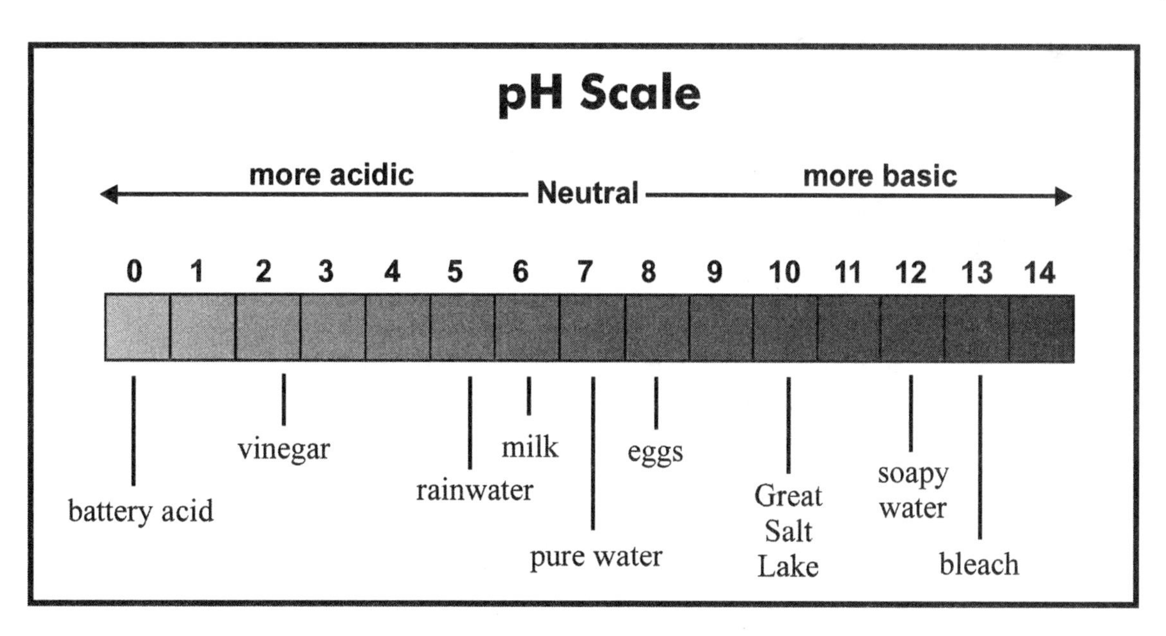

Figure 9.11 pH Scale

The pH of a substance can be determined by a pH meter or by an **indicator**, such as litmus. Indicators turn different colors depending on the pH of a solution. For instance, acids will turn blue litmus paper red, and bases will turn red litmus paper blue. However, the range over which indicators change colors is small. Litmus is only an effective indicator for solutions in the pH range from 5.5 – 8.0. Therefore, chemists use many different indicators, such as methyl yellow, bromophenol blue, and thymolphthalein, to test the pH of solutions across the entire pH scale.

Neutralization Reactions

When an acid reacts with a base, both substances neutralize, and the products are water and a **salt**. A salt is a compound composed of the positive ion of a base and the negative ion of an acid. For example, equal amounts of hydrochloric acid (HCl) and sodium hydroxide (NaOH) will react to form table salt (NaCl) and water (H_2O). Sometimes this reaction is very violent; therefore, it is not advisable to mix strong acids and strong bases.

$$HCl\ (aq) + NaOH\ (aq) \rightarrow NaCl\ (aq) + H_2O\ (l)$$ **Equation 9.1**

Another example of a neutralization reaction occurs when you take an antacid to calm an upset stomach. The stomach contains hydrochloric acid to help digest food. Consuming a medicine that has a basic pH, an antacid, helps to neutralize the stomach juices.

Reactions with Metals

Acids will react with reactive metals, such as magnesium and zinc. When an acid reacts with a metal, hydrogen and a salt are produced. The hydrogen causes bubbling during the reaction. Equation 9.2 shows an example of an acid-metal reaction between hydrochloric acid and zinc.

$$2HCl + Zn \rightarrow ZnCl_2 + H_2$$ **Equation 9.2**

Gold will not react with any acid.

Acids and Bases in the Atmosphere

Acid rain forms from the reaction of pollutants from cars and factories with water vapor in the clouds. These pollutants include **sulfur dioxide** released primarily from the burning of fossil fuels, and **nitrogen oxide** emitted from automobiles. These gases combine with water to form sulfuric and nitric acid, both of which were listed in Table 9.2 as strong acids. Acid rain has a pH of less than 5.6, with values as low as 2 occurring on the East Coast of the United States.

$SO_2 + H_2O_2 \rightarrow H_2SO_4$	**Equation 9.3**
NO_x + sunlight + $OH \rightarrow HNO_3$	**Equation 9.4**

Acid rain has disastrous effects on the environment such as slowing the growth of trees and leaching nutrients from the soil. Acid rain alters the chemistry of the water in lakes and other bodies of water, thus killing many organisms. Acids can also combine with other chemicals in the air to form smog, which attacks the lungs. In addition to its effects on living systems, acid rain destroys stone structures such as buildings, statues, and ancient ruins due to the corrosive nature of acids.

Section Review 4: Types of Chemical Reactions

A. Define the following terms.

acid	base	salt
dissociate	alkaline	pH scale
strong acid	weak acid	indicator

B. Choose the best answer.

1. Which of the following is a weak acid?
   A. hydrochloric acid
   B. sulfuric acid
   C. nitric acid
   D. acetic acid

2. Which of the following substances is alkaline?
   A. milk
   B. vinegar
   C. bleach
   D. rainwater

3. Orange juice has a pH of 3, and tomato juice has a pH of 4. Orange juice has a lower pH because
   A. it is more basic than tomato juice.
   B. it is a more concentrated acid than tomato juice.
   C. it is a stronger acid than tomato juice.
   D. it is a weaker acid than tomato juice.

4. All of the following are bases ***except***

   A. HBr

   B. $Al(OH)_3$

   C. NaOH

   D. NH_3

5. Which one of the following statements is true?

   A. Strong acids always have a higher pH than weak acids.

   B. Strong acids partially dissociate in water while weak acids almost completely dissociate in water.

   C. Strong acids almost completely dissociate in water while weak acids partially dissociate in water.

   D. Strong acids produce hydroxide ions in solution, and weak acids produce hydronium ions in solution.

C. Answer the following question.

1. If blue litmus paper turns red upon being submerged in a substance, in what pH range does the substance belong?

CHAPTER 9 REVIEW

A. Choose the best answer.

1. Separating a mixture of alcohol and water commonly requires which of the following separation methods?

   A. filtration B. dialysis C. condensation D. distillation

2. Hydrogen peroxide can be manufactured by the electrolysis of sulfuric acid solutions. Therefore, the production of hydrogen peroxide is

   A. a physical change. C. chemical weathering.

   B. a combustion reaction. D. a chemical change.

3. In the neutralization reaction, $HClO_3 + NaOH \rightarrow H_2O + NaClO_3$ $NaClO_3$ is a(n)

   A. solution. B. salt. C. element. D. reactant.

4. Sodium hydroxide completely dissociates into positive and negative ions and is, therefore, a

   A. strong acid. B. weak acid. C. strong base. D. weak base.

5. Carbon dioxide in the atmosphere reacts with rainwater to form carbonic acid. The pH of rainwater is

   A. less than 7. C. greater than 7.

   B. 7. D. not measurable by ordinary means.

6. A substance that causes red litmus paper to turn blue is

   A. an acid. B. a salt. C. a base. D. pure water.

7. A solution with a pH of 1 is how many more times acidic than a solution with a pH of 5?

   A. 4 B. 20 C. 100 D. 10,000

8. Which of the following is a balanced neutralization equation?

   A. $2Al(OH)_3 + 3H_2SO_4 \rightarrow 6H_2O + Al(SO_4)_3$

   B. $2KOH + HNO_3 \rightarrow 3H_2O + KNO_3$

   C. $4Ca(OH)_2 + H_2SO_4 \rightarrow H_2O + 3CaSO_4$

   D. $NaOH + HCl \longrightarrow H_2O + NaCl$

9. A student made a solution by dissolving sugar in tap water. She wanted to increase the concentration of her sugar solution. Select the best way for her to do that.

   A. add more sugar to the existing solution

   B. add more water to the existing solution

   C. warm the solution

   D. filter the solution

10. Identify the liquid that is the best conductor of electricity.

   A. concentrated sugar solution
   B. molten candle wax
   C. pure water
   D. saltwater

11. Identify the property of water that makes water an excellent solvent.

   A. low freezing point
   B. high specific heat
   C. polar molecules
   D. translucent

12. A given solid and a given gas both dissolve in a given liquid. Identify the result that vigorously shaking the liquid will have on the amount of the solid and the amount of gas that can dissolve in the liquid.

   A. More solid and more gas can dissolve.

   B. More solid but less gas can dissolve.

   C. Less solid but more gas can dissolve.

   D. Less solid and less gas can dissolve.,

13. A given solid and a given gas both dissolve in a given liquid. Identify the result that cooling the liquid will have on the amount of the solid and the amount of gas that can dissolve in the liquid.

   A. More solid and more gas can dissolve.

   B. More solid but less gas can dissolve.

   C. Less solid but more gas can dissolve.

   D. Less solid and less gas can dissolve.

14. A group of students boiled saltwater and condensed the vapor given off while the saltwater was boiling. They collected the condensate. Select the best description of that condensate.

   A. pure water
   B. deionized water
   C. saltwater
   D. salt

15. The concentration of citric acid would be highest in which of the following fruits?

   A. apple
   B. pear
   C. orange
   D. grape

Chapter 10
Forces and Motion

PHYSICAL SCIENCE STANDARDS COVERED IN THIS CHAPTER INCLUDE:

SPS8 Students will determine the relationships among force, mass and motion.

MOTION

DISPLACEMENT VS. DISTANCE

Displacement is a term that describes the distance an object moves in a specific direction. The terms displacement and **distance** are similar, but displacement always includes a **direction**. A person traveling 50 miles due north is an example of displacement. The distance is 50 miles, but the **direction** of due north makes it a **displacement** value.

Example 1: Claude rides his scooter 50 yards down the sidewalk from his house. Then, he turns around and rides his scooter the 50 yards back to his house. What is the distance that he traveled? What is the total displacement? Look at Figure 10.1 to the right. Claude travels a total distance of 100 yards, but his displacement is zero since his ending point is the same as his starting point. Think about his travel as if he were on a number line. He travels 50 yards in a positive direction, and when he turns around and travels the other way, his distance is a negative 50 yards. Total displacement is zero, since 50 + (–50) = 0.**

Figure 10.1 Displacement Example 1

**Note that if you end up in the same place as you start, regardless of the path taken, displacement will equal zero.

Example 2: What if a car travels 30 miles due north and then turns and travels 40 miles due east? What distance does the car travel? What is the car's total displacement? The total distance is easy. Just add 30 miles and 40 miles to get a total of 70 miles traveled. The total displacement is the straight line distance between the starting point and the ending point and the resulting angle. Use graph paper and a protractor to find the displacement distance and the angle. Look at Figure 10.2. Each unit on the graph measures 1/4 of an inch and represents 10 miles. The 30 mile leg of travel measures 3/4 of an inch. The second 40 mile leg of travel measures 1 inch. The line connecting the starting point to the ending point is $1\frac{1}{4}$ inches, which equals 50 miles. The measure of the angle is 53° in the direction of northeast. So, the displacement is 50 miles at an angle of 53° northeast.

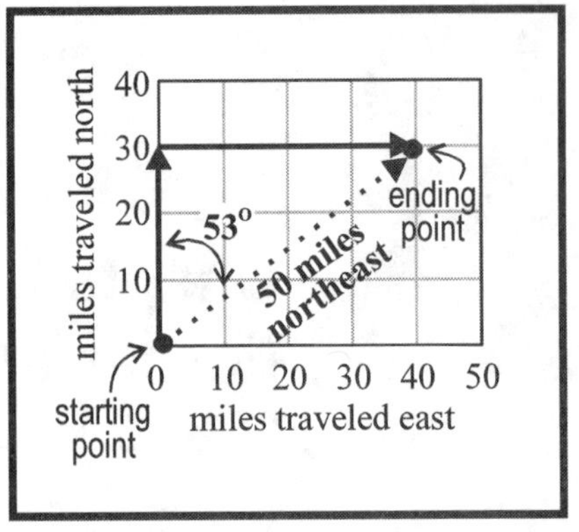

Figure 10.2 Displacement Example 2

Practice Exercise 1: Finding Distance and Displacement

Find the distance and displacement for the following problems. Use graph paper and a protractor when needed, and use a convenient scale.

1. LaKisha sprints 100 meters down a track and then walks back to the starting line.
2. Shawn runs 4 times around a 400-meter oval track and stops at the same point he started.
3. A truck driver travels 50 miles north and then turns west and drives another 50 miles.
4. A basketball player jumps straight up 20 inches and lands in the same spot he started.
5. Dupree Park has a perfectly circular walking track that has a circumference of 200 feet and a diameter of about 64 feet. Dale starts at the westernmost point and walks to the easternmost point.
6. A father runs errands with his children. He leaves his home, drives five miles south to the gas station, 10 miles southeast to the dry cleaners, 5 miles east to the grocery store, 10 miles northeast to the eye doctor, 5 miles north to the park, and 19 miles west to return home.

VELOCITY VS. SPEED

The rate of displacement is called velocity. **Velocity** is the distance traveled in a specified direction (displacement) per unit of time. For example, the velocity of a car might be 55 miles per hour west. Speed is similar to velocity in the same way that displacement and distance are similar. **Speed** is the rate without the direction. So, "55 mph" indicates speed. The terms velocity and speed are often used interchangeably, although they are not the same. To calculate speed, divide the distance traveled by the time as shown in Equation 10.1.

$$\text{speed} = \frac{\text{distance}}{\text{time}}$$

$$\text{or, } s = \frac{d}{t} \qquad \textbf{Equation 10.1}$$

Similarly, the velocity is equal to the displacement divided by time. The SI unit for velocity is meter per sec or **m/s**.

Example: Nancy runs 100 meters in 12.5 seconds. What is her speed?

$$s = \frac{100 \text{ meters}}{12.5 \text{ seconds}} = 8 \text{ m/s}$$

Nancy's speed (rate of motion) is 8 meters per second.

Velocity or speed can also be shown graphically with distance on the *y*-axis and time on the *x*-axis. Figure 10.3 shows the distance that two cars travel versus time. You can use a graph like this to determine speed. Remember, speed is distance divided by time. Since the line for car 1 is straight, the speed is constant. From the graph, you can see that car 1 travels 15 meters every second. Between any two points, the distance divided by the time is 15 meters per second. Therefore, the speed of the car is 15 m/s. The slope of the line (the change in distance divided by the change in time) represents the speed. Recall from math class that the slope of a line equals rise over run, or the difference in the *y*-value divided by the difference in the *x*-value. The graph for the second car is not a straight line. Therefore, the speed is not constant. Car 2 travels 30 meters in the first second, and then 15 m in the next second. Finally, the car only travels 15 meters in the last 2 seconds. The speed of the car is slowing down from 30 m/s initially to 15 m/s, and finally to 7.5 m/s.

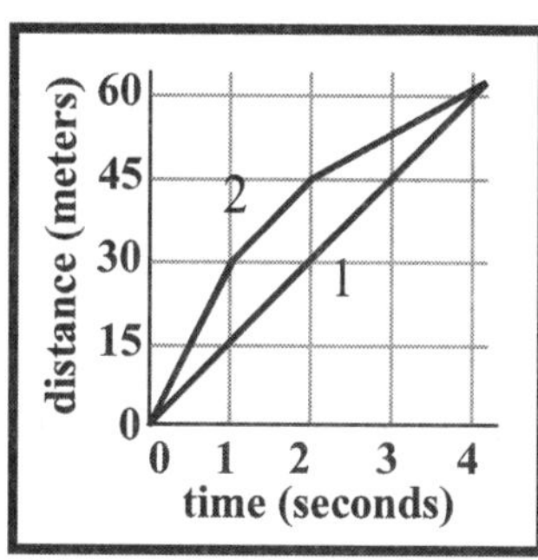

Figure 10.3 Graphical Representation of Car Speed

Practice Exercise 2: Calculating Speed

Calculate the speed for the following problems.

1. James drives 400 km in 5 hours.
2. Two Frenchmen hang glide down a 3,048 m mountain in 20 minutes.
3. The racehorse travels 40 km in 20 minutes.
4. The cyclist travels 170 km in 3 hours.
5. The turtle crawls 100 m in 40 minutes.
6. A gazelle runs 40 km in 30 minutes.
7. Michelle swims 100 m in 50 seconds.
8. Joe drives 180 km in 3 hours.

ACCELERATION

Acceleration is the change in velocity over time. An object can accelerate if either the speed or direction of motion changes with time. Thus, an object in motion that is not traveling at a constant speed is accelerating. When acceleration is a positive number, the object increases in speed. When acceleration is a negative number, the object decreases in speed. Negative acceleration is also called **deceleration**. Equation 10.2 gives the formula to calculate acceleration.

$$\text{acceleration} = \frac{\text{final velocity} - \text{initial velocity}}{\text{change in time}}$$

$$\text{or, } a = \frac{\Delta v}{\Delta t} = \frac{v_f - v_i}{\Delta t} \qquad \textbf{Equation 10.2}$$

where the symbol Δ means "the change in."

Example: A car accelerates from 10 m/s to 22 m/s in 6 seconds. What is the car's acceleration?

$$a = \frac{22 \text{ m/s} - 10 \text{ m/s}}{6 \text{ s}} = 2 \text{ m/s}^2$$

Notice that the units for acceleration are distance per time squared. In this example, the car accelerates 2 meters per second each second, or **m/s²**, which is the SI unit for acceleration.

$$a = \frac{v_f - v_i}{\Delta t}$$

$$\text{unit of a} = \frac{\text{m/s} - \text{m/s}}{\text{s} - \text{s}} = \frac{\text{m/s}}{\text{s}} = \frac{\text{m}}{\text{s}^2}$$

Practice Exercise 3: Calculating Acceleration

Calculate the acceleration in the following problems. Use a negative quantity to indicate deceleration.

1. A sky diver falls from an airplane and achieves a speed of 98 m/s after 10 seconds. (The starting speed is 0 m/s.)
2. A fifty-car train going 25 meters per second takes 150 seconds to stop.
3. A Boeing 747 was flying at 150 m/s and then slows to 110 m/s in 10 minutes as it circles the airport. (Note: Time is given in minutes. Convert minutes to seconds to make your units consistent.)
4. A motorcycle starts from a standstill and reaches a speed of 80 m/s in 20 seconds.
5. A cyclist accelerates from 4 m/s to 8 m/s in 100 seconds.
6. A runner speeds up from 4 m/s to 6 m/s in the last 10 seconds of a race.
7. A car traveling 24 m/s comes to a stop in 8 seconds.

Acceleration can also be shown graphically, with speed on the y-axis and time on the x-axis. Figure 10.4 shows the acceleration of a car by graphing speed versus time. The line in this graph is straight; therefore, the acceleration is constant. Since acceleration is the change in speed over time, the slope of this graph represents the acceleration of the car.

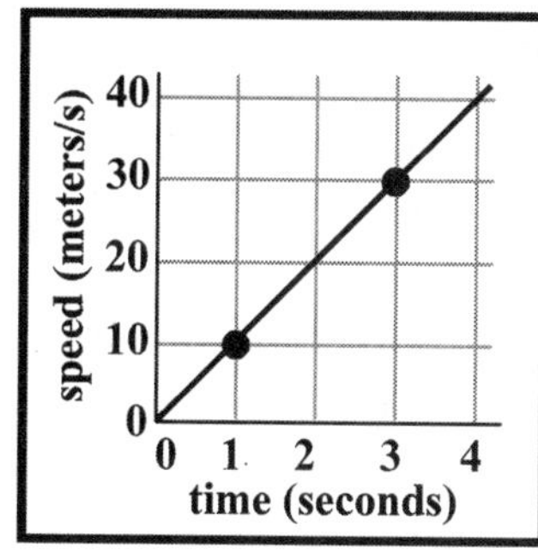

Figure 10.4 Graphical Representation of Car Acceleration

Example: Use the graph to determine the car's acceleration. Pick any two points to calculate the change in speed and the change in time.

$$a = \frac{30\text{ m/s} - 10\text{ m/s}}{3\text{ s} - 1\text{ s}} = \frac{20\text{ m/s}}{2\text{ s}} = 10\text{ m/s}^2$$

The acceleration of the car is 10 m/s^2.

Now, look at Figure 10.5. Let's interpret the meaning of this graph. At time equal to zero, the car had no speed, meaning that it was at rest. Between 0 and 20 seconds, the speed changes from 0 m/s to 10 m/s. During this time interval, the car's speed is changing, which means that the car is accelerating. Between 20 and 30 seconds, the speed remains at 10 m/s and does not increase or decrease during this time interval. Acceleration during this interval is zero, and the car travels at a constant speed of 10 m/s. From 30 seconds to 60 seconds, the car's speed changes again, but this time the speed decreases. Therefore, between 30 and 60 seconds the car decelerates. At 60 seconds, the speed is zero, meaning the car has completely stopped.

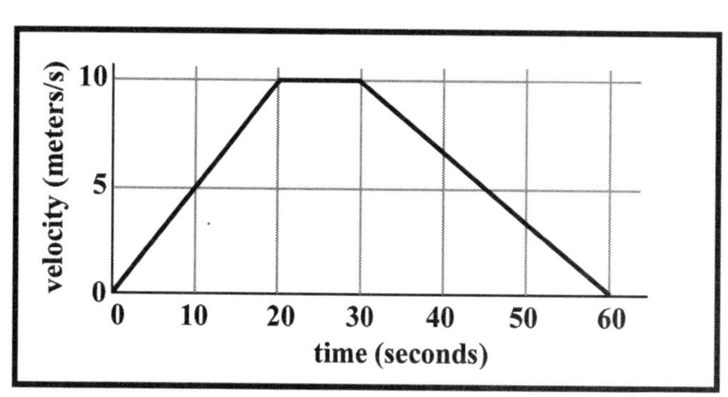

Figure 10.5 Motion of a Car

Practice Exercise 4: Interpreting Graphs of Motion

Use Figure 10.5 above to answer the following question.

1. Calculate the acceleration of the car between 0 and 20 seconds.
2. At 25 seconds, what is the car's acceleration?
3. Calculate the acceleration of the car between 30 and 60 seconds.
4. At 45 seconds, what is the car's speed?

Section Review 1: Motion

A. Define the following terms.

displacement	speed	deceleration
velocity	acceleration	

B. Choose the best answer.

1. Calculate the average speed of a bicyclist who travels 21 miles in 90 minutes.

   A. .25 miles per hour
   B. 4 miles per hour
   C. 10 miles per hour
   D. 15 miles per hour

2. Calculate the acceleration of a race car driver if he speeds up from 50 meters per second to 60 meters per second over a period of 5 seconds.

   A. 2 m/s^2
   B. 5 m/s^2
   C. 60 m/s^2
   D. 50 m/s^2

Answer questions 3 and 4 based on the following information:

Janice drives 120 miles to visit her grandchildren. On the way home, she takes a shortcut and drives 105 miles.

3. What is the total distance Janice traveled?

   A. 120 miles
   B. 105 miles
   C. 225 miles
   D. 15 miles

4. What is Janice's total displacement?

   A. 120 miles
   B. 105 miles
   C. 15 miles
   D. 0 miles

C. Use the graph below to answer the following questions.

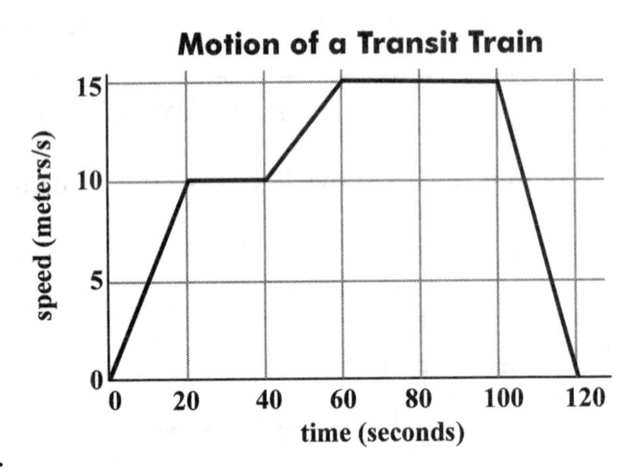

1. What is the acceleration of the train between 0 and 20 seconds?
2. What is the acceleration of the train between 20 and 40 seconds?
3. What is the acceleration between 100 and 120 seconds?
4. At 80 seconds, what is the train's speed?
5. Calculate how far the train traveled during the period of time when its speed was 10 m/s. Hint: speed = distance/time, so distance = speed × time.

FORCES AND MOTION

Force is a push or pull on matter. Force can sometimes cause matter to move. For example, you use force to pull a door open or to push a shopping cart. Force can also cause matter to slow down, to stop, or to change direction. For example, applying force to the brakes of a bicycle causes the bicycle to slow down. When a rock falls and hits the ground, the force of the ground against the rock stops the rock's motion. The batter swings a bat connecting with a baseball. The force of the hit changes the direction of the baseball. Recall that a change in speed or direction of motion with time results in acceleration of the object. Therefore, force can change the acceleration of an object.

Forces also occur at a level that we cannot see. All matter is held together by forces between the particles that compose it. For example, an atom is held together by an electric force between the protons of the nucleus and the electrons surrounding the nucleus. Most observable forces, such as the force of friction that opposes motion, are a result of electric forces acting between atoms and molecules. On a much larger scale, the moon orbits the earth as a result of forces that we cannot see but are there nevertheless.

When the forces that act on an object at rest are in balance, the object remains at rest. When the forces on an object are unbalanced, the object moves. **Sir Isaac Newton** (1642 –1727) formulated three laws of motion that describe how forces affect the motion of objects. Newton's laws and their consequences are often referred to as **Newtonian**, or **classical mechanics**. Newton built heavily on the physics foundation laid by Galileo in forming his laws of motion.

Galileo Galilei (1564 –1642) contributed greatly to the study of mechanics. He proposed the basis for what would become Newton's first law, or the **Law of Inertia**. His proposed theories were based on a series of experiments involving objects sliding down inclined planes, although it is a popular myth that Galileo formulated his theories of motion by dropping balls out of the Leaning Tower of Pisa. Galileo based his ideas on experiments and precise measurements. Galileo asserted that if frictional forces were reduced to zero, then an object moving with constant speed would remain at that speed until another force acted upon it.

We now know that Newtonian, or classical mechanics is a special case of quantum and relativistic mechanics. **Quantum mechanics** applies to objects at small scale distances, and **relativistic mechanics** describes the motion of objects moving at very high speeds (approaching the speed of light). The concept of relativity or relativistic mechanics was first introduced by Einstein, as explained in the previous paragraph. Quantum mechanics led to the discovery that electrons did not orbit the nucleus the way the planets orbit the sun, but they actually form electron clouds. While Newton's laws and classical mechanics do have some inconsistencies, they still hold true for everyday objects and thus, are still in use.

NEWTON'S FIRST LAW OF MOTION

Newton's first law of motion states that an object at rest will remain at rest, and an object in motion will remain in motion unless an outside force acts on the object. This law is also referred to as the **law of inertia**. The tendency of matter to remain at rest or in motion is called **inertia**. You feel inertia when you are in a car that starts suddenly, stops suddenly, or goes around a sharp curve. When a car starts suddenly, the inertia of your body keeps you at rest even though the car moves forward. The result is that you feel pushed back into the seat even though actually, the seat is being pushed into you! The opposite occurs when you are in a car that stops suddenly. The inertia of your body is going forward, but the car is stopping. The result is that you feel like your body is being thrown forward. When you are riding in a car that goes around a sharp curve, the inertia of your body keeps you moving in a straight line, but the car's motion is in the opposite direction. You feel pushed in the opposite direction. These are the forces that seatbelts are designed to counteract. The seatbelt stops at the same rate as the vehicle, and because it surrounds your body, it exerts a stopping force on your body.

Figure 10.6 Newton's 1st Law: Examples of Inertia

FRICTION

So, if Newton's first law is true, why does a ball slow down and eventually stop when you roll it down a long hallway? It slows down and stops because **frictional forces** are resisting the forward motion of the ball. The forces of friction occur because of the interaction of an object with its surroundings. An example of a frictional force is **air resistance**, which is the resistance an object encounters when moving through a viscous (resistant to flow) medium such as air or water. Friction, the resistance to motion due to a rough surface, is another common frictional force. To visualize friction, think of trying to walk on an icy sidewalk during the winter. The ice provides less friction than the bare sidewalk because it is smoother; therefore, it is much easier to glide (and sometimes fall) on an icy sidewalk. When you throw a ball straight up into the air, the force of gravity (Earth's attractive force) as well as air resistance oppose the upward motion, and push or pull the ball back down toward Earth.

As mentioned in the previous paragraph, **friction** is resistance to motion. Friction occurs between two surfaces in contact because the irregularities on the surfaces rub against one another. Friction occurs between any two surfaces whether they are solids, liquids, or gases. Let's take a closer look at types of friction and how to decrease friction.

1. **Static friction** is the force required to overcome inertia of a stationary object. In other words, it is the force required to start a stationary object in motion. This kind of friction is the hardest to overcome. Static friction is greater between rough surfaces than smooth surfaces. For example, moving an object on a highly polished glass surface will require less force than moving an object on rough concrete.
2. **Kinetic friction** is the force required to keep an object moving at a constant speed. Kinetic friction is less than static friction because the object is already in motion. Kinetic friction is greater between rough surfaces than smooth surfaces.
3. **Rolling friction** is the force required to keep an object rolling at a constant speed. Rolling friction is the easiest to overcome. For example, ball bearings are used in roller skate wheels to decrease friction between the moving parts.
4. To further decrease friction, surfaces can be lubricated with a liquid such as oil or even water. Friction between a liquid and a solid is less than friction between two solids. Friction between a gas and a solid is even less.

Figure 10.7 Newton's 1st Law: Examples of Frictional Forces

NEWTON'S SECOND LAW OF MOTION

Newton's second law of motion states the mathematical relationship between force, mass, and acceleration. Equation 10.3 relates force, mass, and acceleration. The mass of an object multiplied by the acceleration of an object determines the force of the object.

$$\text{Force} = \text{mass} \times \text{acceleration}$$

$$\text{or, } F = ma \qquad \textbf{Equation 10.3}$$

Recall that **mass** is the amount of matter making up an object. Mass is constant and does not change. Remember, force is the effort it takes to put an object into motion or to change an object's motion. In the SI system of measurement, force is measured in **newtons** (N). One newton of force is equal to one kilogram-meter per second per second (1 kg·m/s^2).

$$F = m \cdot a$$

$$1 \text{ newton} = kg \cdot \frac{m}{s^2}$$

Look at Figure 10.8 on the following page. If you throw a bowling ball with a high acceleration, the bowling ball hits the pins with a large force, and you get a strike. But, if a small child throws the same bowling ball with a low acceleration, then the ball will not hit the pins with as much force and only a few pins will be knocked down. Similarly, if you bowl with a basketball, which has a much lower mass than a bowling ball, then the force may not be enough to knock down any of the pins.

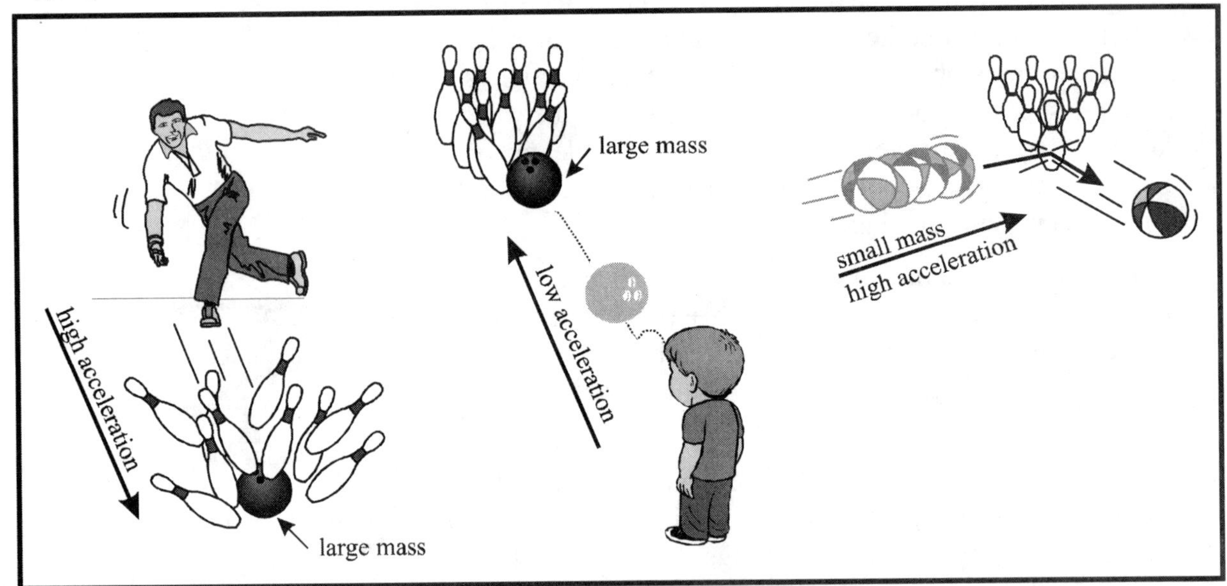

Figure 10.8 Newton's 2nd Law: The Relationship Between Force, Mass, and Acceleration

Rearranging the terms of Equation 10.3 describes how mass and force affect acceleration. Equation 10.4 is equivalent to Equation 10.3, but it shows that acceleration can be determined by dividing the force of an object by its mass.

$$\text{acceleration} = \frac{\text{Force}}{\text{mass}}$$

$$\text{or, } a = \frac{F}{m} \qquad \textbf{Equation 10.4}$$

Newton's Third Law of Motion

Newton's third law of motion is the law of action and reaction. It states that for every force or action, there is an equal and opposite force or reaction. Your book lying on your desk exerts a force on the desk. The desk exerts an equal and opposite force on the book. The force of the desk on the book is called the **normal force**. The force is given the name "normal" because it always acts at a 90° angle, or perpendicular, to the object. You might recall from math class that when objects are perdendicular to each other, they are said to be normal to each other. In a jet engine, air is forced out in one direction, which then drives an airplane forward in the opposite direction. Although they are not shown in Figure 10.9, the plane also has forces exerted on it in the vertical direction. The weight of the plane acts as a downward force and air resistance acts as the upward, normal force. When describing the motion of an object, it is helpful to visualize the motion by drawing a diagram of all the forces acting on the object. This type of diagram is called a **free body diagram**. Figure 10.9 shows some example free body diagrams for the motion of different objects.

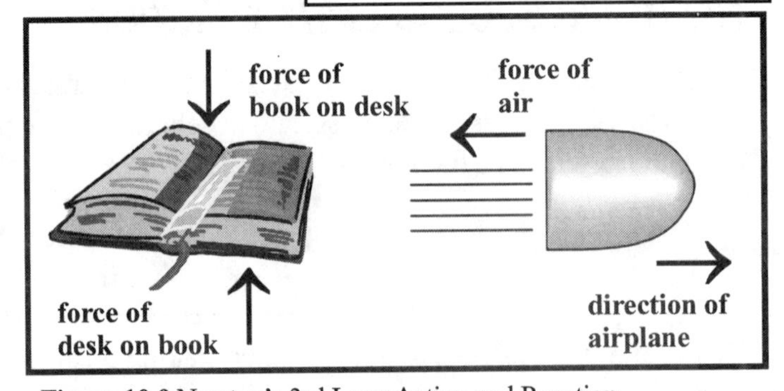

Figure 10.9 Newton's 3rd Law: Action and Reaction

Practice Exercise 5: Calculating Force

Calculate the amount of force on the mass in each problem below.

1. Ryan hits a baseball with an acceleration of 40 m/s^2. The mass of the baseball is 0.5 kg. What is the force of the swing?
2. Richard hits a golf ball that has a mass of 0.45 kg. The acceleration of the ball is 41 m/s^2. What is the force of Richard's swing?
3. How much force is required to give an object with a mass of 20 kg an acceleration of 12 m/s^2?
4. How much force is required to give a 500 kg crate an acceleration of 2 m/s^2 on an icy surface without friction?
5. Greyhounds can accelerate at 7 m/s^2 at the start of a race. How much force does it take a 35 kg dog to produce this acceleration?
6. A 2,000 kg car accelerates at the rate of 2.6 m/s^2. How much force is required to achieve this acceleration?

Section Review 2: Forces and Motion

A. Define the following terms.

force	inertia	friction	Newton's third law
Newtonian mechanics	frictional forces	Newton's second law	normal force
Newton's first law	air resistance	mass	free body diagram

B. Choose the best answer.

1. Which of the following is a force that can oppose or change motion?
   A. gravity
   B. air resistance
   C. friction
   D. all of the above

2. A passenger in a car that suddenly stops will
   A. lean forward.
   B. lean backward.
   C. lean to the right.
   D. feel no motion.

GRAVITATIONAL FORCE

There are four major forces in nature. They are gravitational force, electromagnetic force, weak nuclear force, and strong nuclear force. The strong nuclear force was discussed in Chapter 5. The electromagnetic force, which will be introduced as separate electric and magnetic forces, will be discussed in Chapter 11.

Sir Isaac Newton also formulated the **Universal Law of Gravity**. This law states the following:

- Every object in the universe pulls on every other object;
- The more mass an object has, the greater its gravitational force (pull);
- The greater the distance between two objects, the less attraction they have for each other.

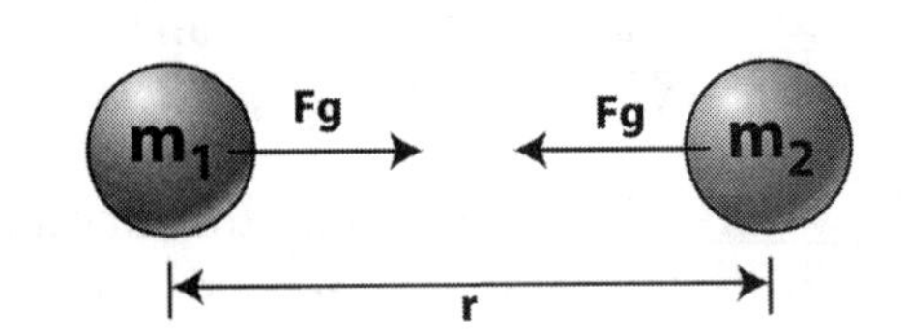

Figure 10.10 Attractive Gravitational Force Between Two Masses

Newton's law of gravitation can be expressed by the following equation:

$$\text{Force of gravity} = \text{constant} \cdot \frac{\text{mass of object 1} \cdot \text{mass of object 2}}{(\text{distance between objects})^2}$$

$$\text{or, } F_g = G \cdot \frac{m_1 \cdot m_2}{r^2} \qquad \textbf{Equation 10.5}$$

where G is the universal gravitational constant and has a value of 6.67×10^{-11} N·m²/kg². This relationship is often called the **inverse square law** because the gravitational force is proportional to the inverse square of the distance between the objects. So, gravitational force increases with increased mass and decreases as distance between masses becomes greater.

$$F_g = G \cdot \frac{m_1 \cdot m_2}{r^2}$$

$$1 \text{ newton} = \frac{N \cdot m^2}{kg^2} \cdot \frac{kg \cdot kg}{m^2} = N$$

Gravity gives the mass of an object its weight. Many confuse the terms "mass" and "weight." Mass is <u>not</u> the same as weight. As we know, mass measures the amount of matter in an object. Weight is a measure of the force of gravity exerted on an object by the earth. Weight depends on the mass of the object and its distance from the earth. In the SI measurement system, weight is measured in newtons, the same unit as force. Weight is calculated by using the same equation as given in Equation 10.3, Newton's second law.

Galileo discovered that objects accelerate toward the earth at a constant rate of 9.8 m/s², which is referred to as the **free fall acceleration** or the **acceleration due to gravity.** If you drop a ball, the earth's gravity will cause that ball to accelerate towards the earth's surface at 9.8 meters per second each second. This value for acceleration can be substituted into Equation 10.3 and multiplied by mass to calculate weight. Since acceleration due to gravity on the earth is different than the gravity on the moon, you do not weigh the same on the earth as you would on the moon. Your mass, however, is constant. Equation 10.6 is the formula to calculate weight. It replaces "force" with "weight," and "acceleration" with "acceleration due to gravity." Using Newton's second law, we can express weight with the following equation where **w** is the weight and g is the acceleration due to gravity.

$$\text{weight} = \text{mass} \times \text{acceleration due to gravity}$$
$$\text{or, } w = mg \qquad \textbf{Equation 10.6}$$

Section Review 3: Gravitational Force

A. Define the following terms.

Universal Law of Gravity | weight | free fall acceleration
inverse square law

B. Choose the best answer.

1. What is weight?
   A. the force of an object due to gravity
   B. the mass of an object
   C. the acceleration of gravity
   D. all of the above

C. Answer the following questions.

1. The gravitational pull of Mars is less than the gravitational pull of Earth. Would you weigh more or less on Mars than you do on Earth? Explain why you think so.
2. If an object weighs 49 N on Earth, how much does it weigh on the moon if the gravitational acceleration of the moon is approximately 1.63 m/s^2? HINT: Solve for mass first. Then calculate the weight on the moon.

CHAPTER 10 REVIEW

Choose the best answer.

1. A (n) ____________ is a unit of force.

   A. newton
   B. joule
   C. Kelvin
   D. ampere

2. Velocity includes both speed and

   A. distance.
   B. rate of change.
   C. time.
   D. direction.

The graph to the right shows the motion of a roller coaster from the beginning of the ride to the end. Use the graph to answer questions 3 and 4.

Roller Coaster Motion

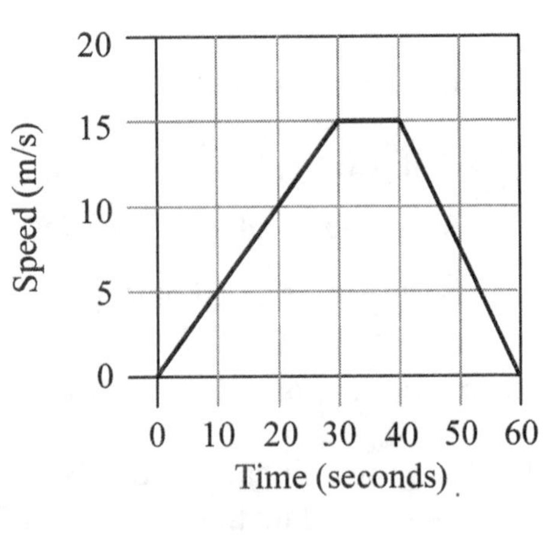

3. Calculate the acceleration of the roller coaster for the first 30 seconds of the ride.

   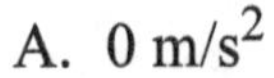

   A. 0 m/s^2
   B. 15 m/s^2
   C. 2 m/s^2
   D. 0.5 m/s^2

4. Identify the motion of the roller coaster during the first 30 seconds, the middle 10 seconds, and the final 20 seconds of the ride.

   A. acceleration, constant speed, negative acceleration
   B. acceleration, stopped, acceleration back to starting point
   C. constant speed up hill, stopped at top of hill, acceleration down hill
   D. constant speed up hill, constant speed at top, constant speed down hill

5. Identify what a person will feel when they are in a car that is accelerating very rapidly in a straight line on a flat road.

   A. pushed up out of seat
   B. pushed back into seat
   C. pushed down into seat
   D. pushed forward out of seat

6. Identify the changes in force and mass that together produce an increase in acceleration.

   A. increased force and increased mass
   B. increased force and decreased mass
   C. decreased force and increased mass
   D. decreased force and decreased mass

7. Identify the changes in mass and distance between two objects that act together to produce an increase in the gravitational force between those two objects.
   A. increased mass and increased distance
   B. increased mass and decreased distance
   C. decreased mass and increased distance
   D. decreased mass and increased distance

8. An unbalanced force acts on a body. Identify the change or changes in motion the unbalanced force can produce.
   A. increased speed only
   B. increased or decreased speed only
   C. increased speed and direction change only
   D. increased speed, decreased speed, or direction change

9. An object is taken from the earth to the moon. Identify the statement that describes the mass and weight of the object on the moon compared to its mass and weight on Earth.
   A. Mass is the same and weight is the same on the moon.
   B. Mass is greater and weight is greater on the moon.
   C. Mass is the same and weight is less on the moon.
   D. Mass is less and weight is less on the moon.

B. Answer the following questions.

10. Barbara accelerates from 0 to 55 mi/h (miles per hour) in 10s (seconds). Calculate her acceleration. Show all your work, and include the units for each number.
11. Using the information in problem number 10, graph Barbara's motion during her trip. Make sure you label the graph with appropriate numbers and units.
12. A beach ball with a mass of 0.2 kg and a bowling ball having a mass of 6 kg are rolled down a bowling alley with the same acceleration, and both strike the bowling pins in the same location. The beach ball bounces off the pins, but the bowling ball knocks the pins over. Explain why in terms of force.
13. Using Newton's first law of motion, explain why a cup of water sitting in a car might tip over as the car goes around a curve.

Chapter 11
Energy, Work, and Power

PHYSICAL SCIENCE STANDARDS COVERED IN THIS CHAPTER INCLUDE:

SPS8 Students will determine the relationships among force, mass and motion.

TYPES OF ENERGY

Energy is the ability to do work, and **work** is the process of moving matter. Energy falls into two broad categories: potential energy and kinetic energy.

Figure 11.1 Potential Energy

Potential energy is stored energy due to the object's position or state of matter. Examples of potential energy are water behind a dam, the chemical energy stored in a lump of coal or a match, the electrical potential of a battery, and the elastic potential energy of a set mouse trap.

Kinetic energy is energy of motion as an object moves from one position to another. Examples of kinetic energy are a moving car, a rock rolling down a hill, falling water, and electrons moving through a circuit. The increased movement of particles as a result of increased temperature is another example of kinetic energy.

Figure 11.2 Kinetic Energy

Energy, whether potential or kinetic, can have many forms. Table 11.1 below lists forms and examples of energy.

Table 11.1 Energy Sources

Type of Energy	Example of Energy
Thermal	fire, friction
Sound	thunder, doorbell
Electromagnetic	sunlight, microwave, ultraviolet light, x-rays
Chemical (potential)	battery, wood, match, coal, gasoline
Electrical	lightning, generator
Mechanical	gasoline engine, windmill, simple machines
Nuclear (potential)	radioactive elements, sun and stars

The Law of Conservation of Energy

Recall from Chapter 8 that the amount of energy, like matter, is always conserved. The **law of conservation of energy** states that energy is never destroyed, but it can be transferred or converted from one form into another. Energy is constantly changing forms. Energy transfer can occur by doing work or heat transfer.

For example, a match has stored, potential, chemical energy. Once the match is struck, the chemical energy is converted to light and heat energy. As another example, water behind a dam has potential energy due to its position. Once the water is released over the dam, its potential energy is converted to kinetic energy. As the water turns the turbines of an electric generator, the water's kinetic energy is converted to mechanical energy. The mechanical energy of the turning turbine is converted to electrical energy by the generator. As the generator turns, some heat is created and dispersed into the environment. The main product of the generator is electrical energy. Electrical energy may power a light bulb and be converted to light energy. The light energy also produces heat energy. Table 11.2 gives common energy changes from one form to another.

Table 11.2 Common Energy Changes

Use of Energy	Resultant Change in Energy	Energy Lost As
turning on a battery-powered flashlight	chemical to electrical to light	heat from flashlight bulb
turning the turbine in an electric generator	mechanical to electrical	heat from friction within the generator
turning on a light bulb	electrical to light	heat from bulb
using a nuclear reaction to produce heat	nuclear to thermal	heat from reaction
rock rolling down a hill	potential to kinetic	heat from friction of rock against earth

Conversion of Energy: How Batteries Work

The battery is an example of how one form of energy can be converted into another. Batteries generate electrical energy through a chemical reaction. Basically, a battery is a closed container with chemicals that react with each other. This chemical reaction produces electrons that flow out of the battery producing electrical energy.

Look at a flashlight battery. Notice that it has two terminals, one on each end of the battery. The one on the top of the battery is called the **positive terminal**, and the one on the bottom of the battery is called the **negative terminal**.

A wet cell is a battery in which the electrolyte (solution that conducts electricity) is a liquid. In a dry cell battery, the electrolyte is a paste. As electrons build up in a wet cell at the **anode** (negative terminal), a force is created which causes the electrons to flow through the conductor to the **cathode** (positive terminal). In the dry cell, a chemical reaction causes a flow of electrons from the negative zinc container to the positive carbon rod.

In order to use the battery's electrical energy, you can connect a load to the battery. The **load** is anything that requires electrical energy to operate. When the device is switched on, the electrons move from the negative terminal to the load, which then uses the electrical energy to operate the device.

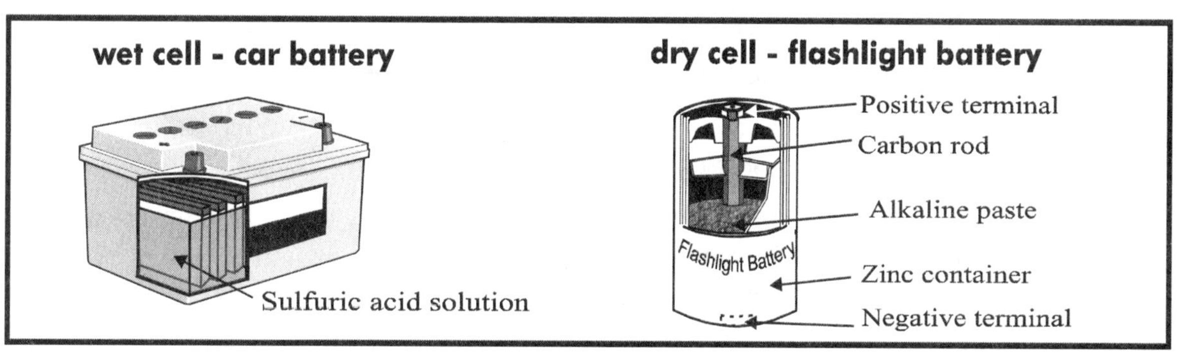

Figure 11.3 Examples of Batteries

POTENTIAL ENERGY

Recall from the previous chapter, that all objects exert a gravitational force on other objects. Objects on Earth have a **potential energy** related to the distance between the object and either the ground or a body of water. This gravitational potential energy is equal to the object's mass measured in kilograms × the acceleration due to gravity (9.8 m/s^2) × height from the ground in meters. For all objects on Earth, Equation 10.1 below applies:

$$\text{potential energy} = \text{mass} \times \text{acceleration} \times \text{height}$$

$$\text{or, } PE = mgh \qquad \textbf{Equation 11.1}$$

The SI unit for potential energy is the **joule** (J), which is also called a newton-meter.

$$PE = m \cdot g \cdot h$$

$$1 \text{ joule} = kg \cdot \frac{m}{s^2} \cdot m = N \cdot m$$

Example: Suppose a rock weighing 2.2 kilograms is about to be dropped off a cliff that is 300 meters above the ground. What is the potential energy of this rock?

Step 1. Set up the equation: PE = mgh

Step 2. Insert the known information. In this problem, we know the mass of the rock (2.2 kg), the acceleration due to gravity (9.8 m/s^2), and the height above ground (300 m).

Therefore, the equation becomes: PE = (2.2 kg)(9.8 m/s^2)(300 m)

Step 3. Solve: PE = 6,468 joules = 6.5×10^3 J

Practice Exercise 1: Potential Energy

1. If a student drops a book having a gravitational potential energy of 5 joules from a height of 1.1 meters, what is the mass of the book?
2. How many meters high on a table is a bottle of correction fluid for paper having a mass of 0.1 kilograms and a gravitational potential energy of 1 joule?
3. What is the gravitational potential energy of a microscope with a mass of 0.49 kilograms on a shelf 1.6 meters from the floor?
4. What is the mass of a stereo sound system located on a shelf that is 1.7 meters above the floor and that has a gravitational potential energy of 83.3 joules?
5. A 1,000 kg roller coaster car sits at the top of a 30 meter drop. What is the gravitational potential energy of the roller coaster car?
6. A 3.1 kg cat sitting on a tree branch has a potential energy of 153.7 joules. How far above ground is the tree branch?

KINETIC ENERGY

Moving objects also possess energy related to the object's mass and velocity. **Kinetic energy** is the energy that a moving object possesses. The formula for calculating kinetic energy is shown in Equation 11.2 below.

$$\text{kinetic energy} = \frac{1}{2} \times \text{mass} \times \text{velocity}^2$$

$$\text{or, KE} = \frac{1}{2}mv^2$$ **Equation 11.2**

The joule is also the SI unit for kinetic energy as shown to the right:

$$KE = \frac{1}{2}\cdot m\cdot v^2$$

$$1\text{ joule} = kg\cdot\left(\frac{m}{s}\right)^2 = kg\cdot\frac{m}{s^2}\cdot m = N\cdot m$$

Example: Suppose that a bowling ball has a mass of 4.0 kg and travels at a speed of 5.0 meters/second in a southward direction. What is the kinetic energy of the bowling ball?

Step 1. Set up the equation: $KE = \frac{1}{2}mv^2$

Step 2. Insert the known information. In this problem, the mass is 4.0 kg and the velocity is 5.0 m/s. Therefore, the equation becomes: $KE = 1/2 \cdot (4.0\ kg) \cdot (5.0\ m/s)^2$

Step 3. Solve: KE = 50 J

Practice Exercise 2: Kinetic Energy

1. A sport utility vehicle (SUV) with a mass of 2,450 kilograms is traveling at 21 meters/second southeast. What is the kinetic energy of the SUV?
2. A jet skier with a total mass (Jet Ski and person) of 250 kilograms is traveling south with a kinetic energy of 40,500 J. What is the velocity of the jet skier?
3. A cat with a kinetic energy of 160 J is traveling up a tree at a velocity of 8 meters per second. What is the cat's mass?
4. A 173 gram baseball is thrown with a velocity of 21 m/s. What is the kinetic energy of the baseball?
5. A satellite with a kinetic energy of 2.45×10^{10} joules orbits the earth at 7×10^3 m/s. What is the mass of the satellite?
6. A 50 kg cheetah runs in the African plains with a kinetic energy of 24,000 joules. What is the velocity of the cheetah?

Section Review 1: Types of Energy

A. Define the following terms.

energy	potential energy	law of conservation of energy
work	kinetic energy	joule

B. Choose the best answer.

1. The energy in a battery is

A. chemical potential energy.
B. mechanical potential energy.
C. electromagnetic kinetic energy.
D. thermal kinetic energy.

2. Which of the following is true concerning the law of conservation of energy?

   A. Energy is conserved when converting from potential to kinetic energy, but is destroyed by friction.

   B. Energy cannot be destroyed, but it can be transferred from one form to another.

   C. A negligible amount of energy is destroyed when converting from one form of energy to another.

   D. Energy is conserved as long as mass is also conserved, but when mass is destroyed, energy is also destroyed.

3. Which of the following is an example of potential energy?

   A. a rock rolling down a hill
   B. a rock at the bottom of a hill
   C. a rock at the top of a hill
   D. a rock bouncing down a hill

4. Which of the following is an example of kinetic energy?

   A. a baseball flying through the air
   B. a baseball in a catcher's mitt
   C. a baseball in a locker
   D. a baseball stuck in a house gutter

5. An engine converts 95% of energy input into useful work output. What happens to the remaining 5% of the energy?

   A. It is converted to heat or to some other form of unusable energy.

   B. It is destroyed in the process of converting from one type of energy to another.

   C. It is stored in the engine for later use.

   D. It is lost along with the mass of the fuel.

C. Answer the following question.

1. What is the gravitational potential energy of an object with a mass of 2.85 kilograms dropped from a height of 5.71 meters?

WORK

Work occurs when the force applied to an object moves the object a certain distance. Work, like energy, is measured in joules. To calculate work, multiply the force applied to an object by the distance that it moves the object as shown in Equation 11.3 below. Remember, you can push on a heavy box all day, but unless the box moves, no work is done.

$$\text{Work} = \text{force} \times \text{distance}$$

$$\text{or, } W = Fd \qquad \textbf{Equation 11.3}$$

In the SI system of measurement, work is measured in joules (J), force is measured in newtons (N), and distance is measured in meters (m).

$$W = F \cdot d$$

$$1 \text{ joule} = N \cdot m$$

Example: Jason moved a chair 2 meters using 10 newtons of force. How much work did he do?

Step 1. Set up the equation: $W = Fd$

Step 2. Insert the known information. In this problem, the force is 10 N and the distance is 2 m.

Therefore, the equation becomes $W = (10\text{ N}) \cdot (2\text{ m})$

Step 3. Solve: $W = 20\text{ J}$

Practice Exercise 3: Work

Calculate the work done in the following problems.

1. It takes 30 newtons of force to move a chair. It is lifted 0.5 meters.
2. Bill uses 15 N of force to move the ladder 30 m.
3. Mike uses a force of 50 newtons to move a box of books 6 m.
4. Sara uses 40 N of force to pick up her dog 0.8 m off the floor.
5. Cedrick uses 90 N of force to pull a table 11 meters.
6. Andrew lifts a log 1.5 meters with 60 newtons of force.
7. Amy lifts her book bag 1.2 meters. Her book bag weighs 12 N.
   (Remember, weight is a force.)

Machines make work easier by changing the speed, the direction, or the amount of effort needed to move an object. **Effort force** is the force exerted by a person or a machine to move the object. The resistance force is the force exerted by the object that opposes movement (equals the weight of the object in newtons). A machine can change the amount of effort force needed to overcome the resistance force of an object. Figure 11.4 below shows the six types of simple machines: pulley, wheel and axle, screw, inclined plane, wedge, and lever.

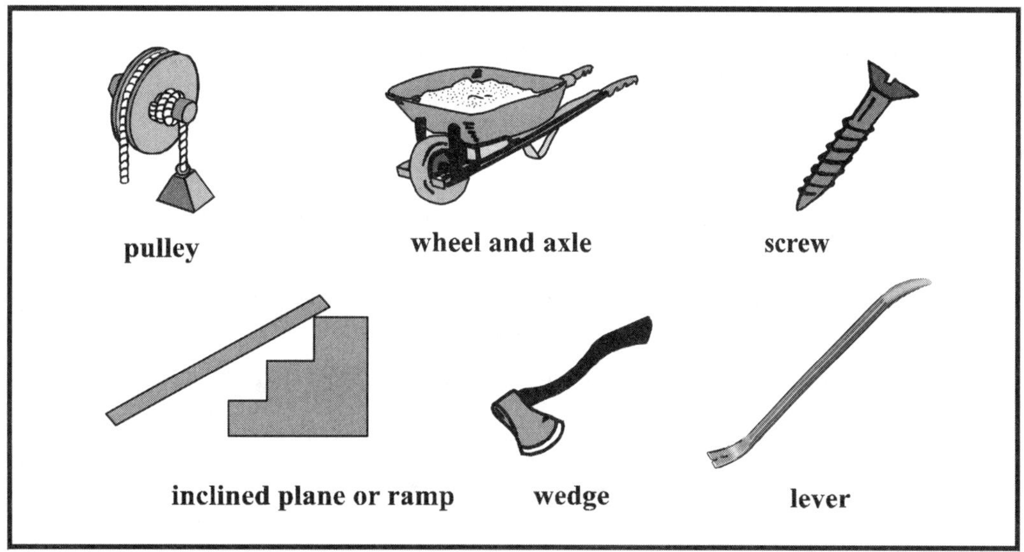

Figure 11.4 Examples of Machines

The inclined plane, or ramp, allows you to overcome a large resistance force by applying a smaller effort force over a longer distance. The mechanical advantage comes from the ratio of the length of the ramp (L) to the height of the ramp (h), as in:

$$\frac{L}{h} = \frac{F_r}{F_e} \qquad \textbf{Equation 11.6}$$

To illustrate, let's look at two scenarios that confront Pete, who wants to life a 100N box a vertical distance of 1 meter into the back of a moving van.

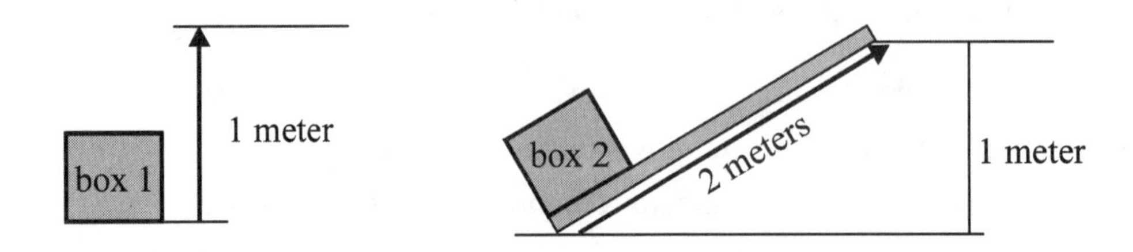

In scenario A, Peter lifts the box straight up. The work that he does is:

$$W = F \times d = 100\text{ N} \times 1\text{m} = 100\text{ J}$$

In scenario B, Pete chooses to push the box up a 2 meter ramp to get it into the van. The work that he does is:

$$W = F \times d = 100\text{ N} \times 2\text{m} = 200\text{ J}$$

Wait a minute, that's no good — he has done more work! Huh? Well, let's find out the mechanical advantage of using the ramp over a straight lift. From Equation 11.6, you can see that the mechanical advantage of the inclined plane equals the length of the plane divided by the height of the plane's terminal end. In our example this is:

$$MA = \frac{L}{h} = \frac{\textbf{2 meters}}{\textbf{1 meter}} = \textbf{2}$$

That means that Pete's effort force, when using the plane, is

$$F_e = \frac{F_r}{MA} = \frac{100\text{ N}}{2} = 50\text{ N}$$

So, Pete did more work, but the work was *easier.* You know it was easier, because it required a lower effort force to move the box up the ramp than it did to lift it straight up.

In real life, an inclined plane will have friction that opposes motion. Remember, friction decreases mechanical advantage. What would be the mechanical advantage of the inclined plane if Pete had to apply an additional 10 N of force to overcome friction as he pushed the box up the inclined plane?

$$MA = \frac{F_r}{F_e} = \frac{\textbf{100N}}{\textbf{50 N} + 10\text{ N}} = \frac{100\text{ N}}{60\text{ N}} = 1\tfrac{2}{3}$$

The force necessary to move the box up the inclined plane is still less than lifting it vertically, but friction increases the effort force and, therefore, decreases the mechanical advantage.

***Test hint**: The mechanical advantage of a frictionless inclined plane will always be the length of the plane (in our example, 2 meters) divided by the height of the plane's terminal end (in our example, 1 meter).

Practice Exercise 4: Machines

Answer the following questions on mechanical advantage and efficiency.

1. A knife is an example of a wedge. Why would a sharp knife give you more mechanical advantage than a dull knife?
2. Which machine do you think would be more efficient: a rope on a rusty pulley or a rope on a highly polished, well-greased pulley? Why?
3. Taylor applied 20 N of force to turn an ice cream crank. The ice cream's resistance was 60 N. What was the mechanical advantage of the crank?
4. A machine uses 500 J of energy to produce 300 J of work. What is the efficiency of the machine?
5. Jan uses a crowbar to open a crate. She applies 10 N of force to the end of a 30 cm crowbar. The resistance of the crate lid is 40 N. The crate opens 6 cm. What is the mechanical advantage of the crowbar? What is the efficiency of the crowbar?

POWER

Power is the rate at which work is completed or, in other words, the amount of work completed over time. Power can also be calculated using a force or weight multiplied by a rate (distance divided by time). The formula for power is given in Equation 11.6.

$$\text{Power} = \frac{\text{Work}}{\text{time}} = \frac{\text{Force} \times \text{distance}}{\text{time}}$$

$$\text{or, } P = \frac{W}{t} = \frac{Fd}{t}$$

Equation 11.6

Note: The W in this formula stands for work, while the unit for power is **watts** (W). One watt equals one joule of work per second. W is work in joules, t is time in seconds, F is force in newtons, and d/t is the rate in meters per second.

$$P = \frac{W}{t} = \frac{F \cdot d}{t}$$

$$1 \text{ watt} = \frac{J}{s} = \frac{N \cdot m}{s}$$

Example: A forklift picks up a crate that weighs 400 N. It moves the crate 20 meters in 50 seconds. How much power is used?

Step 1. Set up the equation: $P = \frac{W}{t}$. Therefore, we need to calculate the amount of work done before we can solve for power.

Step 2. Set up and solve the equation for work: $W = F \cdot d$

$$W = (400\ N) \cdot (20\ m)$$

$$W = 8000\ J$$

Step 3. Next, insert the values for work and time into the power equation.

$$P = \frac{8000\ J}{50\ s}$$

Step 4. Solve: $P = 160\ W$ (watts)

Efficiency can also be calculated using power calculations as seen in Equation 11.7.

$$\text{efficiency} = \frac{\text{Power out}}{\text{Power in}} \times 100\%$$

$$\text{or, efficiency} = \frac{P_{out}}{P_{in}} \times 100\%$$ **Equation 11.7**

Example 2: An engine has a power input of 800 watts. It can move a 100 N object at a speed of 6 m/s. What is the efficiency of the engine?

Step 1. Set up the equation: $\text{efficiency} = \frac{P_{out}}{P_{in}} \times 100\%$. First, we need to calculate the power output.

Step 2. Set up and solve the equation for power output: $P = \frac{W}{t} = \frac{F \cdot d}{t}$

$$P_{out} = \frac{100\ N \cdot 6\ m}{1\ s}$$

$$P_{out} = 600\ W$$

Step 3. Next, insert the value for power output into the efficiency equation. $\text{efficiency} = \frac{600\ W}{800\ W} \times 100\%$

Step 4. Solve: $\text{efficiency} = 75\%$ The engine is 75% efficient.

Practice Exercise 5: Power

Calculate the power used in the following problems.

1. Angela loads her wheelbarrow with dirt. She uses 10 N of force to push the wheelbarrow 50 meters to her garden in 4 minutes. How much power does she use? (Note: here, time is given in minutes. Make sure you convert minutes to seconds before continuing this problem.)
2. An 800 watt engine produces 650 watts of usable power. What is the efficiency of the engine?
3. Neil runs 50 meters in 20 seconds. How much power does he use if he has a mass of 100 kg? (Hint: Calculate the force, then solve for the power.)
4. A power boat engine has a power rating of 5000 watts. The 250 N boat can move at a speed of 13 m/s. What is the efficiency of the engine?

Section Review 2: Work and Power

A. Define the following terms.

work power watt

B. Choose the best answer.

1. Which of the following is a true statement regarding the relationship between energy and work?
   A. Without energy, work could not occur.
   B. Work is calculated as the rate of energy input per unit of time.
   C. Work output is always greater than energy input.
   D. Without work, energy could not be produced.

2. A horse pulls a cart weighing 450 newtons for a distance of 150 meters. How much work did the horse do?
   A. 3,000 J B. 67,500 J C. 450,000 J D. 150,000 J

3. The weight of a rock is 100 newtons. Using a lever, the rock was lifted using 80 newtons of force. What was the mechanical advantage of the lever?
   A. 1.25 B. 2 C. 0.8 D. 180

C. Answer the following questions.

1. How much work would it take to lift a 50 N box straight up to a height of 1.5 meters?
2. Explain how work and power are related.

CHAPTER 11 REVIEW

A. Choose the best answer.

1. Beating a drum represents what kind of energy conversion?
   A. electrical to mechanical
   B. mechanical to sound
   C. chemical to electrical
   D. sound to heat

2. What is the relationship between work and power?
   A. Work and power are the same.
   B. Power is the rate at which work is done.
   C. Work is the amount of power used over a certain distance.
   D. Power is work multiplied by force.

3. Which of the following is true regarding the relationship between work and efficiency of a machine?
   A. Since work output is always less than work input, efficiency is always less than 100%.
   B. Since friction increases work output, friction increases the efficiency of machines.
   C. Since work input and work output are always equal, these two quantities do not affect the efficiency of a machine.
   D. Since efficiency of a machine is determined by the ratio of work output to work input, the greater difference in these numbers results in greater efficiency.

4. A pulley system is 100% efficient. The system is used to lift a 99N weight through 3m. An applied force of 33N is needed. Identify the true statement.
   A. The pulley system does not change the work needed to lift the weight.
   B. The pulley system reduces the work needed to lift the weight.
   C. The effort force will be applied through a distance of 3m.
   D. The effort force will be apploed through a distance of 1m..

5. A watt is a unit of
   A. electricity.
   B. resistance.
   C. motion.
   D. power.

6. Gravitational potential energy is measured in
   A. newtons.
   B. joules.
   C. ohms.
   D. amperes.

7. Assume the acceleration due to gravity (g) equals $10 m/s^2$ (meters per second squared). What is the grvitational potential energy of a book that has a mass of 2.0kg (kilograms) that is sitting 1.5m (meters) above the floor on a table?

   A. 0.3 joules B. 20 joules C. 30 joules D. 200 joules

8. 1,000J (Joules) of work is needed to lift a box onto a shelf. Identify the gravitational potential energy (PE) the box had on the shelf and the kinetic energy (KE) it had the instant before it hit the floor after it fell off the shelf.

   A. 1,000J PE and 1,000J KE

   B. less than 1,000J PE and less than 1,000J KE

   C. less than 1,000J PE and more than 1,000J KE

   D. more than 1,000J PE and more than 1,000J KE

9. Calculate the kinetic energy of a football player if his mass is 100 kg and he is running at a speed of $10 m/s^2$.

   A. 5000 joules B. 500 joules C. 1000 joules D. 50,000 joules

10. How much power is needed to lift a 10,000 N roller coaster car to the top of the first hill in 50 seconds if the hill is 40 meters high?

    A. 12,500 watts B. 8,000 watts C. 5 watts D. 5,000 watts

11. A car uses gasoline for fuel. Which of the following describes the energy conversion from gasoline to the movement of the car?

    A. mechanical to electrical

    B. heat to light

    C. electrical to nuclear

    D. chemical to mechanical

12. Which of the following is an example of electrical energy being converted to light energy?

    A. ringing a doorbell

    B. striking a match

    C. turning on a computer monitor

    D. water falling over a dam

13. Identify what happens to the kinetic energy (KE) and gravitational potential energy (PE) of a rock as it falls.

    A. KE and PE both increase

    B. KE and PE both decrease.

    C. KE increases and PE decreases

    D. KE decreases and PE increases

14. A large boulder is set at the top of a cliff. What kind of energy does the boulder have?

    A. kinetic B. potential C. chemical D. electrical

Use the diagram below to answer questions 15 and 16.

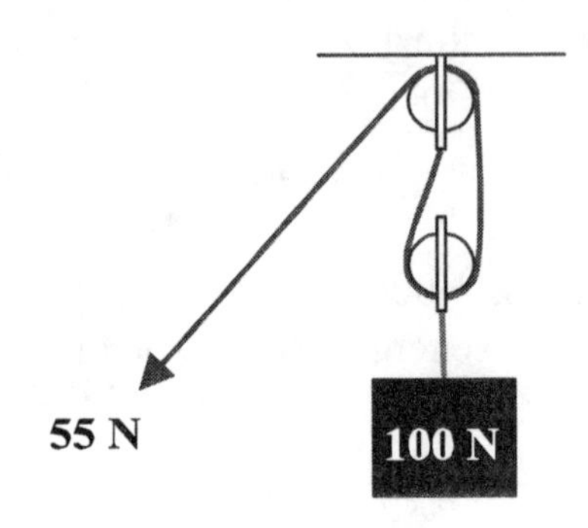

15. Which of the following is the resistance force for the diagram above?

    A. 55N   B. 100 N   C. 45 N   D. 155 N

16. Calculate the mechanical advantage of this pulley system.

    A. 0.55   B. 2.2   C. 1.8   D. 3.4

B. Answer the following questions.

17. Give an example of an object that has potential energy and an example of an object that has kinetic energy. Be sure to identify which is which.
18. As a ball rolls across the floor, it has kinetic energy. As it slows down, it's kinetic energy decreases. What is happening to the kinetic energy from the rolling ball? Is it lost, destroyed, or converted? Explain.
19. A power generation plant burns coal to heat water and produce steam. The steam turns a turbine. The turning turbine produces electricity. Identify the types of energy mentioned in this example and record how each energy source is converted to another.
20. 150 joules of work were done on a bowling ball to lift it. If the bowling ball is 75 N, how far in centimeters was it lifted? Explain in a paragraph how to solve this problem.

Chapter 12
Waves

PHYSICAL SCIENCE STANDARDS COVERED IN THIS CHAPTER INCLUDE:

SPS9	Students will investigate the properties of waves.

TYPES OF WAVES

Waves are visible or invisible evidence of energy being transferred through matter or space. Waves can be placed into one of two categories: mechanical waves and electromagnetic waves. **Mechanical waves** require a medium through which to travel. A **medium** is any material, liquid, solid, or gas, having molecules that can transport the wave's energy. Mechanical waves disturb and move through the medium. Sound waves, water waves, and seismic waves are examples of mechanical waves. Mechanical waves cannot move without their corresponding mediums. For sound waves, air is the medium, water is the medium for ocean waves, and the earth's crust is the medium for earthquakes. No mechanical waves can travel in the vacuum of space, as there is no medium through which they can move.

Electromagnetic waves do not need a medium to travel. They can travel through a vacuum or empty space. Visible light, radio waves, and ultraviolet light are all examples of electromagnetic waves. We receive light from the sun because electromagnetic waves can travel without a medium.

It is important to understand that waves transport energy and *not* matter. For example, think of an ocean wave. Consider a person floating on his back in the ocean. The person will move up and down and sideways about his original position, but will not move towards the shore. The water wave moves toward the shore, but the person is not carried with it. Similarly, the water in the ocean does not move along with the wave. It is simply disturbed by the wave as it passes toward shore.

There are two main types of waves. In **transverse waves,** the particles of the medium move at right angles, or perpendicular, to the direction of the wave. **Longitudinal waves** consist of particles that move parallel to the motion of the wave.

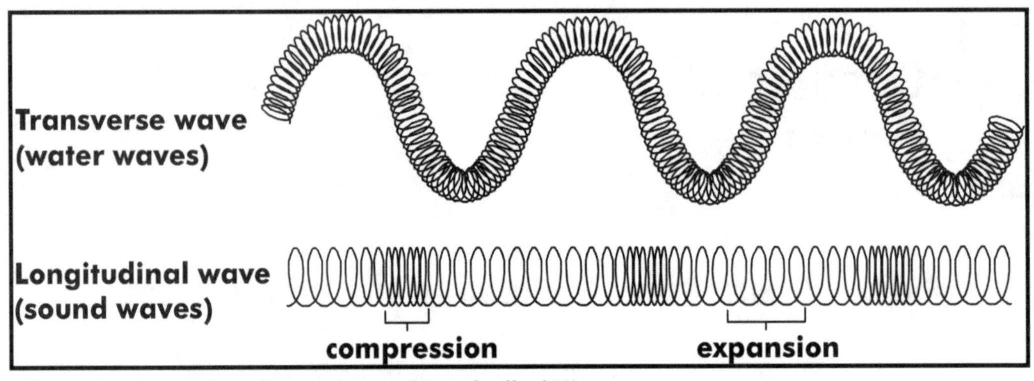

Figure 12.1 Examples of Transverse and Longitudinal Waves

Lab Activity 1: Wave Motion in a Slinky

Obtain a long slinky. Get a friend to hold one end of the slinky on the floor about 5 –10 feet away from you. While holding the other end of the slinky, repeatedly move your hand to the right and left along the floor to introduce a wave into the slinky. Notice that the slinky is moving left and right with the movement of your hand. However, the slinky wave is traveling toward your friend. What type of wave is this?

Now, repeatedly move your hand holding the slinky toward and away from your friend. Notice that both the slinky and the slinky wave are moving toward your friend. What type of wave is this?

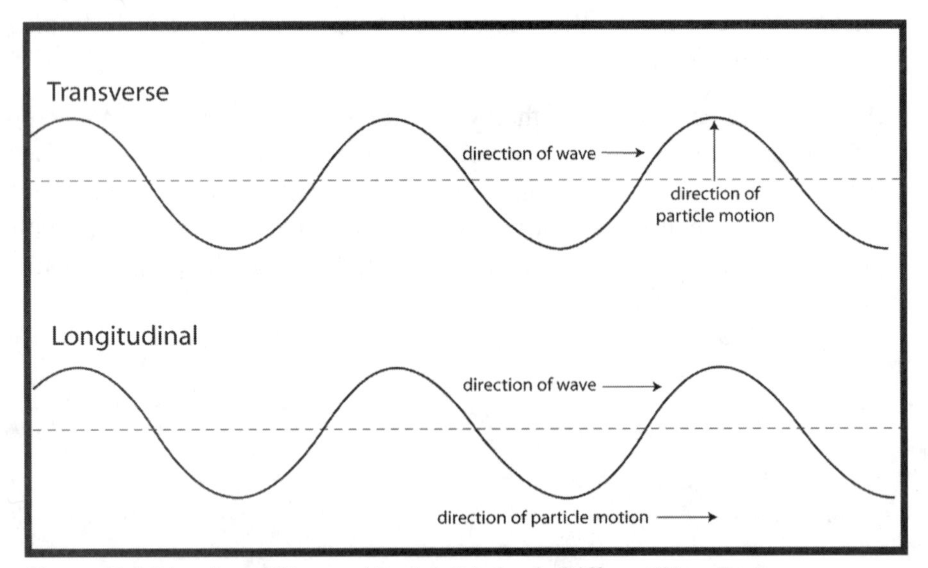

Figure 12.2 Direction of Wave and Particle Motion in Different Wave Types

PROPERTIES OF WAVES

The distance between any two identical points on a wave is called the **wavelength**. For example, the distance between two crests is the wavelength. Since wavelength is a measure of distance, its SI unit is the meter. However, wavelength is often given in nanometers (1 nm = 1×10^{-9} m) because visible light waves have wavelengths on that scale. The **amplitude** is the maximum displacement of a wave particle from its starting position. In other words, the amplitude is the height of the wave. The **period** is the amount of time required for a wave particle to complete one full cycle of its motion. Period is measured in seconds. The number of wave crests that occur in a unit of time is called the **frequency**. Frequency is measured in hertz. One **hertz** (Hz) is equal to one peak (or cycle) per second, 1/sec. Since the period and frequency are inversely related, the mathematical relationship between the two quantities is known as a **reciprocal relationship**:

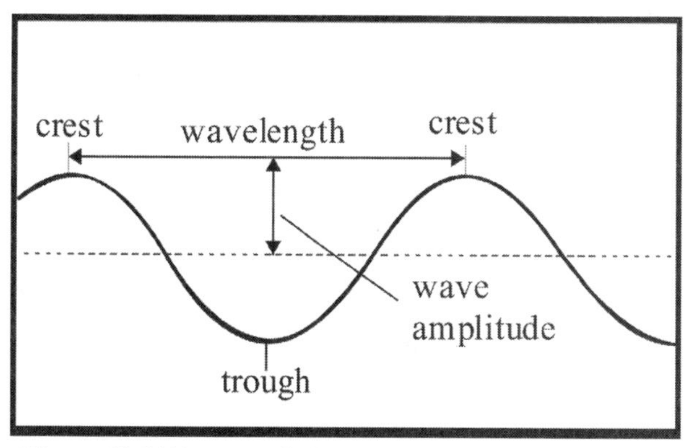

Figure 12.3 Diagram of a Wave

$$T = \frac{1}{f} \quad \text{and} \quad f = \frac{1}{T}$$

The **velocity** of a wave, or the rate at which a wave moves through a medium, is given by Equation 12.1, where λ (pronounced lambda) is the symbol for wavelength. This equation is known as the **wave equation**. As wavelength increases, the wave frequency decreases (in the same medium). The frequency or wavelength of a wave can be determined by rearranging the terms of the equation.

$$\text{velocity} = \text{frequency} \times \text{wavelength}$$

$$\text{or, } v = f\lambda \qquad \textbf{Equation 12.1}$$

A long wave has a low frequency, and a short wave has a high frequency.

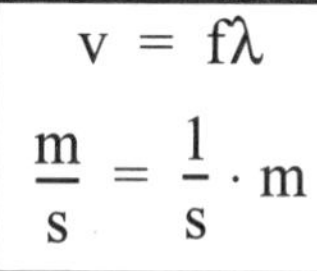

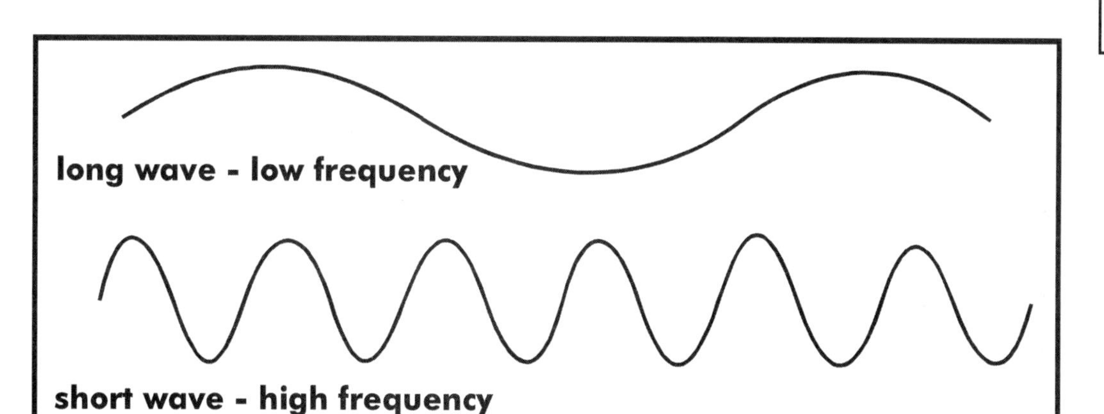

Figure 12.4 Relationship Between Wavelength and Frequency

As mentioned earlier, waves transport energy through a medium without transporting matter. The energy transported by a wave (**wave energy**) is proportional to the amplitude squared, as shown in Equation 12.2. Recall that the symbol $\propto$ means "proportional to."

$$\text{wave energy} \propto \text{amplitude}^2$$

$$\text{or, } E \propto A^2 \qquad \textbf{Equation 12.2}$$

Section Review 1: Types and Properties of Waves

A. Define the following terms.

wave	transverse wave	amplitude	hertz
mechanical wave	longitudinal wave	period	velocity
medium	wavelength	frequency	wave equation
electromagnetic wave		reciprocal relationship	wave energy

B. Choose the best answer.

1. Which of the following is true regarding mechanical waves?
   A. Mechanical waves must travel through matter.
   B. Mechanical waves can travel through matter and space.
   C. Mechanical waves can only travel through a vacuum.
   D. Mechanical waves can change matter.

2. The height of a wave is its
   A. amplitude. B. wavelength. C. period. D. crest.

3. Which of the following is *not* an example of a mechanical wave?
   A. sunlight
   B. vibrations of a guitar string
   C. ocean waves
   D. sound waves

4. Which of the following is *not* an example of an electromagnetic wave?
   A. sunlight B. X-rays C. radar D. sound waves

C. Answer the following question.

1. Label the wavelength and amplitude of the following wave.

BEHAVIOR OF WAVES

When a wave hits a surface, it can be reflected or transmitted. Reflection of the wave when it hits the surface can be partial or complete. Transmitted waves can then be refracted, diffracted, or absorbed as shown in the diagram to the right. The diagram in Figure 12.5 shows the waves as straight lines called **rays** for the sake of simplicity. The type of behavior the wave shows depends on the medium it is traveling in, the material it is entering, and the energy of the wave itself. Table 12.1 describes the possible responses of a wave when it hits a surface.

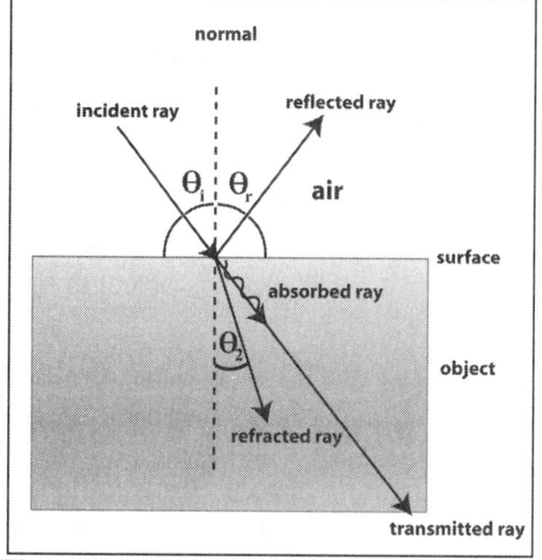

Figure 12.5 Interaction of Waves With a Surface

Table 12.1 Possible Interactions of a Wave with an Object

Behavior	Description of Wave Motion
Reflection	bounces off the surface at the same angle it hit with
Transmission	travels through the material at the same angle it entered with
Refraction	travels through the material, but at an altered angle
Diffraction	travels through the material until it encounters an obstacle, which it then bends around
Absorption	cannot travel all the way through the material

REFLECTION

When a wave hits a surface, it can bounce back. This bouncing off an object is known as **reflection**. The surface can be a solid, liquid, or gas. We'll illustrate reflection using light waves. When you turn on a light bulb, the light travels through the room until it hits an object such as the wall. The light wave then bounces off the wall and continues to travel as reflected light. The **Law of Reflection** states that the angle of reflection equals the angle of incidence. In other words, the angle the light hits the surface with equals the angle at which it bounces off the surface.

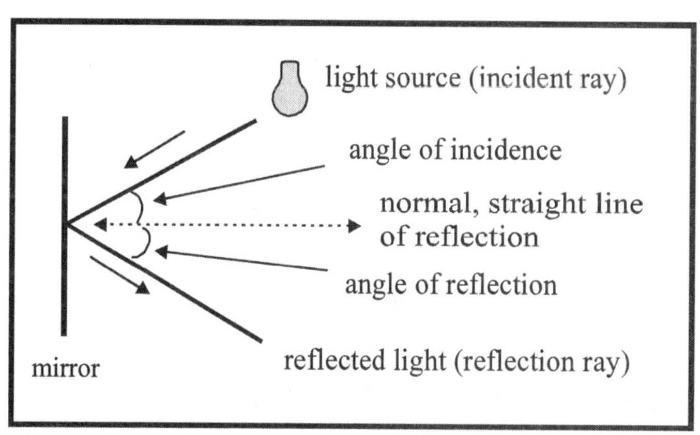

Figure 12.6 Reflection from a Mirror

In the figure above:

- A flat mirror is a smooth shiny surface that reflects light.
- Light travels in a straight line as it is reflected from a surface.
- The **incident ray** is the light ray that strikes the surface.
- The **reflection ray** is the light ray that reflects off the surface.

If a wave hits an object straight on, it will bounce back straight. If a wave approaches an object from the left, it will bounce off the object toward the right at the same angle.

When light waves hit a mirror, they reflect, and we can see ourselves in the reflection. When you shout in a cave or an empty room, you hear an echo. The **echo** is the reflection of the sound waves you created when you shouted. Bats use these principles of waves bouncing off objects to navigate at night. Whales, porpoises, and dolphins use sound echoes to find their way in the ocean. A reflecting telescope uses reflected light to illuminate the image of objects in space.

REFRACTION

Refraction is the bending of a wave by the change in density of the medium. The bending of the wave is a consequence of the reduced velocity of the wave as it enters a medium of higher density. To visualize this phenomenon, consider an army of soldiers marching along a concrete surface toward a muddy clearing on the right. The concrete surface is easy to walk on and does not impede the movement of the soldiers. When the soldiers reach the muddy clearing, they are not able to march as fast. This results in a turn to the right because the right side of the formation reaches the muddy area first, and they are slowed down. For example, Figure 12.7 shows a pencil placed in a clear glass of water viewed through the glass. Since the water is more dense than the air, the light rays passing through the water will bend, causing the pencil to look broken and disconnected.

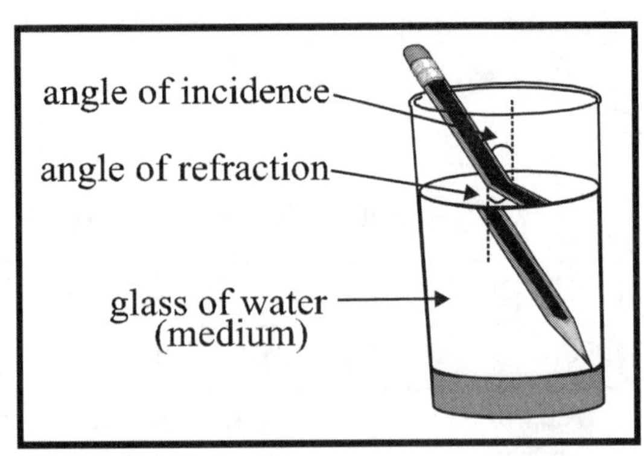

Figure 12.7 Refraction

The amount the wave bends is determined by the **index of refraction** (n) of the two materials. Index of refraction is also referred to as refractive index. The amount the wave is bent, called the **angle of refraction** (θ_2), is determined using **Snell's law**. A schematic representation of Snell's law is shown in Figure 12.8. Notice that in the figure the wave is bent toward the normal (the dashed line perpendicular to the surface) as it moves from air into the glass. This occurs because the refractive index of glass is greater than the refractive index of air. The refractive index of a material is related to the density and atomic structure of the material. The higher the index of refraction, the more the material will bend the incoming wave. A wave moving from glass into air will bend away from the normal.

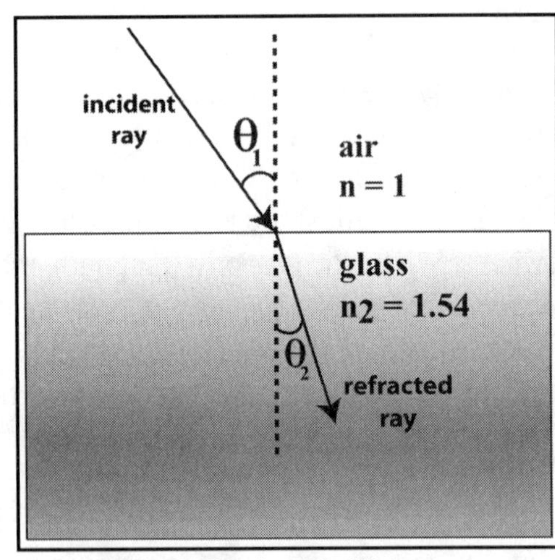

Figure 12.8 Refraction of a Wave

So far we have only discussed the refraction of electromagnetic light waves. Refraction of sound waves also occurs. Sound travels in all directions from its source. The listener can usually only hear the sound that is directed toward him. However, refraction of the sound waves bends some of the waves downward, toward the listener, in effect amplifying the sound. Refraction of sound waves can occur if the air above the earth is warmer than the air at the surface. This effect can be observed over cool lakes early in the morning. The cold water of the lake keeps the air at the surface cool, but the rising sun starts to heat the air that is higher up. This effect is called thermal inversion and results in the refraction of sound.

Diffraction

Another property of waves, called **diffraction**, relates to the ability of a wave to bend around obstacles or through small openings. Waves tend to spread out after going through an opening, which results in a shadow region. Diffraction depends on the size of the obstacle and the wavelength of the wave. The amount a wave bends, or diffracts, increases with increasing wavelength. Therefore, the diffraction angle is greater for waves with longer wavelengths. Waves with wavelengths smaller than the size of the obstacle or opening will not diffract.

Sound waves can diffract around objects or through very small holes. This is why we can hear someone speak even when they are around a corner, in a different room. Sound waves with long wavelengths are efficient at diffraction. Therefore, longer wavelength sounds can be heard at a greater distance from the source than shorter wavelength sounds. In addition, sound waves that have longer wavelengths become less distorted when they bend around objects. If a marching band were approaching, the first sounds that would be heard would be the long wavelength, low pitch, bass sounds. Elephants use this property of sound waves to communicate across the African planes using very long wavelength, low pitch sounds. Elephants travel in large herds, and it is easy for them to get separated from each other. Since they are sometimes out of visible range, they communicate using subsonic sound waves that are able to diffract around any obstacles present.

Figure 12.9 Diffraction of Sound Waves

Light waves diffract differently than sound waves. The type of light diffraction that you are probably familiar with is called **scattering**. In order for scattering to occur, the obstacle must be on the *same order* of size as the wavelength of the wave. Light scattering is responsible for the corona we sometimes see around the sun or moon on cloudy days. The water droplets in the clouds act as obstacles to the light from these objects. The light is then bent and spread out. Therefore, the light from the object appears larger than the actual source and we see a "crown" around the object. *In light scattering, waves with a shorter wavelength are bent more than waves with a longer wavelength.* **Light diffraction** is a special case of light scattering that occurs when a light wave encounters an obstacle with a regularly repeating pattern resulting in a diffraction pattern. The amount of diffraction of a light wave depends on the size of the opening and the wavelength of the light.

Absorption

Materials selectively **absorb** and **transmit** waves depending on the frequency of the wave and the atoms in the material. When a material absorbs a wave, the wave is no longer able to travel. It basically disappears. When a material transmits a wave, the wave travels all the way through the material and eventually exits the material.

The absorption and transmission of electromagnetic waves has consequences in the color of visible light. **Visible light**, the light that humans can see with the naked eye, has wavelengths between 400 and 750 nm. Blue light has a wavelength of approximately 440 nm. (nm is the abbreviation for nanometer. 1 nm is one billionth of a meter.) What we see as the **color** of an object is actually a result of the light frequencies reflected, absorbed, and transmitted by the object. Objects do not have color within themselves. For example, if a material strongly absorbs all wavelengths except those around 440 nm, the object appears to be blue, because it absorbs all wavelengths of visible light except the blue wavelengths. Objects that reflect all wavelengths of visible light appear white, whereas objects that absorb all wavelengths of visible light are black. The sky appears to be blue because the atmosphere selectively absorbs all wavelengths of visible light but the wavelengths that correspond to the color blue. **Chlorophyll** is responsible for the green color of plants. The chlorophyll absorbs the wavelengths corresponding to red and blue, while it reflects the wavelengths corresponding to the color green.

INTERFERENCE

When waves coming from two different sources meet, they affect each other. How they affect each other is known as **interference.** When two waves meet, and the high point (crest) of one wave meets the crest of the other wave, the resultant wave has the sum of the amplitude of the two waves. These waves are said to be **in phase** with one another. The interaction of waves that are in phase is called **constructive interference.** The two waves come together to *construct* a new wave.

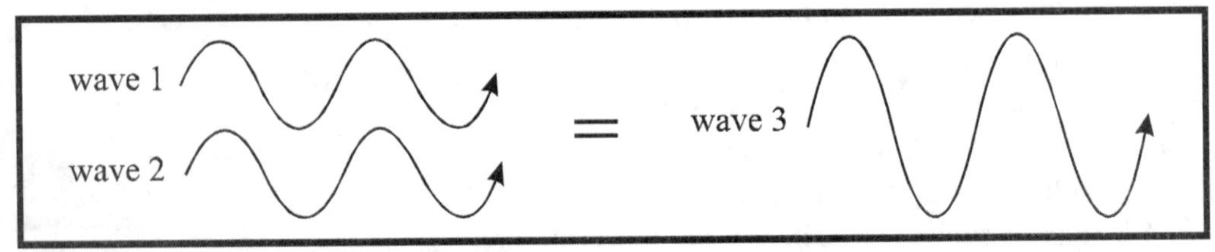

Figure 12.10 Constructive Interference

If, however, the low point (trough) of one wave meets the crest of another wave, then the waves cancel each other, and the wave becomes still. The waves are said to be **out of phase.** The interaction of out of phase waves is called **destructive interference.** The two waves meet and *destroy* each other.

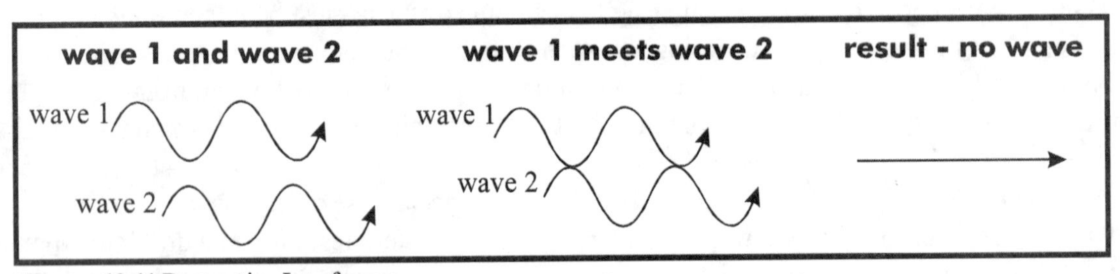

Figure 12.11 Destructive Interference

When waves interfere somewhere between these two extremes, there is some **distortion,** which results in a wave with an irregular pattern. To visualize distortion, think of what happens to the ripples made in a pond when two rocks are thrown into the water close to each other.

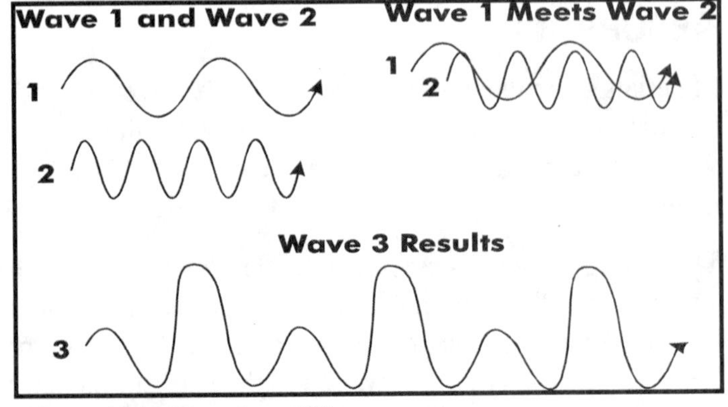

Figure 12.12 Distortion of Waves

Resonance

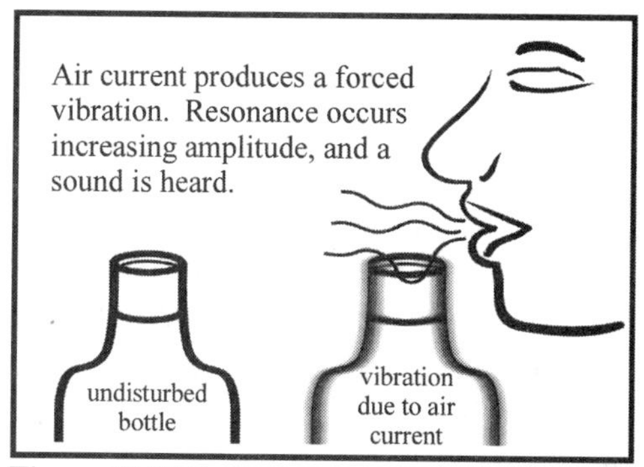

Figure 12.13 Resonance

All matter has its own frequency of vibration, called the natural frequency. However, we rarely hear these natural frequencies unless forced vibrations are applied to an object. Forced vibrations occur when an object is forced into vibrational motion by an adjoining vibrating body. For example, a factory floor vibrates from the motion of the heavy machinery.

When the frequency of forced vibrations on an object matches the object's natural frequency, **resonance** occurs. The resulting vibration has a high amplitude.

Consider an empty glass soft drink bottle for a moment. The bottle contains inherent natural frequencies because its molecules are slowly moving and bumping into each other making sounds. The human ear can't hear these sounds. However, if you blow across the top of the bottle, it will vibrate at its natural frequency. If the bottle and the air molecules within the bottle vibrate at the same frequency, resonance occurs, and you hear a sound. Amplitude of sound waves corresponds to volume. The greater the amplitude, the louder the sound. Therefore, when resonance occurs, the sound is amplified.

Just as amplitude of a sound wave and volume directly correspond, so does frequency of a sound wave and pitch. The greater the frequency of a wave, the higher the pitch. If you blow across a shorter bottle, the wavelength of the waves traveling through the bottle is decreased. As wavelength decreases, frequency increases and results in a higher pitch. A taller bottle will produce a longer wavelength, decreased frequency, and lower pitch.

Section Review 2: Behavior of Waves

A. Define the following terms.

reflection	Snell's Law	constructive interference	forced vibrations
Law of Reflection	diffraction	destructive interference	resonance
incident ray	scattering	distortion	refraction
reflected ray	absorption	natural frequency	index of refraction
echo	transmission		interference

B. Choose the best answer.

1. Which of the following wave characteristics allows us to see objects?
   A. reflection B. refraction C. amplitude D. pitch

2. The indices of refraction for four materials are listed below. Which material will bend incoming light the most?
   A. vacuum n=1.00
   B. air n=1.000277
   C. ice n=1.31
   D. diamond n=2.417

C. Answer the following question.

1. Explain why an object at the bottom of a swimming pool appears in a different location from viewing it outside the water than it does when viewed under water.

SOUND WAVES

A **sound wave** is a mechanical wave produced by a vibrating object. The wave results from the compression and expansion of the molecules surrounding the vibrating object. Sound cannot travel through empty space or a vacuum. Sound travels faster through solids than through liquids and gases because the molecules are packed together more tightly in solids. When temperature increases, the speed of sound increases. The speed of sound also increases when the air becomes more humid (or moist).

Most people hear compression waves of the frequency 20 Hz to 20,000 Hz.

Example: A dog whistle is higher than 20,000 Hz. Elephants make a sound lower than 20 Hz. Therefore, humans cannot hear these sounds.

Sound waves of different frequencies have different wavelengths in the same medium. As frequency increases, wavelength decreases. Frequency of sound waves determines pitch. **Pitch** describes how high or low a sound is. Sounds with higher pitch have higher frequencies.

Example: A police siren has a high pitch. The growl of a large dog has a low pitch.

The intensity or volume of the sound is measured in **decibels**. The amplitude of the sound wave determines the volume. The higher the amplitude, the louder the sound and the higher the decibel value. Lower amplitudes produce softer sounds with lower decibel values.

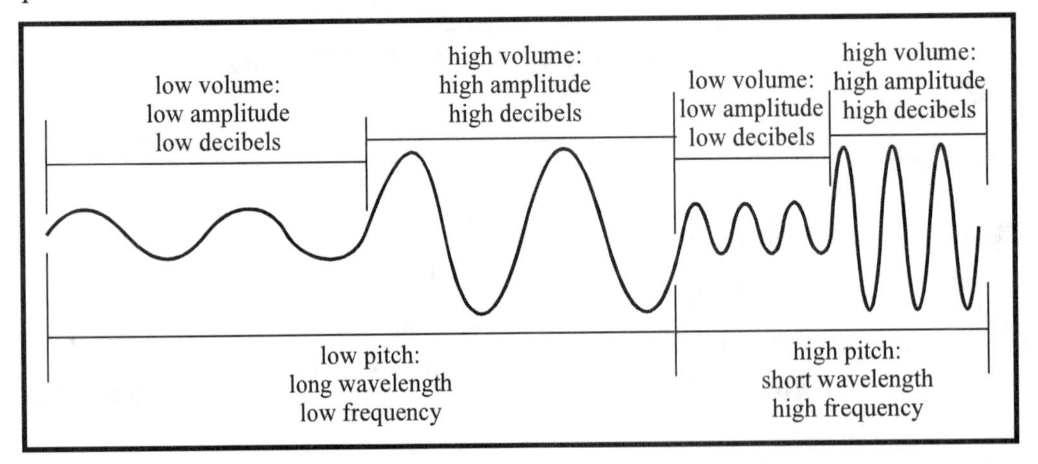

Figure 12.14 Relationship Between Volume and Pitch

THE DOPPLER EFFECT

When a sound source moves toward a listener, the pitch of the sound appears to increase. The reason is this: the movement of the sound emitter has the effect of increasing the frequency of the sound waves that the listener hears. It is important to realize that the frequency of the sounds that the source *emits* does not actually change. Next time you hear a siren while out walking, stop and listen to how the sound changes as it moves past you.

Section Review 3: Sound Waves and Seismic Waves

A. Define the following terms.

sound wave | pitch | decibel | seismic wave

B. Choose the best answer.

1. Increasing the frequency of a sound wave has which of the following effects?
   A. increases wavelength
   B. increases amplitude
   C. increases pitch
   D. increases decibel level

2. Through which of the following would sound travel the fastest?
   A. a vacuum
   B. warm, humid air
   C. warm, dry air
   D. cold, dry air

C. Answer the following question.

1. Relate the speed of sound waves to temperature and medium. Through what type of medium does sound travel fastest? How is the speed of sound affected by temperature?

ELECTROMAGNETIC WAVES

Electromagnetic waves are transverse waves that do not need a medium through which to travel. However, unlike a Slinky where the wave travels in one plane, electromagnetic waves travel in more than one plane. Electromagnetic waves are produced by the acceleration or deceleration of electrons or other charged particles. The **electromagnetic spectrum** is made up of invisible and visible waves, ranging from low frequency to very high frequency, which travel at the speed of light in a vacuum.

The wave equation given in Equation 12.1 can be rewritten for electromagnetic waves by substituting c, the speed of light, for the velocity, v. The **speed of light** in a vacuum is 3×10^{8} m/s.

As stated earlier, as the length of the wave increases, frequency decreases and as the length of the wave decreases, frequency increases. Notice in Figure 12.15 on the next page that radio waves have very long wavelengths, and gamma waves have very short wavelengths. Therefore, radio waves will have low frequencies, and gamma waves will have high frequencies.

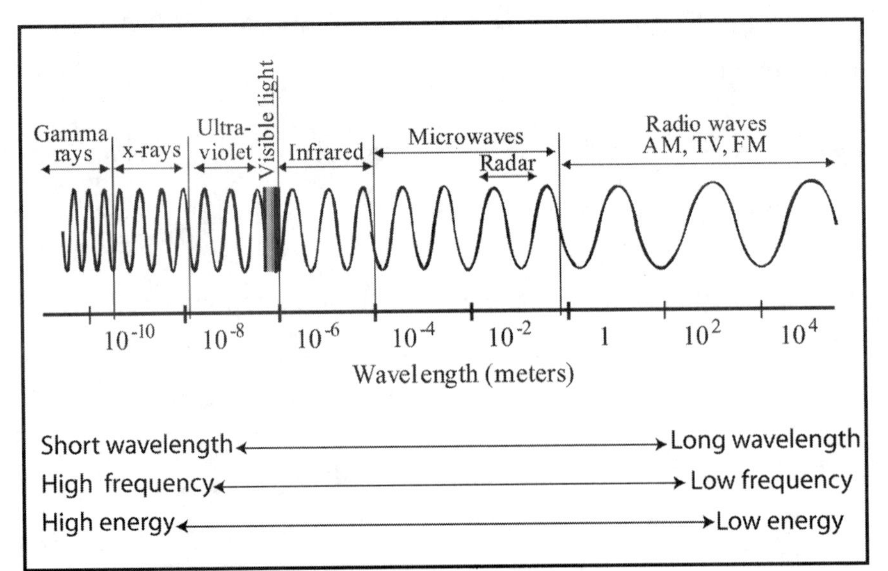

Figure 12.15 The Electromagnetic Spectrum

Light can be described as waves or particles. This is referred to as the **wave-particle duality** of light. Some characteristics of light are better described by wave theory, while others can be described by particle theory. Electromagnetic waves travel through space in particle-like units called **photons**. Photons have no mass. They are pure energy. These particles have an energy (E) proportional to their frequency (f) as expressed by the following equation:

$$E = hf \qquad \textbf{Equation 12.3}$$

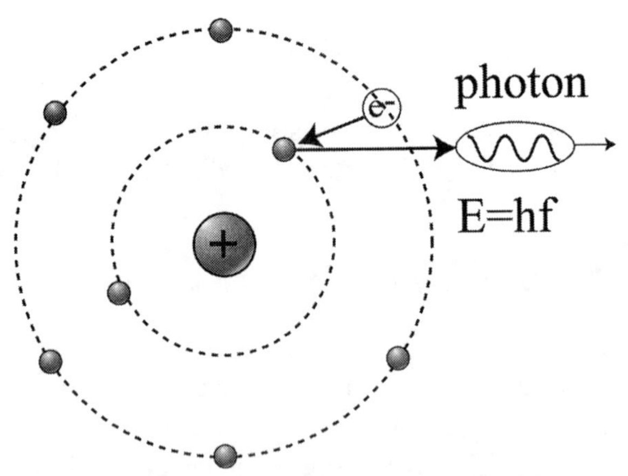

Figure 12.16 Transition of Electron and Resulting Photon

where h is a constant called Plank's constant. Recall from Chapter 5 that atoms consist of electron shells, each with a specific energy level. When energy is absorbed by an atom, an electron can be promoted from its natural (ground) state to a higher energy level. The excited electron only remains at the higher energy level for a short amount of time. When an electron falls back to a lower energy shell as shown in Figure 12.16, a photon is emitted. The wavelength of the emitted photon is a characteristic of the atom that contains the excited electron, and varies depending on which shell it occupied when it was excited and which shell it falls to.

GAMMA RAYS

Gamma rays have the shortest wavelengths, and therefore, the highest energy. Gamma rays from the sun and outer space are absorbed by the atmosphere, but they are generated on the earth by radioactive atoms and nuclear reactions. Because these waves have such high energy, they can penetrate objects more easily than any other type of wave. Gamma rays damage or destroy living cells. For this reason, gamma rays are used to destroy harmful bacteria and cancerous growths. However, exposure to gamma rays can also destroy healthy cells, so, even for medical purposes, it should be kept to a minimum.

X-RAYS

X-ray waves have very short wavelengths. Some are smaller than an atom. Shorter wavelengths mean higher energy, so these waves can also damage living tissue. X-rays coming from outer space, however, do not penetrate to the earth's surface. We generate and use x-rays to penetrate soft tissue in order to "take a picture" of bones. Since bones and metal stop the X-rays, these structures cast a silhouette on the film, but the skin and other soft tissue appear transparent. X-rays are also used in airport security to check people as well as baggage for weapons or explosives. Large doses of X-rays can damage a fetus (baby) in the first trimester of pregnancy. This exposure can cause mental retardation, skeletal deformities, and in some cases a greater risk of certain types of cancer in the newborn. Exposure to X-rays, especially long term exposure, can cause tissue damage, cancers, and diseases of various organs. It can also damage the inheritable DNA so that the genetic damage can be passed on to offspring. Overexposure to X-rays can be very harmful, and for all these reasons should be used carefully.

ULTRAVIOLET LIGHT

Ultraviolet waves are shorter than visible light waves and again, cannot be seen by the unaided human eye. Some insects like the bumblebee can see ultraviolet light. Ultraviolet waves from the sun are responsible for sunburns to the skin. Atmospheric gases such as ozone block most of the ultraviolet waves coming from the sun (which is why the depletion of the ozone layer is of great concern). Ultraviolet waves can be harmful to human health. Long term exposure to ultraviolet waves (sun bathing) and severe or repeated sunburns are a primary cause of premature aging of skin, wrinkling of skin, and skin cancer.

VISIBLE LIGHT

Visible light waves are the only waves that we can see with our unaided eyes. Specialized cells in our eyes called **cones** make it possible for us to discern color from the light reflected off of or transmitted through objects.

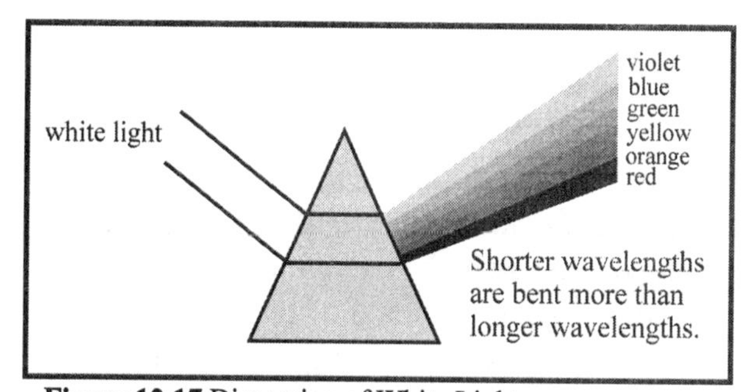

Figure 12.17 Dispersion of White Light

Light waves are part of the electromagnetic spectrum. Each color of light has its own range of wavelengths and frequencies. The frequency of a light wave determines its color. Sunlight and lamp light are a mixture of light waves with many frequencies, which is why they appear to be white in color. A prism breaks up white light into its various wavelengths of color as shown in Figure 12.17. This separation of white light into its constituent colors is called **dispersion** and is a consequence of the refraction of the incident white light. Shorter wavelengths (i.e., blue light) are bent more than longer wavelengths (i.e., red light).

You can remember the order of visible light dispersion by using the pneumonic device **Roy G. Biv**, where the letters in the name correspond to the colors of the visible light spectrum from longest to shortest wavelength.

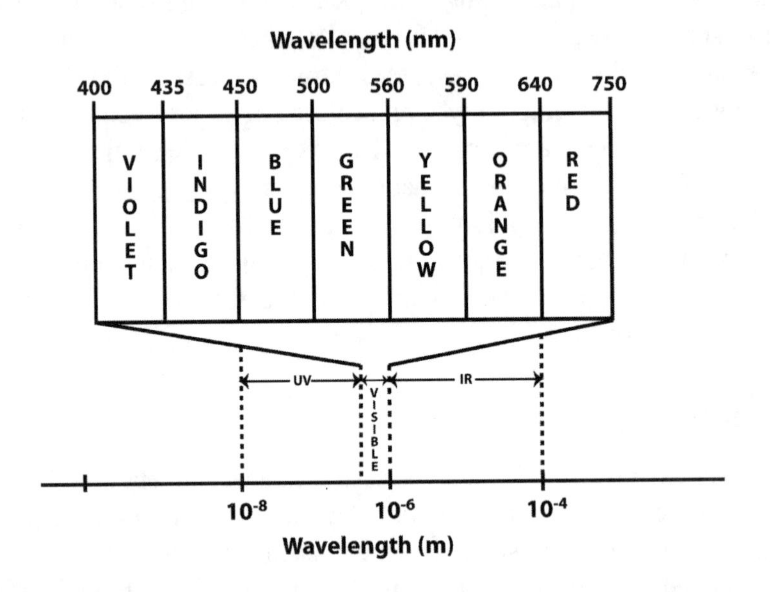

Figure 12.18 Roy G. Biv

Production of Visible Light

Many devices exist that produce visible light. These include incandescent and halogen light bulbs, neon lights, fluorescent lights, and light sticks. The method of producing light varies for the different devices. The simplest device for producing light is the **incandescent light bulb**. The light bulb contains a metal filament which is heated by electrical current. The heat excites the atoms within the filament, so that some of the electrons temporarily move to higher energy levels. When these electrons fall back to their original energy level, they release a photon. The wavelength of the photon depends on the type of atom that was excited. The filament in an incandescent light bulb is made of tungsten metal. **Halogen lights** are another type of incandescent light, which means that it uses heat to produce light. Halogen lights also contain a tungsten filament, but differ in the type of gas that surrounds the filament. The filament in a light bulb is enclosed in a glass envelope. This glass envelope usually contains an inert gas such as argon or nitrogen to prevent combustion (burning) of the filament. In halogen light bulbs, the tungsten filament is surrounded by a gas from the halogen group (i.e., fluorine, chlorine, bromine, etc.). The benefit of using a halogen gas is that it will combine with the tungsten vapor and redeposit on the filament. For this reason, halogen lights last longer than regular incandescent bulbs.

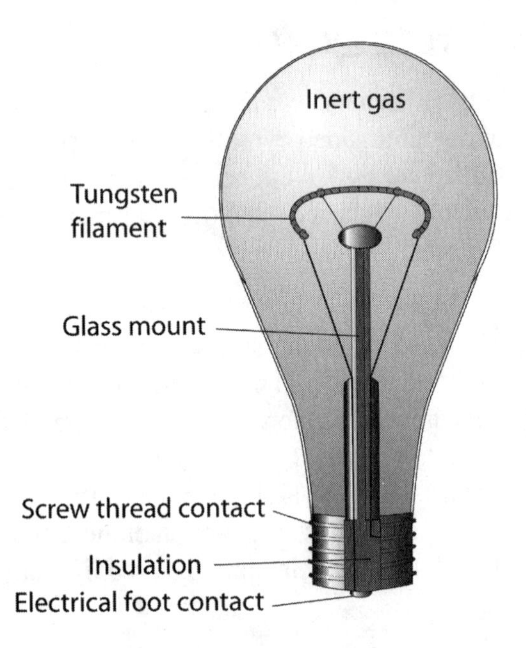

Figure 12.19 Incandescent Light Bulb

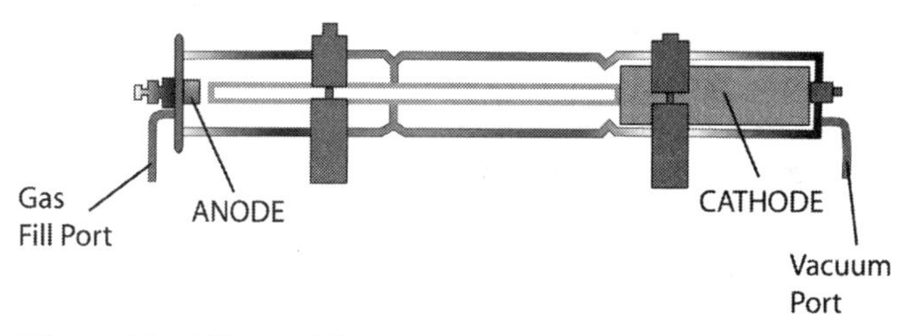

Figure 12.20 Neon Light

Neon lights use a different technique for exciting the electrons within an atom. Electricity is used to directly excite the electrons, instead of using heat. Neon lights consist of a glass tube that contains a gas such as neon, argon, or krypton. There are also electrodes at both ends of the glass tube. When a high voltage is run through the electrodes, electrons flow through the neon gas exciting the neon atoms. As the neon atom returns to its natural state, it emits red light. "Neon" lights that emit colors other than red actually contain a different gas. Neon mixed with argon or xenon will produce a blueish light and neon mixed with krypton will produce a greenish light.

Like neon lights, **fluorescent lights** use electricity to directly excite atoms. However, the process for emitting visible light is slightly more complicated in fluorescent lights. A fluorescent bulb contains a filament and a small amount of mercury liquid surrounded by a glass tube. The inside of the tube is coated with a phosphor powder. The emission of visible light occurs in a two step process. First, electricity is passed through the filament to energize electrons within the tube. The energized electrons excite some of the mercury liquid. When these excited mercury electrons return to their natural energy level, an ultraviolet photon is emitted. However, UV photons cannot be seen by the unaided human eye. So, the light from the mercury then excites electrons in the phosphorus powder. The excited phosphorus electrons emit photons in the visible spectrum when they return to their natural state. Fluorescent bulbs are more efficient than incandescent bulbs because they put the invisible UV light photons to work.

Light can also be produced by exciting electrons through chemical reactions, such as in light sticks.

Infrared Waves

Infrared waves are shorter than microwaves but longer than visible light waves. They vary in wavelength from the size of a pin head to the size of a microscopic cell. The infrared waves that we get from the sun are the part of the sun's energy that gives us warmth. You may have noticed that the temperature inside a building during the winter does not feel as warm as the same temperature would feel outside during the summer. The presence of infrared waves provides warmth. Although we cannot see infrared light, some animals can. Pit viper snakes such as the rattlesnake can detect warm blooded animals even in the dark by sensing the infrared energy. Some night vision goggles work, in part, by allowing the wearer to see the infrared light emitted from objects.

Microwaves

Microwaves are shorter in wavelength than radio waves and vary from 1 millimeter to 10 centimeters. You are probably familiar with microwaves because they are the type of waves used in microwave ovens to heat your food. Microwaves are also used to transmit information across long distances because they can penetrate clouds, smog, and precipitation. For example, microwaves are used to transmit satellite signals from outer space to provide images of the earth.

Radio Waves

Radio waves are waves in the electromagnetic spectrum that have the longest wavelengths and the lowest frequencies. They range in wavelength from less than a meter to around a mile long. The sun, other stars, and other objects in space emit radio waves. Despite what you might think, you cannot hear radio waves. We use radio waves to transmit signals that are then converted by a radio, television, or cell phone into a sound or light wave that we can hear or see. Radio waves are not generally considered harmful to human health. There is some concern that the use of cell phones may be linked to an increased risk of certain brain tumors, but, to date, ongoing scientific research has not been able to verify that link.

Section Review 4: Electromagnetic Waves

A. Define the following terms.

electromagnetic wave	Roy G. Biv	fluorescent light
electromagnetic spectrum	incandescent light	photon
speed of light	halogen light	visible light
wave-particle duality	neon light	dispersion

B. Choose the best answer.

1. Energy from the sun is transported by
   A. mechanical waves.
   B. electromagnetic waves.
   C. sound waves.
   D. compression waves.

2. How are microwaves different from gamma ray waves?
   A. Gamma rays have a higher frequency than microwaves.
   B. Gamma rays have a higher amplitude than microwaves.
   C. Gamma rays have a longer wavelength than microwaves.
   D. Gamma rays are electromagnetic, but microwaves are mechanical.

3. Which of the following statements about sunlight is ***false***?
   A. Sunlight travels through space as photons.
   B. Infrared rays come from the sun.
   C. The sun does not give off radio waves.
   D. Some of the waves coming from the sun are invisible to us.

4. Which of the following color groups correctly shows increasing wavelength?
   A. green, blue, red
   B. blue, yellow, violet
   C. yellow, orange, red
   D. orange, yellow, green

5. The rainbow of light from a prism
   A. shows that white light is the absence of color.
   B. proves that various wavelengths of light bend differently.
   C. works only if the light penetrates the face of the prism perpendicularly.
   D. would not be visible in a vacuum.

C. Answer the following questions.

1. How are electromagnetic waves different from mechanical waves?
2. Give two ways in which electromagnetic waves are similar to mechanical waves.
3. Name the colors of the visible light spectrum, beginning with the shortest wavelength.

CHAPTER 12 REVIEW

CHAPTER REVIEW

A. Choose the best answer.

1. What are mechanical waves?

   A. the means by which energy moves through a medium

   B. photons of energy transported through space

   C. anything that moves energy from one place to another

   D. all of the above

2. Identify the property of electromagnetic waves that is NOT also a property of mechanical waves.

   A. can be reflected

   B. can cause matter to vibrate

   C. can travel through a vacuum

   D. can transfer energy but not matter

3. Which of the following types of electromagnetic waves has the shortest wavelength?

   A. radio wave

   B. visible light wave

   C. ultraviolet wave

   D. gamma ray wave

4. Identify the combination of frequency and amplitude that would maximize the energy transferred by a wave.

   A. high frequency and high amplitude

   B. high frequency and low amplitude

   C. low frequency and high amplitude

   D. low frequency and low amplitude

5. Identify the statement that correctly identifies the units of frequency or wavelength and the relationship between frequency and wavelength.

   A. Frequency, measured in hertz, increases as wavelength increases.

   B. Frequency, measured in hertz, decreases as wavelength increases.

   C. Wavelength, measured in hertz, increases as frequency increases.

   D. Wavelength, measured in hertz, decreases as frequency increases.

6. Identify which of the following relies on refraction.

   A. using echoes to measure distance

   B. using a mirror to see what is behind you

   C. using contact lenses to improve eyesight.

   D. using soundproofing to create a quiet room

7. As you hold Ma large seashell up to your ear, you hear what sounds like the ocean. In fact, sound waves, most of which you cannot hear, from the surrounding area fill the seashell causing the air within the shell to vibrate. If a frequency from the surrounding air causes air within the shell to vibrate at its natural frequency, the sounds are amplified, and you hear something that resembles the splashing of ocean waves. This phenomenon is called

   A. refraction.
   B. pitch.
   C. destructive interference.
   D. resonance.

8. Which of the following is ***not*** a transverse wave?

   A. sound
   B. X-rays
   C. vibrations of a violin string
   D. water waves

9. Identify the relationship between the pitch, frequency, and wavelength of sound.

   A. High pitch equals low frequency and short wavelength.
   B. High pitch equals high frequency and long wavelength.
   C. Low pitch equals low frequency and long wavelength.
   D. Low pitch equals high frequency and short wavelength.

10. Which of the following lists electromagnetic radiations from lowest to highest energies?

   A. radio waves, microwaves, ultraviolet radiation, visible light
   B. microwaves, radio waves, visible light, x-rays
   C. radio waves, infrared radiation, visible light, ultraviolet radiation
   D. gamma radiation, infrared radiation, visible light, X-rays

11. The speed of sound in air at sea level and a temperature of 20 degrees Celsius is 343 meters per second. The musical note A has a frequency of 440 Hz. What would be its wavelength?

   A. 0.78 meters
   B. 1.3 meters
   C. 0.83 meters
   D. 0.75 meters

12. Which statement is true about electromagnetic radiation?

   A. Electromagnetic waves require a medium to travel through.
   B. Electromagnetic waves are produced by vibrating matter.
   C. Electromagnetic waves travel through matter as compressional waves.
   D. Electromagnetic waves travel faster through a vacuum than through matter.

B. Answer the following questions.

13. AM radio signals can usually be heard behind a hill, but FM signals cannot. That is, AM radio signals bend more than FM signals. Use your knowledge of the behavior of waves to explain this phenomenon.

Chapter 13
Electricity and Magnetism

PHYSICAL SCIENCE STANDARDS COVERED IN THIS CHAPTER INCLUDE:

SPS10 Students will investigate the properties of electricity and magnetism.

ELECTRIC FORCE AND ELECTRICITY

ELECTRIC FORCE AND FIELD

There are four major forces of nature. The electric force, which involves charged particles, both positive and negative, is one of these forces. The **electric force** between two charged particles is described by **Coulomb's law**, which is expressed mathematically in Equation 13.1. Although this equation looks confusing, the important points to remember are:

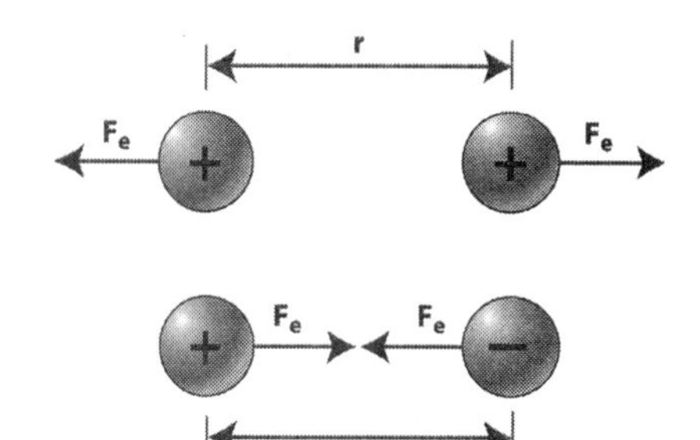

Figure 13.1 Electric Force Between Two Charged Particles

- Charged particles exert forces on each other;
- Like charges repel, opposite charges attract;
- The greater the distance between charges, the less force they will exert on each other.

Coulomb's law states:

$$\text{Electrostatic force} = \text{constant} \cdot \frac{\text{charge of particle 1} \cdot \text{charge of particle 2}}{(\text{distance between particles})^2}$$

$$\text{or, } F_e = k_e \cdot \frac{q_1 \cdot q_2}{r^2}$$ **Equation 13.1**

where k_e is the Coulomb constant and has a value of 8.988×10^9 N·m²/C². When the two charges have the same sign, they repel one another. When they have opposite signs, they attract each other. Notice that the

$$F_e = k_e \cdot \frac{q_1 \cdot q_2}{r^2}$$

$$1 \text{ newton} = \frac{N \cdot m^2}{C^2} \cdot \frac{C \cdot C}{m^2} = N$$

equation for electric force has the same form as the equation for gravitational force, but charge replaces mass, and the constants are different. While the electric force can be attractive or repulsive, the gravitational force can only be attractive.

Recall that atoms are made of a positively charged nucleus surrounded by negatively charged electrons. The attractive electrical force between these charges is what holds the atom together.

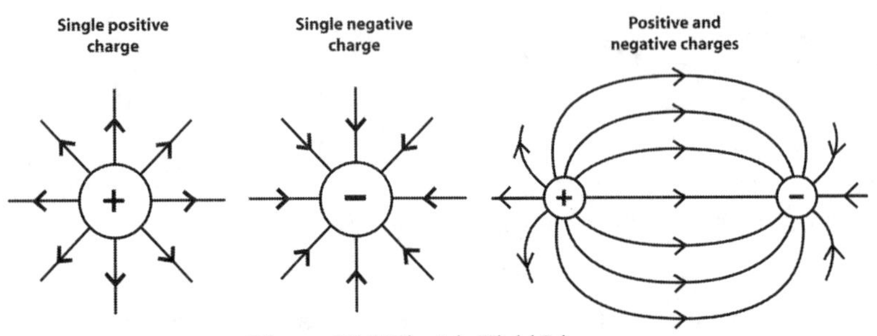

Figure 13.2 Electric Field Lines

The concept of an **electric field** helps to visualize the effects electric charges have on one another. An electric field surrounds every electric charge. If a test charge (a small, charged particle) were placed in the electric field of a charged particle, a force would be exerted upon it. The **electric field lines** or **lines of force** point in the direction that a positive charge would move when in the presence of an electric field. A positively charged particle would be repelled by a positive charge and attracted by a negative charge. Thus, electric field lines always point away from positive source charges and towards negative source charges. Electric field lines do not actually exist in the physical world; they are simply used to illustrate the direction of the electric force exerted on charged particles. The strength of the field surrounding a charged particle is dependent on how charged the particle generating the field is and separation distance between the charged objects.

Practice Exercise 1: Electric Force

Let's calculate the attractive force between the nucleus and an electron in an atom. The charges of the particles and the distance between the particles is indicated in Figure 13.3 to the right.

$$F_e = k_e \cdot \frac{(1.60\times10^{-19}\ C)\cdot(1.60\times10^{-19}\ C)}{(1.0\times10^{-10}\ m)^2}$$

$$F_e = 2.3\times10^{-8}\ N$$

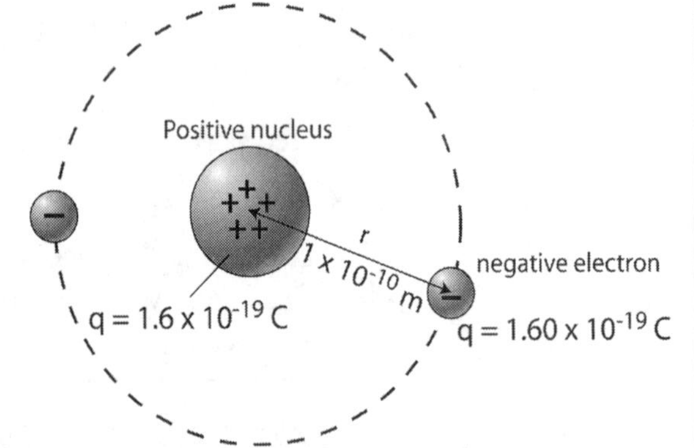

Figure 13.3 Distance Between Nucleus and Electron

Now, let's calculate the gravitational force between these particles given that the mass of a proton is 1.6726×10^{-27} kg and the mass of an electron is 9.11×10^{-31} kg.

$$F_g = G \cdot \frac{(1.6726\times10^{-27}\ kg)\cdot(9.11\times10^{-31}\ kg)}{(1.0\times10^{-10}\ m)^2}$$

$$F_g = 1.0\times10^{-47}\ N$$

Although the electrical force may seem like a very small force, if you compare it to the gravitational force between these particles, you will see it is 39 times greater than the force of gravity.

ELECTRICAL CONDUCTORS AND INSULATORS

Electricity involves the movement of electrons. Electricity results when electrons flow through a material. When this happens, the material is referred to as an **electrical conductor**, meaning that it can conduct electricity. Materials that cannot conduct electricity are **electrical insulators**. **Conductivity** is an intrinsic material property that is a measure of how easily electricity flows through a material. **Resistivity** is the inverse of conductivity and is also an intrinsic material property. In other words, it is the measure of how difficult it is for electricity to travel through a material.

Conductors typically have free electrons that can easily move about in the material. The charge is distributed uniformly in a conductor because the electrons are so mobile. Metals, such as copper and aluminum, are good conductors of electricity because the metallic bonding in most metals results in a sea of electrons that are easily able to move about. However, metals are not the only class of materials that conduct electricity! Some oxides of transition metals, such as Ti and Cr, are also good conductors. This is surprising since most oxides are not conductive because they have very strong ionic bonds. Even some special plastics are good conductors. In addition, water and the human body are good conductors. This is why it is important to keep electrical devices like radios and televisions away from water.

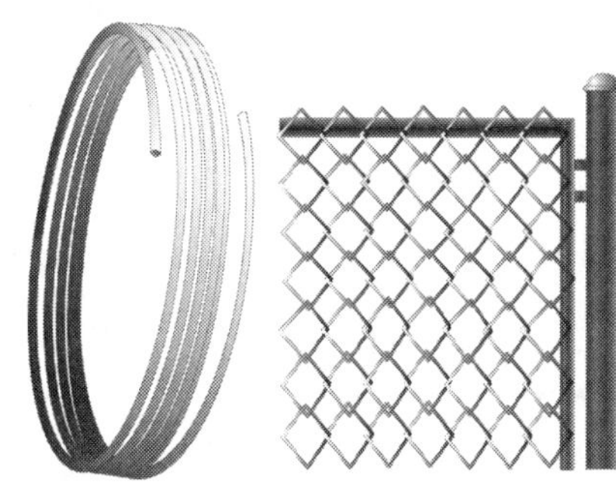

Figure 13.4 Examples of Conductors

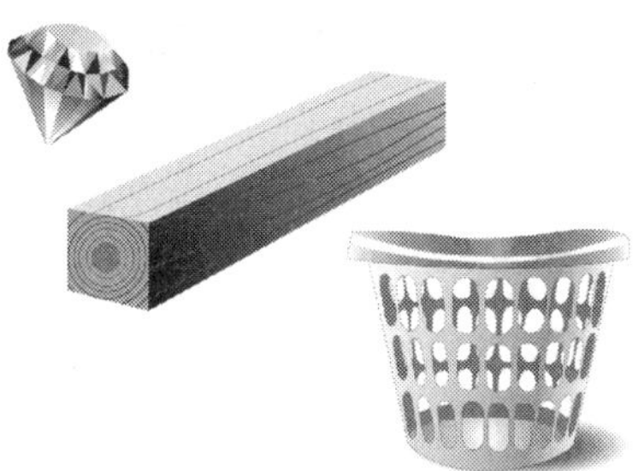

Figure 13.5 Examples of Insulators

In an **insulator,** there are almost no free electrons with which to transfer current. The atoms in insulators have very tightly bound electrons. Thus, if charge is transferred to an insulator at a given location, it will stay at that location. The electrons are not able to move in order to distribute the charge evenly throughout the insulator. Many ionic solids are insulators because the ionic bonds are hard to break and electrons are not free to move about. Some covalent solids are also insulators. Examples of insulators are nonmetals, diamond, wood, glass, rubber, porcelain, dry air, and most plastics and oxides. If you've ever seen a broken electrical cord, then you know that it consists of a conductive wire surrounded by a plastic material. The plastic material is an insulator, thus current cannot flow, and you won't be electrocuted if you touch the electrical cord. This is why it is dangerous to handle equipment with broken electrical cords.

Semiconductors are a third group of materials. These materials have few electrons with which to conduct electricity, however if enough energy (thermal or otherwise) is provided electrons can be freed and allowed to flow. Some elemental semiconductors are silicon and germanium, which are in Group IVA of the periodic table. There are many compound semiconductors formed by elements from groups III and V or groups II and VI of the periodic table. GaAs (gallium arsenide) and CdTe (cadmium telluride) are examples of III-V and II-VI compound semiconductors, respectively. Semiconductors are often used in the computer and optics industries. A semiconducting material forms the base wafer that most computer memory chips are made of.

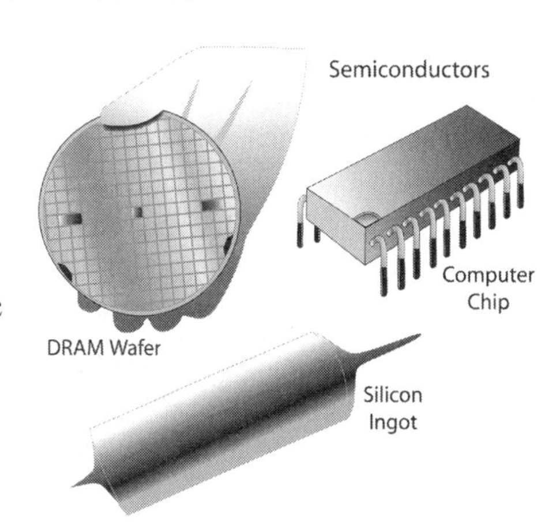

Figure 13.6 Examples of Semiconductors

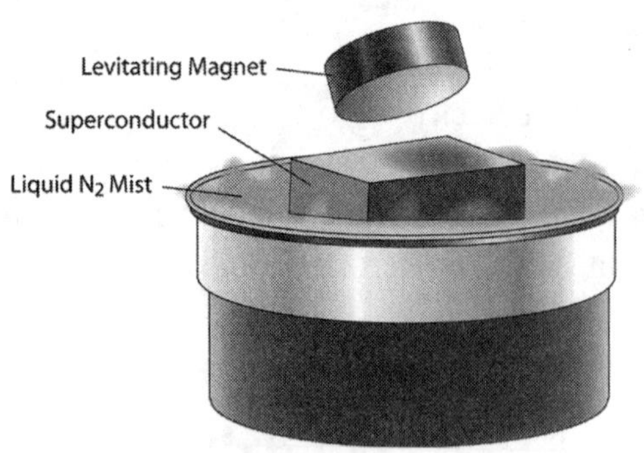

Figure 13.7 Superconducting Material with Permanent Magnetic Field

Superconductors are conductive materials where the resistivity abruptly decreases to zero as the temperature decreases below a critical value, called the **critical temperature**. The value of the critical temperature depends on the chemical composition, pressure, and structure of the material. Mercury was the first material to demonstrate superconductivity below the critical temperature of 4.15K. Recall that the temperature in Celsius equals the temperature in Kelvin minus 273.15. Therefore, the critical temperature for mercury is –269°C, which is *really* cold! Other superconductive elements include niobium, lead, tin, aluminum, and zinc. Some compounds, such as the copper-oxides, exhibit superconductivity at temperatures as high as 134K; however, that is still well below room temperature (22°C).

STATIC ELECTRICITY

As you learned in a previous section**, electricity** involves the movement of electrons. Let's talk about two types of electricity: **static** electricity and **current** electricity. We experience static electricity in everyday life in the form of a shock when we touch a metallic object after dragging our feet along the carpet, or the standing up of our hair when we take off a winter hat. **Static electricity** occurs as a result of excess positive or negative charges on an object's surface. Static electricity is built up in three ways: friction, induction, and conduction.

Figure 13.8 Static Electricity in a Comb

Rubbing two objects together will often generate static electricity through **friction.**Some electrons are held more loosely than others in an atom. The loosely held electrons can be rubbed off and transferred to the other object. Static electricity occurs when an object gains electrons giving the object a negative charge, or an object loses electrons giving it a positive charge. Rubbing a balloon on carpet or combing your hair with a hard plastic comb on a dry day causes static electricity to build up. Like charges repel, and unlike charges attract. Rub two balloons on the carpet and then slowly move the balloons together. Both balloons will have the same negative charge, and you should be able to feel the mild repulsive force. The charged balloon or comb, however, will attract small pieces of paper or other small, light objects having an opposite positive charge. These attractive or repulsive forces are weak forces, but they can overcome the force of gravity for very light objects.

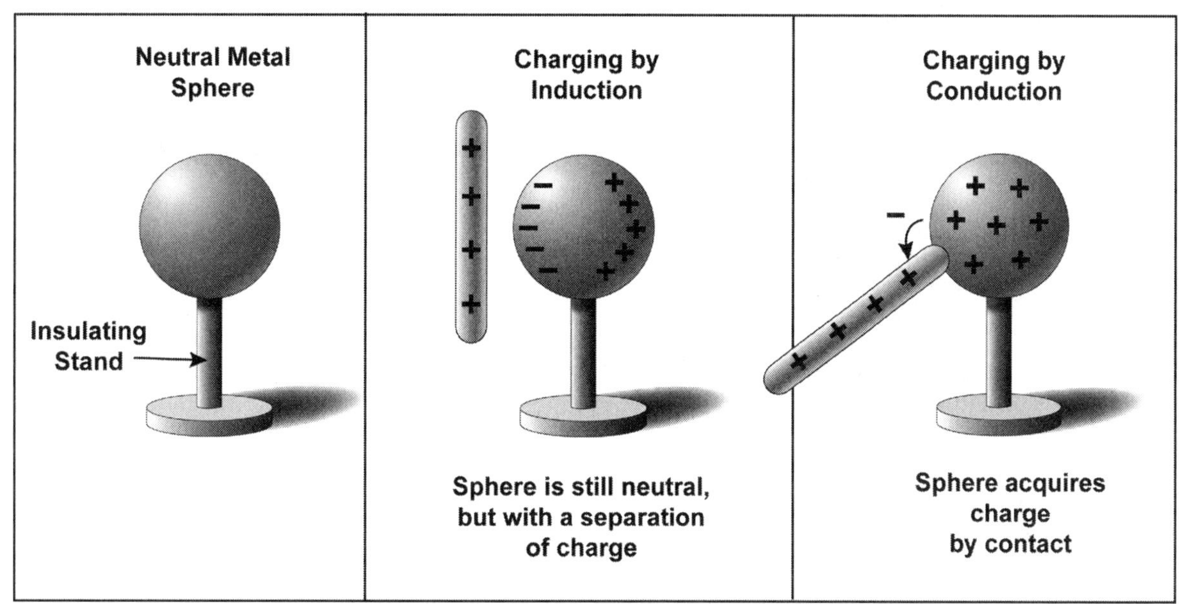

Figure 13.9 Generation of Static Electricity

Electrical charge generated by **induction** occurs when a charged object is brought near to, but not touching an insulator. Molecules within the uncharged object begin to shift with the negative side of the molecule moving closer to the positively charged object.

Electrical charge can also be generated by **conduction**. Conduction occurs when two objects, one charged and one neutral, are brought into contact with one another. The excess charge from the charged object will flow into the neutral object, until the charge of both objects is balanced.

Lightning is another example of static electricity. The actual lightning bolt that we see is a result of electric discharge from clouds that have built up too much excess charge.

CURRENT ELECTRICITY

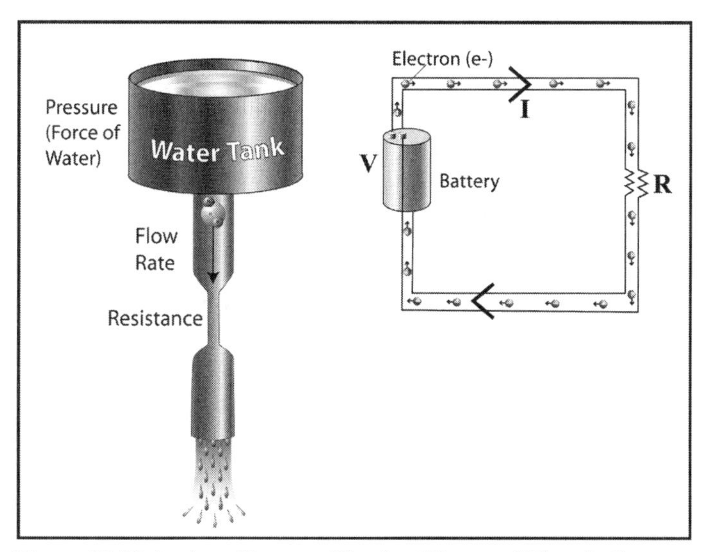

Figure 13.10 Analogy Between Flowing Water and Electric Current

To understand **current electricity**, let's compare electricity to the water flowing through a pipe. The flow rate of water in a pipe might be given in units of gallons per minute. In an electrical circuit, electrons flow through the circuit like water flows through a pipe. **Current (I)** is the flow rate of electrons through the circuit and is measured in **amperes**. As water flowing through a pipe rubs against the walls of the pipe, the water slows down. In the same way, electrons slow down as they move through a circuit. This slowing down of the electrons is called resistance. **Resistance (R)** is the measure of how difficult it is to move electrons through a circuit. Why does water flow through a pipe? A force like gravity or the force of a pump causes water to flow. **Voltage (V)** is the force that moves electrons through a circuit and is measured in **volts**. In other words, voltage drives the current in a circuit. In an electrical circuit, a battery commonly produces this force.

Electrical forces (voltage) found in nature can be very small, or they can be very large. Static electricity that builds up from our shoes as we walk across the carpet is small, and the discharge of that electrical force causes a small spark or shock. Static electricity that builds up in clouds is much larger, and the discharge of that buildup can result in high voltage lightning.

Electrical Units

The unit **ampere** expresses the rate of flow of the electrons past a given point in a given amount of time. One ampere is equal to the flow of one coulomb per second.

$$1 \text{ ampere} = \frac{1 \text{ coulomb}}{\text{second}}$$

$$1 \text{ A} = \frac{\text{C}}{\text{s}}$$

A **coulomb** is the amount of electric charge produced by a current of one ampere flowing for one second, and is equal to 6.3×10^{18} times the charge of an electron.

Resistance is the measure of how difficult it is to move electrons through a conductor.

Potential difference (voltage) is measured in **volts**, or joules of work done per coulomb of charge.

Example: A battery of 6 volts lights a bulb. The potential difference is 6 volts, which is between the 2 terminals of the battery.

$$1 \text{ volt} = \frac{1 \text{ joule}}{\text{coulomb}}$$

$$1 \text{ V} = \frac{\text{J}}{\text{C}} = \frac{\text{kg} \cdot \text{m}^2}{\text{s}^2} \cdot \frac{1}{\text{C}}$$

Ohm's Law

Ohm's Law states that the resistance is equal to the voltage divided by the current as shown in Equation 13.2.

$$\text{resistance} = \frac{\text{voltage}}{\text{current}}$$

$$\text{or, } R = \frac{V}{I} \qquad \textbf{Equation 13.2}$$

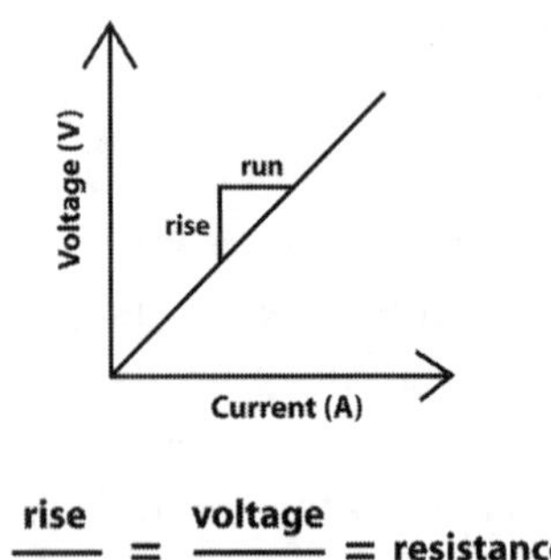

$$\frac{\text{rise}}{\text{run}} = \frac{\text{voltage}}{\text{current}} = \text{resistance}$$

Figure 13.11 Current - Voltage Relationship

where resistance has units of **ohms** (Ω).You may notice that Ohm's law reveals a linear relationship between voltage and current. Given a linear graph of voltage versus current, the slope of the line (i.e. rise over run) is equal to the resistance. Thus, the resistance of a device can be determined experimentally by taking several voltage and current measurements, then plotting the data on a graph. Not all electronic devices have this linear relationship between voltage and current. Those that do have a linear relationship are called ohmic devices.

$$R = \frac{V}{I}$$

$$1 \text{ ohm} = \frac{V}{A}$$

Ohm's law is more frequently written in the form shown in Equation 13.3.

$$V = I \cdot R \qquad \textbf{Equation 13.3}$$

Ohm's law can be used to calculate either resistance, voltage, or current when two of the three quantities are known.

Example: A flashlight bulb with an operating resistance of 50 ohms is connected to a 9.0 V battery. What is the current through the light bulb?

Step 1. Set up the equation: $V = I \cdot R$

Step 2. Insert the known information: $9.0\ V = I \cdot 50\ \Omega$

Step 3. Solve: $I = \frac{9.0\ V}{50\ \Omega} = 0.18\ A$

Practice Exercise 2: Calculations Using Ohm's Law

Use Ohm's Law to calculate resistance, voltage, or current in the problems below.

1. If a potential difference of 15 V is maintained across a wire with a resistance of 7.5 ohms, what is the current in the wire?
2. A radio draws about 5 A when connected to a 120 V source. What is the resistance of the radio?
3. A flashlight bulb with an operating resistance of 2.5 ohms carries a current of 0.9 A. How much voltage must the battery supply to the flashlight bulb?
4. A light bulb has a resistance of 250 ohms when operating at a voltage of 120 V. What is the current through the light bulb?
5. A portable alarm clock draws 1.5 amps from its 9 V battery. What is the operating resistance of the alarm clock?
6. If a current of 0.75 amps flows through a wire with resistance of 100 ohms, what is the potential difference maintained across the wire?

Section Review 1: Electric Force and Electricity

A. Define the following terms.

electric force	electrical insulator	conduction	coulomb
Coulomb's law	semiconductor	current electricity	potential difference
electric field	superconductor	current	volt
electric field lines	static electricity	resistance	Ohm's Law
electricity	friction	voltage	ohms
electrical conductor	induction	ampere	

B. Choose the best answer.

1. What is the force that moves the electrons in an electrical circuit?

   A. ampere B. coulomb C. voltage D. resistance

2. The flow of electricity through a circuit can be compared to the flow of water through a pipe. Using this comparison, the friction caused by the pipe wall would be similar to

   A. the resistance in the circuit. C. the voltage of the circuit.

   B. the amperage of the circuit. D. the coulombs in the circuit.

C. Fill in the blanks.

1. The amount of electric current in a circuit is measured in ________________.
2. The volt is a unit of electrical ________________.
3. The opposition of a conductor to the flow of electrons is called ________________.

D. Answer the following questions.

1. Explain the difference between an electrical conductor and an insulator.
2. Compare and contrast semiconductors and superconductors.

ELECTRICAL CIRCUITS

Electricity is a form of energy caused by moving electrons called electric current. The path through which the electricity is conducted is called a **circuit**. When we draw electrical circuits, we use the symbols shown in Figure 13.12 to represent voltage sources, resistors, and wires. Batteries are commonly used as

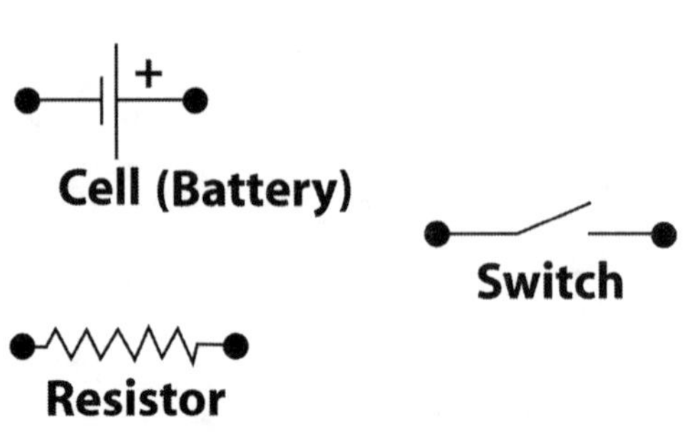

Figure 13.12 Symbols for Circuit Elements

voltage sources. Devices such as radios and televisions provide resistance to the flow of electricity. These devices, or **loads**, are usually represented as a simple resistor in circuit diagrams. There are two types of circuits: series and parallel circuits.

SERIES CIRCUITS

In a **series circuit**, all current is the same through each part or load. If a resistor is broken or damaged, current will no longer be able to flow through a series circuit.

Figure 13.13 Series Circuit

We can also represent a series circuit with a **circuit diagram** (as shown in Figure 13.14 to the right) using the symbols shown in Figure 13.12. The three resistors represent the resistance provided by each light bulb. In a series circuit you can determine the total equivalent resistance of the circuit by adding the individual resistance values. Equation 13.4 illustrates this relationship.

$$R_{eq} = R_1 + R_2 + R_3 \qquad \textbf{Equation 13.4}$$

Thus, adding a resistor in series increases the overall resistance of the circuit. All resistors in a series have the same amount of current, or amperage.

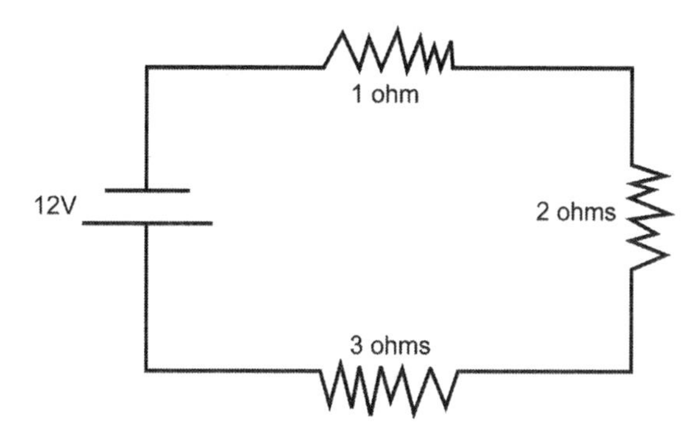

Figure 13.14 Series Circuit

A **switch** may be used to open and close the circuit. When the switch is open, electricity will not flow.

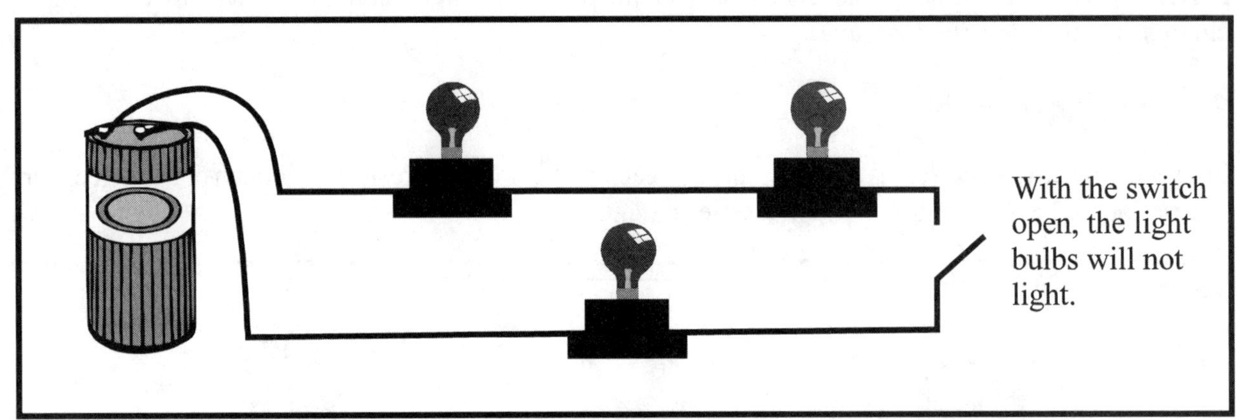

Figure 13.15 Switch in a Series Circuit

Figure 13.15 is shown as a circuit diagram in Figure 13.16 to the right.

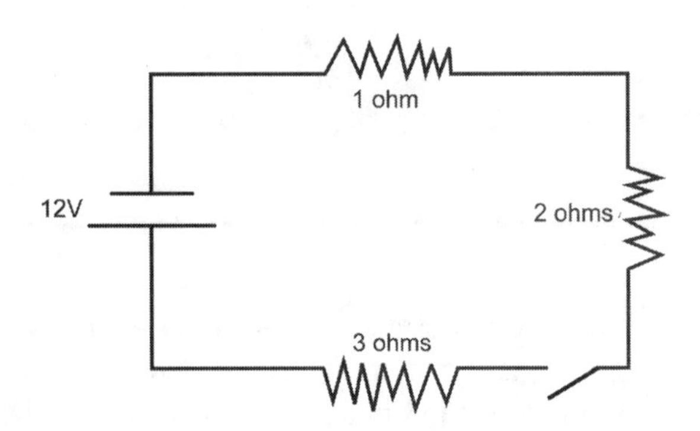

Figure 13.16 Series Circuit with Switch

Parallel Circuits

A **parallel circuit** has more than one path for the electricity to flow. The voltage is the same through all of the resistors in the circuit. If one path is removed or broken, current will still be able to flow in a parallel circuit. Most households are wired with parallel circuits, so that when you turn off a light, the television doesn't turn off as well.

Figure 13.17 Parallel Circuit

We can also represent a parallel circuit with a circuit diagram using the symbols shown in Figure 13.18. The overall resistance of a parallel circuit is reduced as more resistors are added. Thus, more current flows through the circuit. The equivalent resistance of a parallel circuit is expressed by Equation 13.5 below.

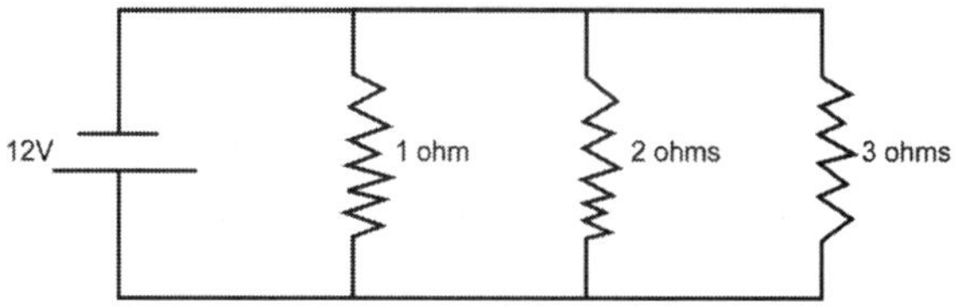

Figure 13.18 Parallel Circuit

$$\frac{1}{R_{eq}} = \frac{1}{R_1} + \frac{1}{R_2} + \frac{1}{R_3} \quad \textbf{Equation 13.5}$$

Ohm's law can be used to determine the voltage, current, or resistance in simple circuits, provided enough information is given.

Practice Exercise 3: Series and Parallel Circuits

Draw the appropriate circuit diagram based on the descriptions given. Be sure to indicate the direction of current flow through the circuit.

1. Two light bulbs, one with a resistance of 100 ohms and one with a resistance of 150 ohms, are connected in series to a 25-V battery.
2. Three resistors, all with a resistance of 50 ohms, are connected in parallel to a 9-V battery.
3. A strand of lights with five bulbs are connected to a 120-V voltage source. When one bulb goes out, the other four bulbs go out as well.
4. A strand of lights with four bulbs are connected to a 210-V voltage source. When one bulb goes out, the remaining bulbs stay lit.

ELECTRICITY IN DAILY LIFE

A **circuit breaker** is used to open circuits when there is too much current in the circuit. **Fuses** are used to protect circuits from overload currents. The conductor in the fuse will melt and break if the current in the circuit is above the allowed level. The break in the fuse stops the flow of current, and thus the device stops working. **Surge protectors** protect against sudden surges, or increases, in voltage. Surges can occur as a result of lightening during a bad thunderstorm, the operation of high power household devices like refrigerators, washers, and dryers, and faulty wiring or problems at the electric company. These surges cause the wires in the device to heat up and burn. If there is a sudden increase in voltage, the surge protector changes the direction of the electricity to go through the ground wire. Most surge protectors have a parallel circuit design, the power strip simply forces the current to travel in a different path. The surge protector only diverts the extra current, thus allowing the normal amount of current to flow to the devices connected to the power strip.

INTEGRATED CIRCUITS

Integrated circuits operate with transistors (devices that act as a switch for electronic signals) and other electronic components, which are created by patterning and depositing layers of conducting and insulating materials on a silicon chip. No wires or separate components are used. The development of the silicon chip made it possible for electric current to flow very short distances. The shorter the distance the electricity has to flow, the greater amount of tasks that can be accomplished in a short period of time. The more electronic components on a circuit, the smaller the components have to be. Otherwise, the silicon chip would be huge. The semiconductor industry is constantly coming up with new technologies to pack more electronic components on the same size chip. The small size of integrated circuits has made it possible to have very small electrical instruments for work, entertainment, and medicine.

Section Review 2: Electrical Circuits

A. Define the following terms.

circuit	circuit diagram	parallel circuit
series circuit	switch	integrated circuit
surge protector	fuse	circuit breaker

B. Choose the best answer.

1. A switch is inserted into a series circuit of Christmas lights. During the night, the switch is left open. The lights will

   A. continue to burn.
   B. be turned off.
   C. become a parallel circuit.
   D. burn brighter.

2. A series and a parallel circuit each have two resistors of two ohms each. A third two ohm resistor is then added. The overall resistance of the __________ circuit __________.

   A. parallel, increases
   B. series, increases
   C. parallel, decreases
   D. B and C only

3. A series circuit has three resistors. R1 = 2 ohms, R2 = 2 ohms and R3 = 3 ohms. What is the total resistance of the circuit, and what will happen if R3 fails?

   A. The total resistance of 1.3 ohms will decrease to 1 ohm when R3 fails.
   B. The total resistance of 6 ohms will decrease to 3 ohms if R3 fails.
   C. The total resistance of 6 ohms will decrease to 0 ohms of R3 fails.
   D. The total resistance of 6 ohms will increase infinitely; current will not flow if R3 fails.

4. A fuse
   A. stops the flow of current in the circuit if it gets too high.
   B. diverts excess current to ground if the current gets too high.
   C. stops the flow of current in the circuits if it drops too low.
   D. diverts current to other paths through the circuit when it breaks.

C. Fill in the blank.

1. A circuit that has more than one path of current flow is a ________________ circuit.

D. Answer the following question.

1. Are the headlights in a car wired in series or parallel? Explain your answer.

MAGNETISM AND MAGNETIC FORCE

A **magnet** is a metallic substance capable of attracting iron and certain other metals. It has a north and south pole that create a **magnetic field** consisting of invisible lines of force around the magnet between the two poles. These invisible lines, called **magnetic field lines**, always point from the north pole to the south pole of a magnet. The earth acts as a giant magnet having a North Pole and a South Pole, and the magnetic field circles the earth longitudinally.

A **compass** contains a small, thin magnet mounted on a pivot point. The end of the magnet that points toward the earth's geographic North Pole is labeled as the north pole of the magnet; correspondingly, the end that points south is the south pole of the magnet.

The earth's current *geographic north* is thus actually its *magnetic south.*

To avoid confusion between geographic and magnetic north and south poles, the terms *positive* and *negative* are sometimes used for the poles of a magnet. The positive pole is that which seeks geographical north.

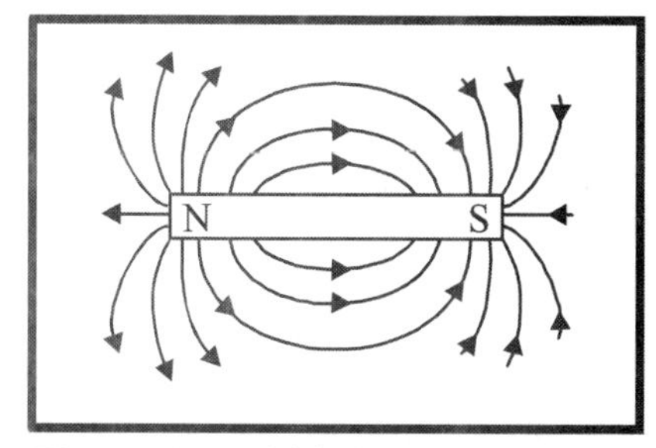

Figure 13.19 Field Lines in a Magnet

The like poles on two magnets exhibit a repulsive (magnetic) force, but two unlike poles exhibit an attractive force. For example, the north pole of one magnet will repel the north pole of another magnet, but the north pole of one magnet will attract the south pole of another magnet.

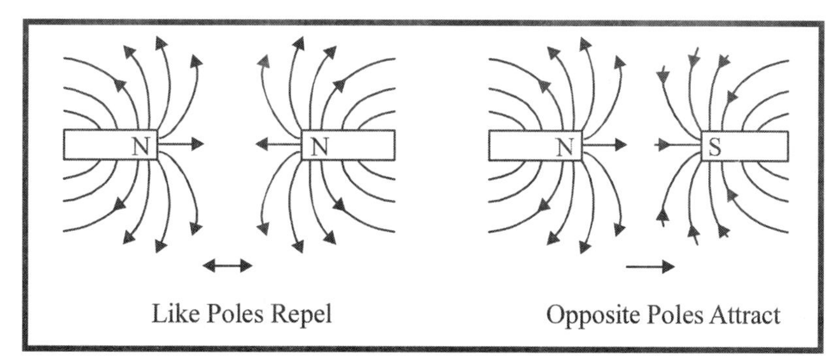

Figure 13.20 Interaction of North and South Poles

Magnets can be naturally occurring or man-made. Magnets found in nature are called lodestones. Man-made magnets are made from a mixture of iron and other metals.

Section Review 3: Magnetism and Magnetic Force

A. Define the following terms.

magnet magnetic field magnetic field lines compass

B. Chose the best answer.

1. Which of the following will attract one another?
   A. the north pole of a magnet and the south pole of another magnet
   B. the north pole of a magnet and the north pole of another magnet
   C. the south pole of a magnet and the south pole of another magnet
   D. all of the above

2. The diagram below shows a compass placed next to a powerful bar magnet.

Identify the arrow that shows the direction of the compass needle.

C. Answer the following questions.

1. In a magnetic compass, explain why the needle points north. What characteristics of Earth cause a magnetic compass to work? What material(s) must the needle be made of?
2. Why do like poles on magnets repel? Draw a diagram to show the magnetic forces around two magnets and use the diagram to explain your answer.

ELECTROMAGNETIC FORCE AND FIELDS

An electric current, as described in a previous section, can produce a magnetic field, and thus, a magnetic force. We know from Newton's third law that for every action there is an equal and opposite reaction. Therefore, it stands to reason that a magnet must exert a force on a wire carrying an electric current. As you can tell from this phenomenon, the electric and magnetic forces are intimately related. They are actually considered to be one force, called the **electromagnetic force**, which is one of the four fundamental forces of nature.

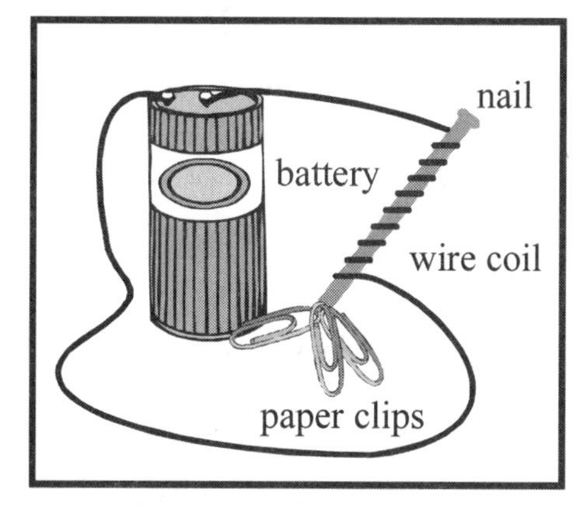

Figure 13.21 Electromagnet

Electrical and magnetic fields are related. For example, a magnetic field can be created by winding a wire around a conducting core and passing electricity through the wire. This type of man-made magnet is called an **electromagnet**. The magnetic field of an electromagnet can be strengthened by the number of turns in the wire coil or by the amount of electric current going through the wire. More coils, more current, or greater voltage equates to larger magnetic force. Note that the current used to create an electromagnet is DC, *direct current,* which we will describe shortly.

When an electromagnet is placed between the poles of a permanent magnet, the poles attract and repel each other as the electromagnet spins. Electrical energy is converted to mechanical energy. Electromagnets become more powerful as the amount of applied current to them is increased. They are often used to lift heavy metal objects, carry them and set them down again by turning off the current.

Not only can an electrical current create a magnetic field, but a magnet can produce an electric current by moving the magnet through a coil. Creating an electric current using a magnet is called **electromagnetic induction**. **Electric generators** are devices that use electromagnetic induction to create electricity. Figure 13.22 is a simple diagram of electromagnetic induction. Note that the magnet or the coils must be in motion in order for an electric current be generated. The direction that the electrons travel depends on the direction that the magnet travels.

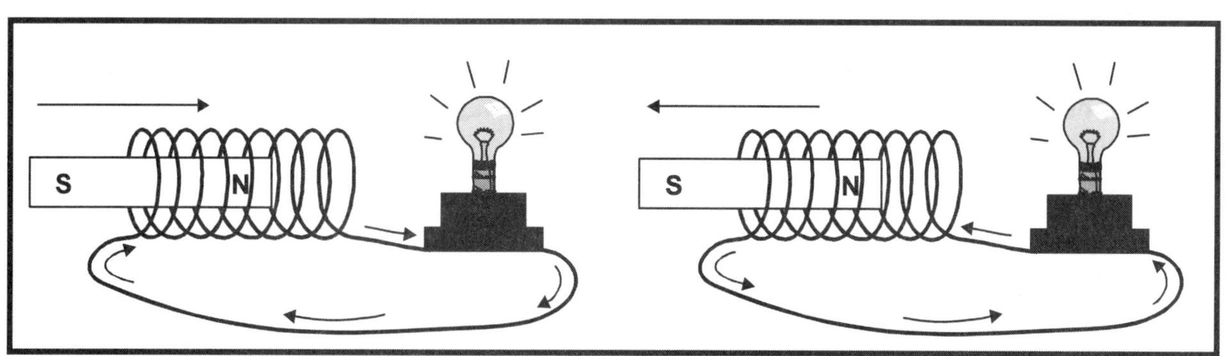

Figure 13.22 Electromagnet Induction

In the United States, electric power generators produce electricity by turning a coil between north and south poles of a magnet. Each time the coil switches from north pole to south pole, the direction of the current changes direction. This type of current is called **alternating current** or **AC**. Some countries use **direct current** or **DC**, which is the kind of current produced by a battery. Direct current flows in only one direction.

Section Review 4: Electromagnetic Force and Fields

A. Define the following terms.

electromagnetic force	electromagnetic induction	alternating current
electromagnet	electric generator	direct current

B. Choose the best answer.

1. Which of the following would strengthen an electromagnet?
   A. increasing the electric current
   B. increasing the number of coils
   C. increasing the voltage of the circuit
   D. all of the above

2. How can a magnet produce an electric current?
   A. by wrapping a wire around the magnet
   B. by moving a magnet through a wire coil
   C. by placing a magnet next to a battery
   D. all of the above

3. Look at the diagram below, and then answer the following question.

   What would happen if the poles of the magnet were reversed?
   A. The direction of the current would be reversed.
   B. The light bulb would not light.
   C. No current would be produced.
   D. The current would increase.

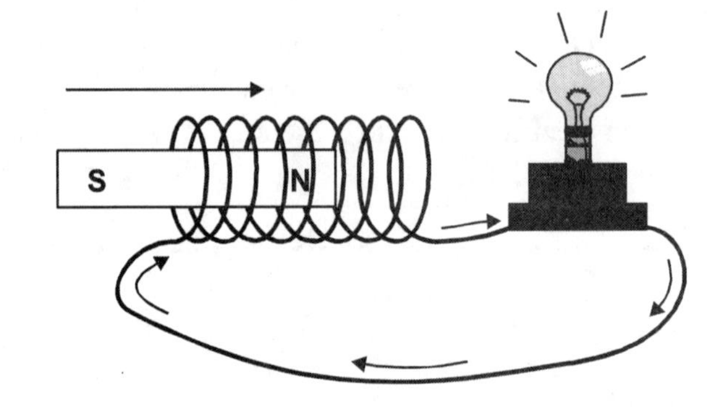

C. Answer the following questions.

1. In physics class, two groups of students experimented with making an electromagnet using a 9-volt battery, an iron nail, and copper wire. The first group made an electromagnet that would pick up 5 paper clips. The second group of students was able to make an electromagnet that picked up 7 paper clips. If the paper clips were all the same size, what other factor could have accounted for the difference? How would you suggest making an electromagnet that would pick up even more paper clips?
2. What is the difference between direct current and alternating current? How is alternating current produced?

CHAPTER 13 REVIEW

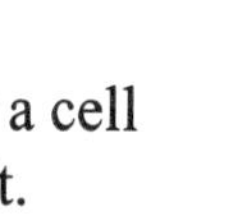

A. Choose the best answer.

1. A current of 0.5 amps flows in a circuit that is powered by a cell that produces 9.0 volts. Identify the resistance of the circuit.
   A. 18.0 ohms B. 9.5 ohms C. 8.5 ohms D. 4.5 ohms

2. A voltage (V) is applied to a circuit with a resistance (R) producing a current (I). Identify the current when a voltage (5V) is applied to a circuit of resistance (R)
   A. 0.2 I B. I C. 5 I D. 10 I

3. A 125 volt battery delivers a current of 2.0 amperes to a portable radio. What is the resistance of the radio?
   A. 0.02 ohms B. 2.0 ohms C. 63 ohms D. 250 ohms

4. A 120 volt line supplies the electricity to a light bulb with an operating resistance of 60 ohms. How many amperes of current will it take to burn the lamp?
   A. 720 amperes C. 20 amperes
   B. 0.5 amperes D. 2 amperes

5. Electrical systems for computers are efficient because they operate with transistors and a silicon chip. The circuits used in computers are
   A. integrated. B. series. C. parallel. D. alternating.

6. Which of the following statements is ***not*** true?
   A. A magnet can produce an electric field.
   B. The flow of electricity can produce a magnetic field.
   C. An electromagnet can be strengthened by decreasing the number of wire coils.
   D. An electromagnet can be strengthened by increasing the number of wire coils.

7. The diagram below shows two bodies, X and Y, that are distance, d, apart. Each body carries a charge of +q. The electrical force exerted on Y by X is equal to F.

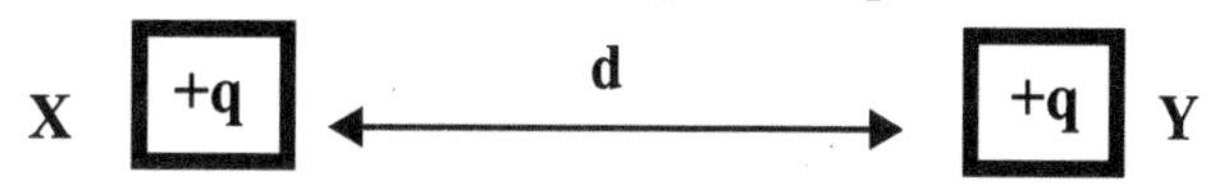

   Identify the change that would result in the biggest increase in the force exerted on Y by X.
   A. Change the charge on Y from +q to –q.
   B. Increase the charge on Y from +q to +3q.
   C. Increase the distance between X and Y from d to 2d.
   D. Decrease the distance between X and Y from d to 0.5d.

8. Identify the best description of an electric current.

   A. a flow of protons
   B. a flow of electrons
   C. a build up of positive charge
   D. a build up of negative charge

9. Identify the type of current used in battery-powered flashlights.

   A. static current
   B. direct current
   C. potential current
   D. alternating current

10. Identify the graph that shows the relationship between the voltage (V) applied across a given resister and the current (I) flowing through that resistor in an ohmic device.

A.

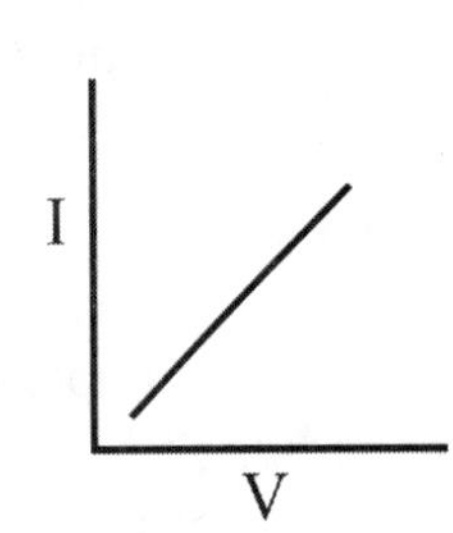

C.

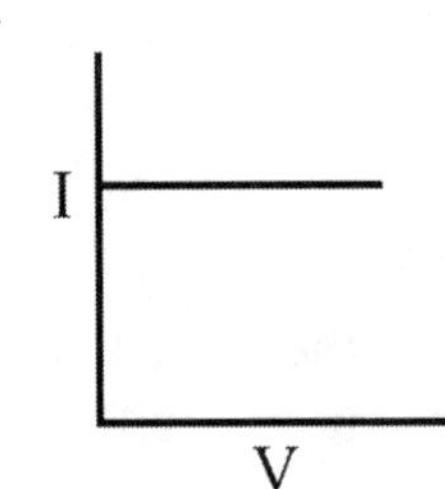

B.

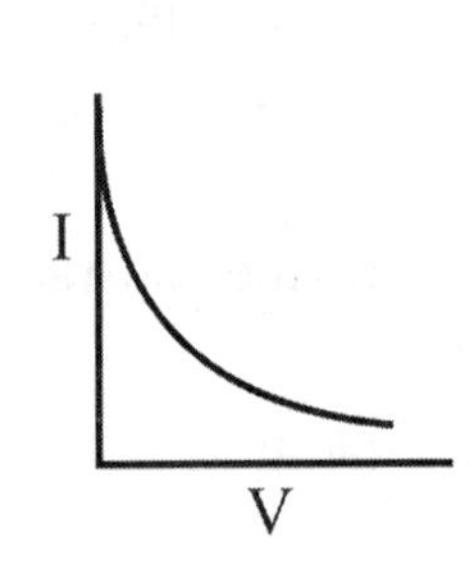

D.

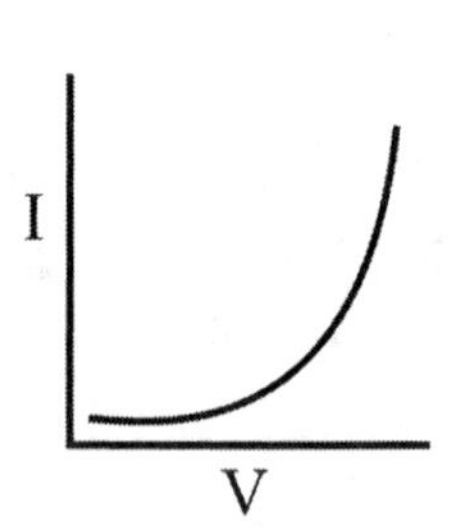

11. Identify the diagram that best represents the electrical field between two positively charged bodies.

A.

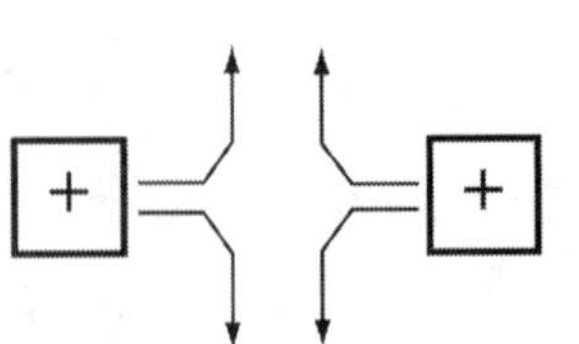

C.

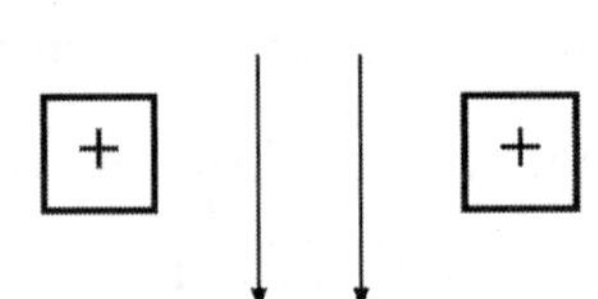

B.

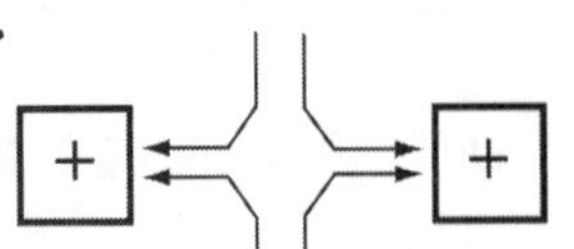

D.

+ +

12. An analogy can be drawn between the work done by an electric current flowing through an electrical appliance and the work done by water flowing over a waterfall. In such an analogy, identify the property of the waterfall that is analogous to the potential difference (voltage) across the electrical appliance.

    A. width of the waterfall
    B. height of the waterfall
    C. rate of flow of the water
    D. temperature of the water

Use the following figure to answer questions 13 and 14.

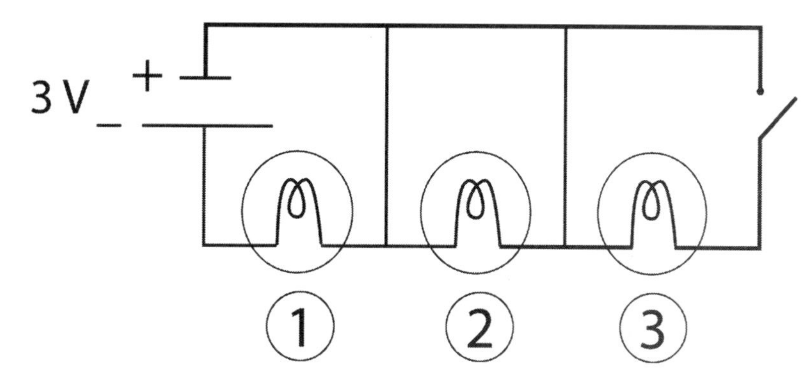

13. Describe the circuit.

    A. One 3V battery connected in series to three loads, with one open switch.
    B. Two 1.5V batteries connected in series to three loads, with one open switch.
    C. One 3V battery connected in parallel to three loads, with one open switch.
    D. An open parallel circuit where no current will flow through any of the three loads.

14. Which bulbs will be lit in the following circuit?

    A. 1 only
    B. 1 and 2 only
    C. 1, 2 and 3
    D. none will be lit

B. Answer the following questions.

15. Bill has a string of lights on a Christmas tree. He wants to know if they are wired in a series or a parallel circuit. What is a simple test to check the kind of circuit?

16. You are assigned the task of creating an electromagnet using copper wire, a battery, and an iron nail. You have your choice of the following materials: a 12-inch piece of copper wire, a 9-inch piece of copper wire, a 6-inch piece of copper wire, a 12-volt battery, a 9-volt battery, a 2-inch long iron nail, and a 4-inch long iron nail. Explain which of these materials you would use to make the strongest electromagnet. Explain why you chose each piece of material and how it contributes to a stronger electromagnet.

17. Identify the following circuits as either parallel or series.

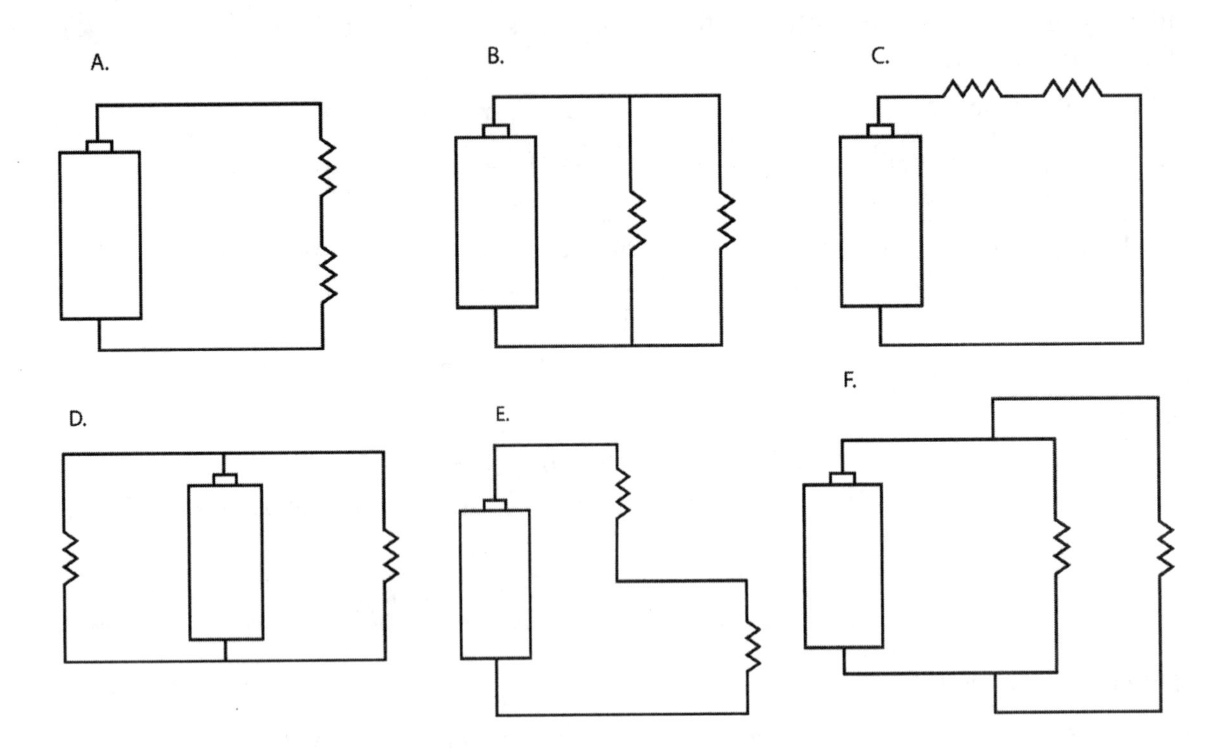

Georgia Physical Science Post Test 1

SECTION 1

Use the information below to answer questions 1and 2.

Bill wants to determine which of two brands of weed killer will kill the most weeds in his yard. To test this, he buys two brands of weed killer, Brand A and Brand B. He marks off three equal-sized squares in his lawn, and applies Brand A to one square, applies Brand B to another square, and does not apply either to the last square. The squares all receive the same amount of sunlight.

1. What is the square that does not receive a weed killer called? SCSh3 Ch 1

   A. the independent variable
   B. the dependent variable
   C. the control group
   D. the hypothesis

2. What type of graph would be best for Bill to use to show the number of weeds growing in each of the patches after a week? SCSh3 Ch 1

   A. line graph
   B. bar graph
   C. pie graph
   D. scatter plot

3. Identify the best scientific hypothesis. SCSh3 Ch 1

   A. Pea plants that are fertilized weekly grow faster than pea plants that are not fertilized.
   B. All plants prefer to be fertilized.
   C. Pea plants that receive fertilizer are better than those that do not.
   D. Plants that are fertilized are healthier than those that are not.

4. Mrs. Morton's biology classes used microscopes to view pond water samples. Which one of the following sets of students used a safe laboratory procedure? SCSh2 Ch 2

   A. James and Bastian carried the microscope by holding the arm of the microscope with one hand and supporting the base with the other hand.

   B. After spilling some pond water on the microscope, Danielle and Jake rinsed the microscope under running water while it was still plugged in.

   C. While preparing their slide, Tara and Kendall dropped it, breaking it in half. Since they only needed a small amount of pond water to view, they decided to just use half of the broken slide during the lab.

   D. The outlet at Pedro and Cierra's lab station was not working. They decided to use an extension cord and plug their microscope into the outlet at the vacant station across the room.

5. In the lab, which piece of equipment would you use to measure 25 mL of glucose solution? SCSh4 Ch 2

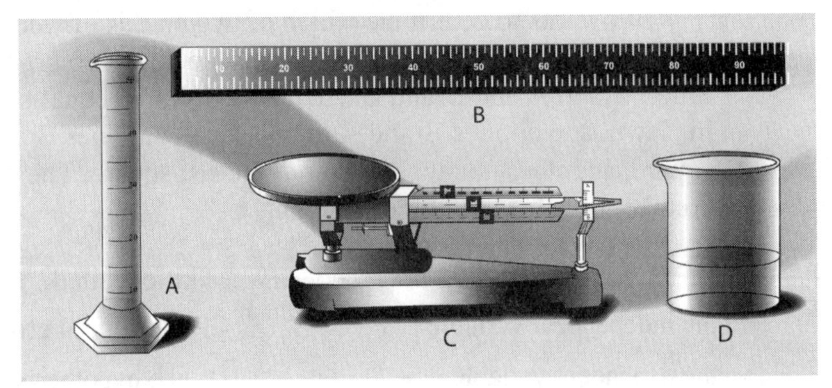

   A. A, graduated cylinder

   B. B, meter stick

   C. C, triple beam balance

   D. D, beaker

Use the following scenario to answer questions 6 and 7.

In lab, Paul and Suzanne measured a small amount of a solution containing an enzyme in a 10 mL graduated cylinder. The enzyme solution was to be added to a test tube containing reactants in order to see if the enzyme catalyzes a particular reaction. While they were pouring the enzyme solution into the test tube, they spilled a couple of drops of the enzyme solution.

6. Which of the following factors will be most affected by the spillage of enzyme? SCSh5 Ch 4

   A. the precision of the measurement

   B. the reliability of the measurement

   C. the accuracy of the measurement

   D. both B and C

7. Which of the following describes the impact of the spill on the experiment? SCSh5 Ch 1

   A. unpredictable impact

   B. predictable impact

   C. correctable impact

   D. little or no impact

Fabio did an experiment on the growth of plant seedlings. He used the same type of seedlings and soil, he gave each seedling the same amount of water and light, and he kept the temperature constant. He planted one seedling in each of three pots and arranged the pots so that one was upright, one was on its side, and one was upside down. He positioned a growlight above each plant and allowed the seedlings to grow. The diagram below shows the set up of Fabio's experiment and the growth of each seedling after 16 weeks.

8. Based on the diagram, identify the statement that is a valid conclusion from Fabio's experiment. SCSh3 Ch 1

   A. All plants need light to grow.

   B. Plants grow towards their light sources.

   C. Plants grow against the force of gravity.

   D. The rate of plant growth depends on the amount of light.

9. Which of the following is the primary purpose of an experimental group? SCSh3 Ch 1

   A. to test the hypothesis by changing one variable

   B. to indicate error in the experiment

   C. to gauge the validity of data taken in an experiment

   D. to generate ideas for future experiments

Use the graph below to answer questions 10 –11.

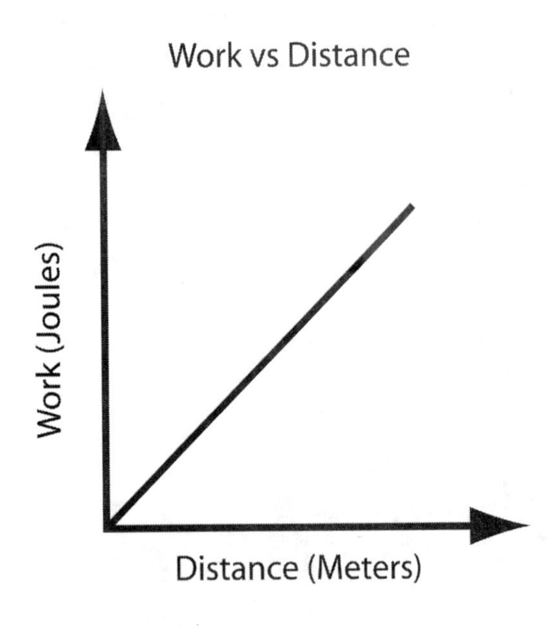

10. What does the slope of this graph represent? SPS8 Ch 11

    A. speed

    B. acceleration

    C. force

    D. energy

11. The slope of the graph would have units of SPS8 Ch 11

    A. newtons.

    B. watts.

    C. joules.

    D. seconds.

12. Joseph measures out 25 mL of distilled water for his chemistry experiment. How many liters of water is this? SCSh4 Ch 1

    A. 0.25 L B. 0.025 L C. 2.5 L D. 2.5×10^{-3} L

13. ACME, Inc. built a factory on the shore of a small lake. The data below was collected over the next five years. Based on the data, identify the most reasonable conclusion. SCSh1 Ch 3

Year	Air Pollution Parts/Liter	Average Water Temperature (°F)	Lake Area (acres)	Plant Species (number)
1	30	70.0	25	42
2	49	71.5	24	41
3	78	71.5	22	39
4	97	73.0	19	35
5	99	75.5	18	32

A. The factory had little environmental impact.

B. The factory negatively impacted the air, water, and plant life.

C. The lake was too small to support over 40 plant species in the long term.

D. The lake ecosystem benefited from the elimination of less resistant plant species.

Birth and death rates for a particular country over a five-year period are given in the table to the right

Year	Birth Rate*	Death Rate*
1995	14	15
1996	10	18
1997	11	19
1998	15	14
1999	16	12

*Birth Rate and Death Rate are given per 1000 people

14. Select the most plausible explanation for the pattern in the birth and death rates. SCSh3 Ch 1

    A. four years of widespread disease

    B. three years of overpopulation followed by starvation due to food shortages

    C. two years of war

    D. five years of economic prosperity leading to a gradual population increase

15. A scientist wants to perform research on how exposure to a certain chemical affects one's likelihood of developing cancer. What should he do before he begins conducting his investigation? SCSh7 Ch 3

    A. document his expected results

    B. enroll participants in his study

    C. review the relevant scientific literature

    D. submit his idea to a journal for peer review

16. The emission of a helium nucleus is called SPS3 Ch 6

    A. alpha particle emission. C. fusion.

    B. beta particle emission. D. fission.

17. On the speed v. time graph to the right, identify the line or curve that represents the motion of a car driven from one stop sign to a second stop sign. SPS8 Ch10

    A. line A  C. curve C

    B. line B D. curve D

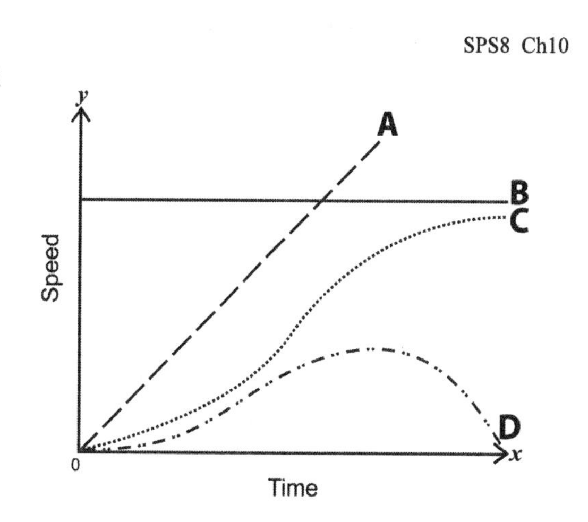

18. Identify the force needed to accelerate a car from 0m/s to 30m/s in 10s if the mass of the car is 2000kg. SPS8 Ch 10

    A. 6.7N B. 667N C. 6,000N D. 600,000N

19. Ross was riding his bike down a hill, and he ran straight into a mailbox. Identify the statement that most closely describes Ross's motion immediately following his collision with the mailbox. SPS8 Ch 10

    A. He is thrown forward over the handlebars.

    B. He is thrown backwards off the bike.

    C. He is thrown sideways off the bike.

    D. He is thrown upward into the air.

20. Gravity is the force of attraction between any two objects that have mass. The gravitational force that a body exerts depends, in part, on its mass. Which of the following factors also affects the amount of gravitational force experienced between two bodies? SPS8 Ch 10

    A. the distance between the bodies

    B. the relative speed of the bodies

    C. the altitude of the bodies

    D. the angle of the bodies

21. Which of the following statements is true concerning the motion of satellites around the earth? SPS8 Ch 10

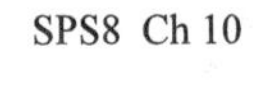

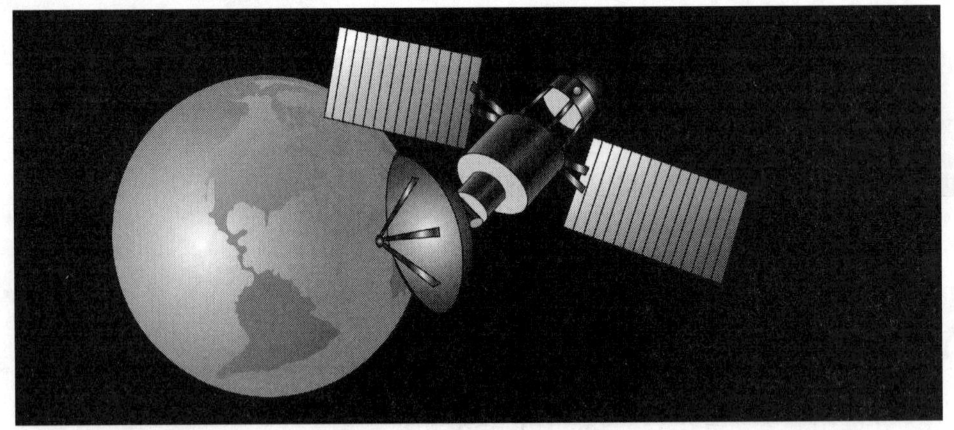

    A. Satellites continue to fall around the earth due to the force of magnetism.

    B. Satellites continue to fall around the earth due to the force of gravity.

    C. When a satellite's supply of energy is depleted, the satellite will fall to the earth.

    D. Satellites experience no acceleration.

22. Identify the result, if any, of moving a conducting wire loop through a magnetic field. SPS10 Ch 13

    A. The wire loop will exert greater gravitational force on metal objects.

    B. Electromagnetic radiation will be emitted by the wire loop.

    C. An electric current will flow in the wire loop.

    D. There will be no detectable result.

23. Identify the property that would make a wire suitable for use as a 20-amp fuse in a 110-volt circuit. SPS10 Ch 13

    A. The wire can conduct at least 2200 amps without overheating.

    B. The wire can conduct at least 20 amps without overheating.

    C. A 20 amp current will cause the wire to melt.

    D. A 2200 amp current will cause the wire to melt.

24. In a parallel circuit the current has more than one path to follow through the circuit. In the circuit below the total resistance is SPS10 Ch 13

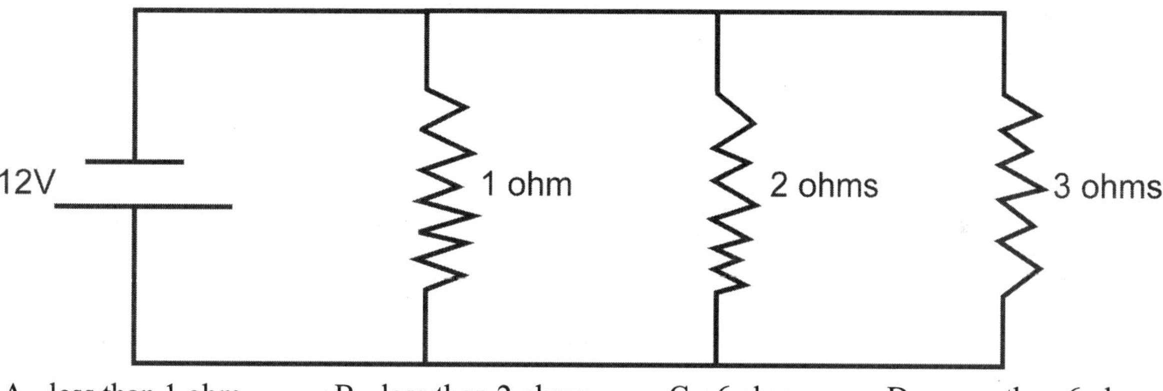

A. less than 1 ohm B. less than 2 ohms C. 6 ohms D. more than 6 ohms

25. An electric iron operating at 120 volts draws 6.0 amps of current. What is the resistance of the iron? SPS10 Ch 13

    A. 0.05 ohms B. 6 ohms C. 20 ohms D. 40 ohms

26. *Kinetic* and *potential* are terms that describe the expression of the energy of an object. Which of the following statements is true? SPS8 Ch 11

    A. All potential energy becomes kinetic energy.

    B. Potential energy is expressed as motion.

    C. All kinetic energy becomes potential energy.

    D. Potential energy is stored.

27. A steel ball is released and allowed to roll down a steep, smooth ramp. Identify the statement that correctly describes the changes in potential energy (PE) and kinetic energy (KE) of the ball. SPS8 Ch 10

 A. PE and KE are both greatest mid-way down the ramp.

 B. PE and KE are both remain constant as the ball rolls down the ramp.

 C. PE is greatest at the top of the ramp; KE is greatest at the bottom.

 D. PE is greatest at the bottom of the ramp; KE is greatest at the top.

28. Which form of electromagnetic radiation is used to image bones? SPS9 Ch 12

 A. ultraviolet B. nuclear C. infrared D. X-ray

29. Identify the three elements that have valence electrons in the *s* orbital. SPS4 Ch 5

 A. magnesium, manganese, and molybdenum

 B. calcium, barium, and sodium

 C. aluminum, silicon, and sulfur

 D. rubidium, strontium, xenon

30. Tritium is the form of hydrogen that contains two neutrons and one proton. Tritium is a/an____________ of hydrogen. SPS3 Ch 6

 A. isotope C. hydronium ion

 B. ion D. hydroxide ion

31. All of the elements in the halogen family SPS4 Ch 5

 A. need to give up one electron to become stable.

 B. need to gain one electron to become stable.

 C. do not react at all with other elements.

 D. do not need to gain one electron to become stable.

32. Consider the portion of the Periodic Table of the Elements shown below. Which of the following elements has the lowest ionization energy?

SPS4 Ch 5

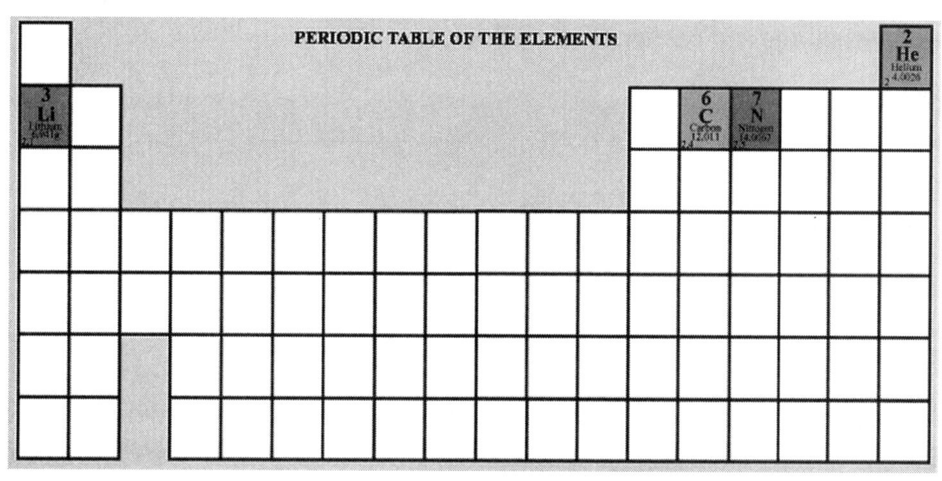

A. nitrogen B. helium C. carbon D. lithium

33. Which of the following types of elements have ductility and malleability?

SPS5 Ch 8

A. metals B. non metals C. metalloids D. gases

34. Sodium chlorate ($NaCIO_3$) and potassium nitrate (KNO_3) are solids at room temperature. The solubility curves for sodium chlorate and potassium nitrate in water are presented in the graph to the right. Based on the data provided, identify the valid conclusion.

SPS6 Ch 9

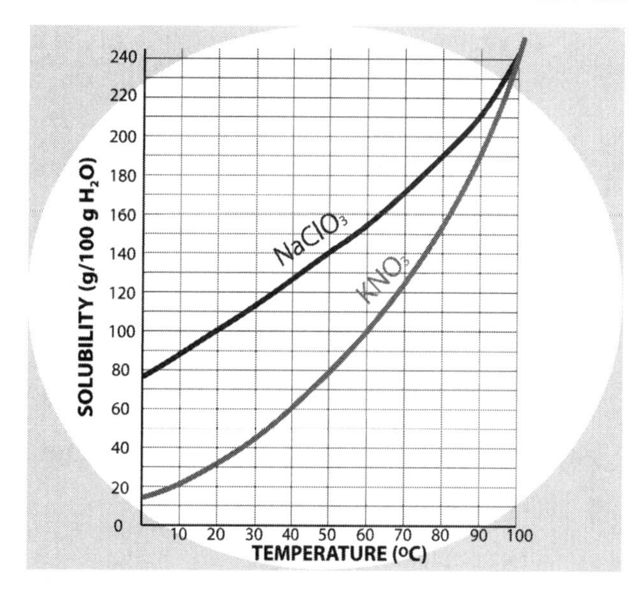

A. The solubility of these solids in water decreases as water temperature increases.

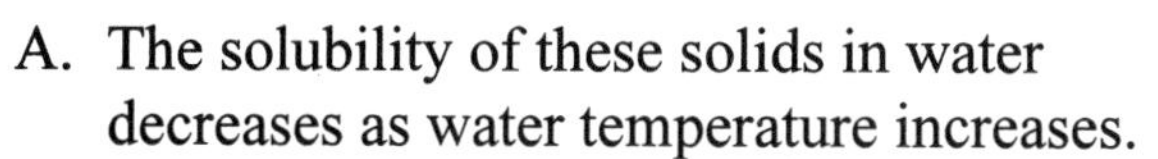

B. The solubility of these solids in water increases as water temperature increases.

C. Nitrate ions are more soluble in water than are chlorate ions.

D. Sodium is more soluble in water than is potassium.

35. By mass, about 80% of the human body is water, and water's ability to surround ions is essential to the proper functioning of many biological processes. Identify the property of water that allows it to surround ions.

SPS6 Ch 9

A. density B. volatility C. buoyancy D. polarity

36. The rusting of iron is represented by the following chemical equation. SPS2 Ch 7

$$4Fe(s) + 30_2(g) \rightarrow 2Fe_20_3(s)$$

The oxidation states of iron and oxygen are as follows:

Elemental iron: 0

Elemental oxygen: 0

Iron in iron (III) oxide: +3

Oxygen in iron (III) oxide: –2

Identify the number of electrons lost by each iron atom that rusts.

A. 2 B. 3 C. 5 D. 12

37. The following reaction is propane burning in oxygen to produce carbon dioxide and steam. SPS2 Ch 7

_____C_3H_8 + _____O_2 → _____CO_2 + _____H_2O

Which of the following sets of numbers will balance this equation?

A. 1, 5, 3, 4 B. 1, 3, 4, 5 C. 5, 4, 3, 1 D. 4, 5, 1, 3

38. Rust forms by the reaction of iron with oxygen to produce iron oxide. An 88 g iron nail rusted. The rusted nail (iron oxide) had a mass of 102 g. Identify the mass of oxygen that reacted in the rusting of the nail. SPS2 Ch 7

A. 14 g B. 88 g C. 102 g D. 200 g

39. The Arrhenius concept of acids and bases is limited in that it SPS6 Ch 9

A. does not consider aqueous reactions.

B. considers H+ the only source of basic character.

C. considers OH– the only source of basic character.

D. considers H_2O the only source of basic character

40. Atoms are held together by _______ forces between the nucleus and the electrons. SPS1 Ch 5

A. gravitational C. electrostatic

B. nuclear D. centripetal

41. The electron configuration of a chlorine atom is 1s2, 2s2, 2p6, 3s2, 3p5. How many electrons must a chlorine atom gain or lose to have a full valence shell? SPS1 Ch 5

A. gain 1 B. lose 1 C. gain 15 D. lose 17

42. A carbon atom is capable of forming up to ______ bond(s) with other atoms. SPS1 Ch 5

A. 2 B. 4 C. 1 D. 3

43. A negatively charged plastic comb is brought close to, but does not touch, a small piece of paper. If the comb and the paper are attracted to each other, the charge on the paper

SPS10 Ch 13

A. may be negative or neutral.
C. must be negative.
B. may be positive or neutral.
D. must be positive.

44. Two spheres, X and Y, each have a charge of Q. The spheres are a distance, d, apart. The electrostatic force between the spheres is F. If the spheres are separated to a distance of $3d$, identify the electrostatic force between the spheres.

SPS10 Ch 13

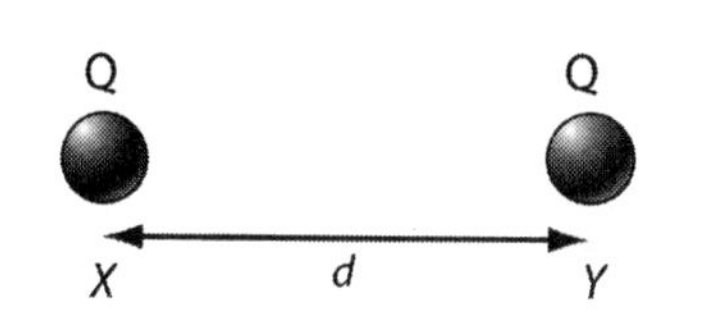

A. F/3 B. F/6 C. F/9 D. F/18

45. The largest component of the air we breathe is

SPS5 Ch 8

A. nitrogen B. oxygen C. carbon dioxide D. argon

This is the halfway point in the test. With the approval of your instructor, you may take a 5-minute stretch break.

SECTION 2

46. Which of these elements is the best conductor of electricity? SPS4 Ch 5

    A. nitrogen B. chlorine C. sulfur D. copper

47. What is the total number of electrons found in an atom of phosphorus? SPS4 Ch 5

    A. 3 B. 5 C. 15 D. 31

48. Which particle diagram represents one pure substance only? SPS5 Ch 8

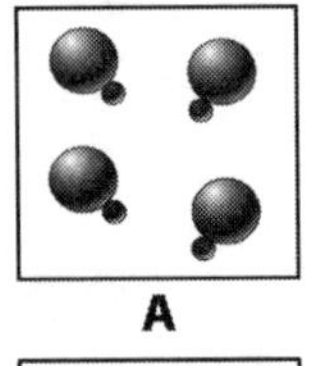

A

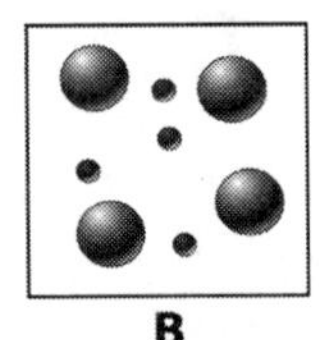

B

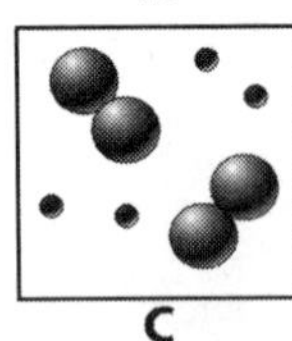

C

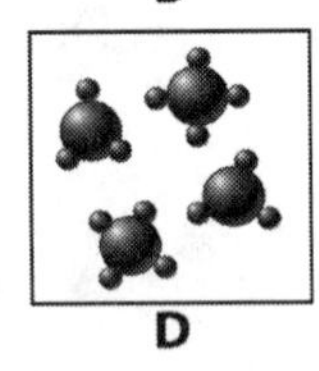

D

49. Which sample contains particles in a rigid, fixed, geometric pattern? SPS5 Ch 8

    A. $CO_2(aq)$ B. HF(g) C. $H_2O_2(l)$ D. $SiO_2(s)$

50. Distillation is a method by which SPS6 Ch 9

    A. liquids are mixed. C. gases are mixed.

    B. liquids are purified. D. gases are purified.

51. At standard temperature and pressure, a 5 gram sample of powdered iron will react more quickly with dilute hydrochloric acid than a 5 gram sheet of hammered iron. Why is this? SPS5 Ch 8

    A. Because the iron sheet is denser than the iron powder.

    B. Because the iron powder has more surface area exposed to the acid than the iron sheet.

    C. Because the iron sheet has more surface area exposed to the acid than the iron powder.

    D. Because the iron sheet is less dense than the iron powder.

52. A gas has a volume of 16 liters at a pressure of 200 kilopascals. If the temperature of the gas is kept constant, identify the volume the gas will have at a pressure of 100 kilopascals. SPS5 Ch 8

    A. 4 liters B. 8 liters C. 32 liters D. 256 liters

53. Which of the following is often a spontaneous process? SPS7 Ch 8

A. endothermic reactions
B. exothermic reactions
C. catalyzed reactions
D. fusion reactions

54. The diagram below shows two neutral metal spheres, x and y, that are in contact and on insulating stands. SPS10 Ch 13

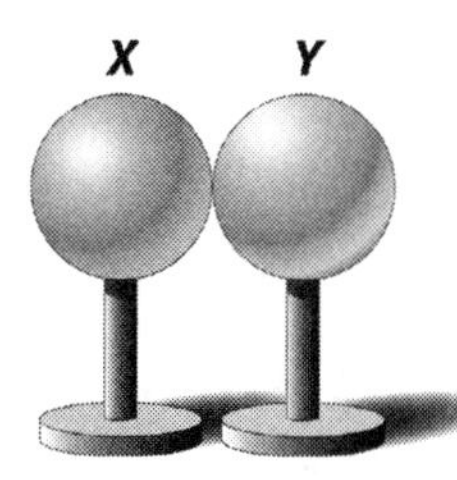

Which diagram best represents the charge distribution on the spheres when a positively charged rod is brought near sphere x, but does not touch it?

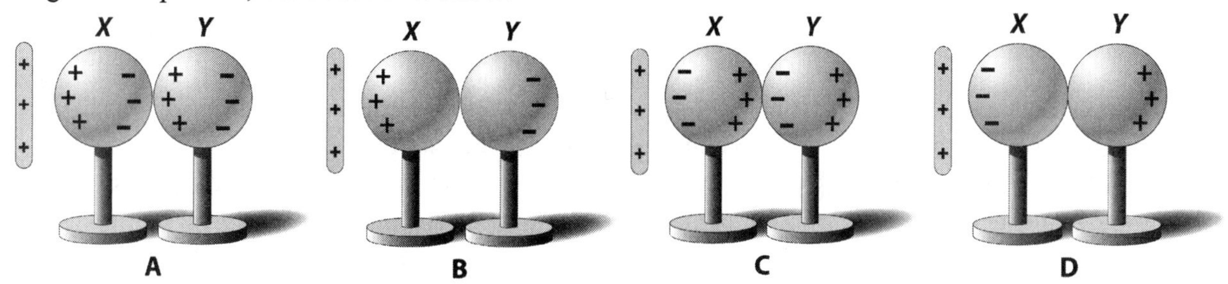

55. Electrons oscillating with a frequency of 1.5×10^8 hertz produce electromagnetic waves. Use the diagram below to classify these waves. SPS9 Ch 12

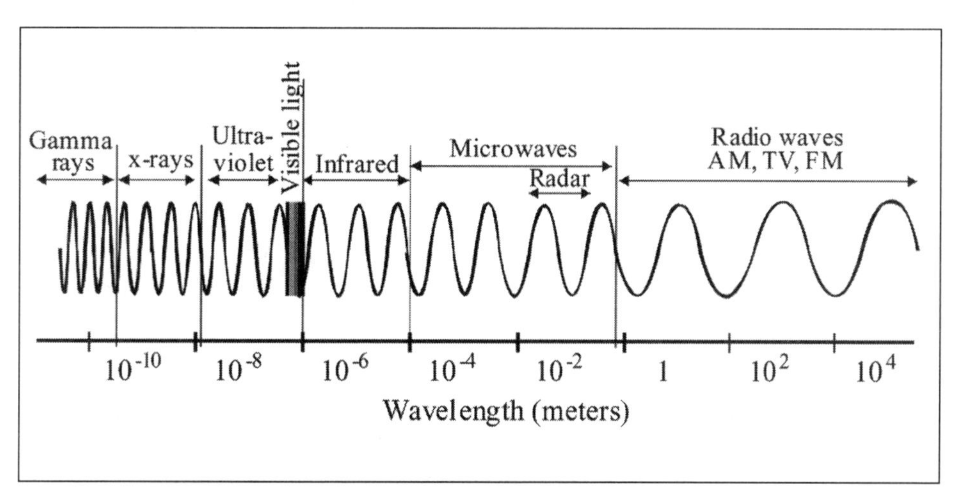

A. ultraviolet
B. visible
C. microwave
D. radio

56. Which of the following is a quantitative data set? SCSh3 Ch 1

A. the number of every player on the football team

B. the color of the jerseys for every team in the league

C. the number of rushing yards gained by each player

D. the weather conditions for every game of the season

57. A student heats a beaker full of water over a low flame. While heating, the student adds 3 grams of salt that has been measured out on a tared beam balance. He allows the solution to boil until its volume has been reduced by 50%. Then he uses metal tongs to transfer the beaker from the flame to an ice bath. What is the likely result of this series of actions? SCSh2 Ch 2

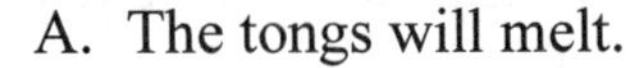

A. The tongs will melt. C. The beaker will crack.

B. The beaker will melt. D. The tongs will crack.

58. The International System of Units (SI) standardizes measurement units, enabling scientists from all over the world to communicate their results in common metric units. Which of the following correctly identifies the metric units for length, mass, time and temperature? SCSh5 Ch 4

A. meter, kilogram, second and Kelvin C. meter, kilogram, second, Celsius

B. kilometer, kilogram, second, Kelvin D. meter, gram, second, Kelvin

59. Which pair of elements is most likely to form an ionic bond? SPS1 Ch 5

A. K and H B. N and C C. K and Cl D. C and H

60. This figure shows SPS1 Ch 5

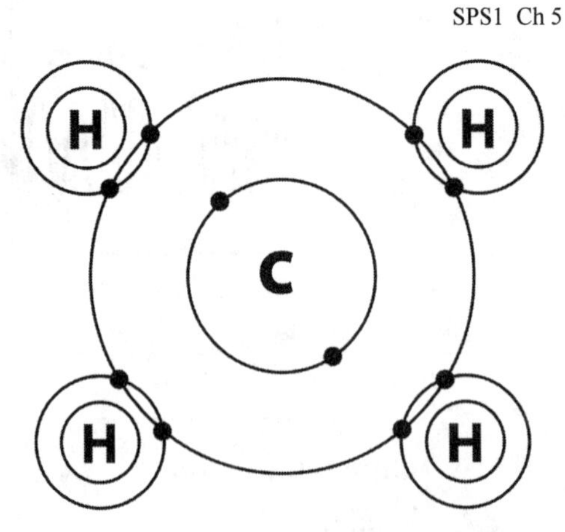

A. 5 atoms and 10 electrons

B. 1 molecule and four ionic bonds

C. 5 molecules and 10 electrons

D. 1 atom and four covalent bonds

61. Which of the following elements forms the most polar bonds? SPS4 Ch 5

A. F B. Cl C. Br D. I

62. Which of the following radioactive emissions is the largest in size? SPS3 Ch 6

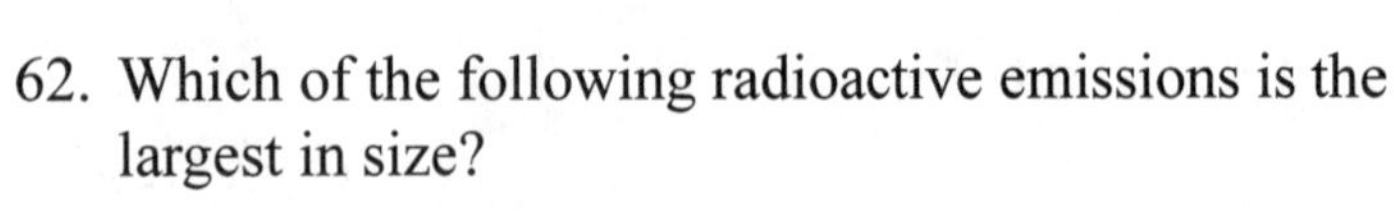

A. alpha particle C. gamma

B. beta particle D. X-ray

63. What process does the following description refer to: "a block of dry ice sitting on a laboratory benchtop has a cloud of gas around it?" SPS5 Ch 8

A. melting B. sublimation C. evaporation D. fumigation

64. Lucas conducted an experiment on human blood, to determine if a patient's gender affected the number of platelets in his or her blood. He drew 5mL of blood from 100 females and 100 males to make his comparison. In his experiment, SCSh3 Ch 1

A. gender is the independent variable and the number of platelets is the dependent variable.

B. the volume of blood is the independent variable and gender is the dependent variable.

C. the volume of blood is the independent variable and the number of platelets is the dependent variable.

D. the number of platelets is the independent variable and gender is the dependent variable.

65. If carbon dioxide is produced when fuel is burned, which of the following atoms were in the fuel? SPS2 Ch 7

A. oxygen B. nitrogen C. helium D. carbon

66. A radioactive isotope, Pu-241, has a half-life of 14.4 years. If you start with 10 g of pure Pu-241, how much will be left in 28.8 years? SPS3 Ch 6

A. 14.4 g B. 5 g C. 20 g D. 2.5 g

67. Which of the following is a correct statement about a periodic trend? SPS4 Ch 5

A. In general, the atomic radius of elements decreases going down the periodic table.

B. In general, an electron is more tightly bound to its atom going from left to right across the periodic table.

C. The reactivity of metals decreases going down the periodic table.

D. In general, the atomic radius of elements increases going from left to right on the periodic table.

68. Which of the following symbols does not represent an element? SPS4 Ch 5

A. Li B. Bt C. Na D. Mg

69. Three chemistry lab students in three different cities simultaneously bring water to boil in a beaker. Student 1 is in Norden, CA, a city that is about 6,900 feet above sea level. Student 2 is in San Francisco, CA, just at sea level. Student 3 is in Forrum, CA at 205 feet below sea level. Which student's results will indicate that the water boiled at a temperature below 100 degrees Centigrade? SPS6 Ch 9

A. Student 1

B. Student 2

C. Student 3

D. None of the students will get that result.

70. Four boys were working out with free weights in gym class. SPS8 Ch 10

Brett was holding 150 pound barbell above his head.

Philip was spotting for Brett.

Bryan was walking toward the teacher carrying a box full of weights.

Doug was holding a squat with a 30 pound barbell in each hand.

The physical science teacher came by and said only one boy was actually doing any work. Who was it?

A. Brett B. Philip C. Bryan D. Doug

71. Lucas threw a softball up in the air to a height of 10 meters. Greg threw a softball up 12 meters. Greg threw the ball higher because SPS8 Ch 10

A. the pull of gravity on the ball was less.

B. he threw with greater force.

C. there was less air friction when he threw it.

D. gravity pulled the ball higher.

72. Which of the following weighs about 1 kilogram? SPS2 Ch 7

A. a cup of water

B. a liter of water

C. a milliliter of water

D. a kiloliter of water

73. A 200 N force, F_1, and a 250 N, F_2, force are applied to the same point at the same time to a large trunk on a frictionless level surface. Which diagram below shows the position of the forces that will give the greatest acceleration to the trunk? SPS8 Ch 10

A.

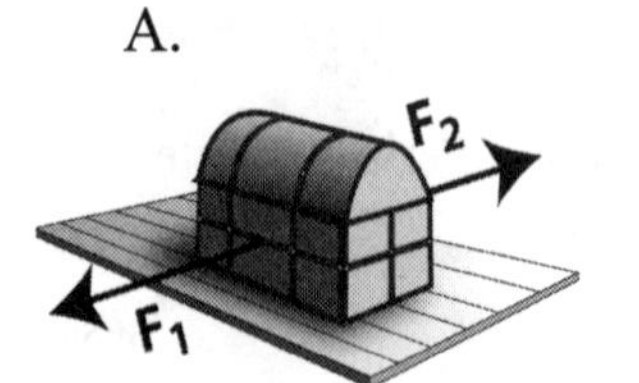

B.

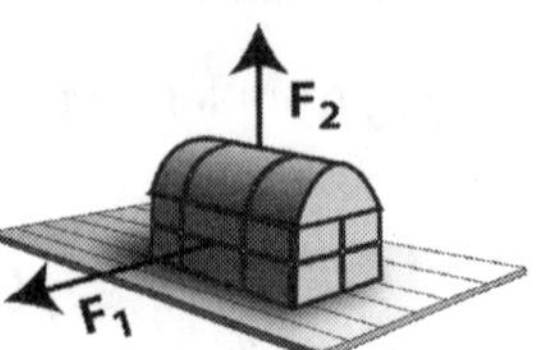

C.

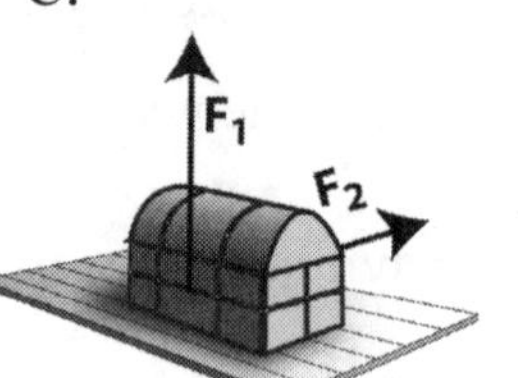

D.

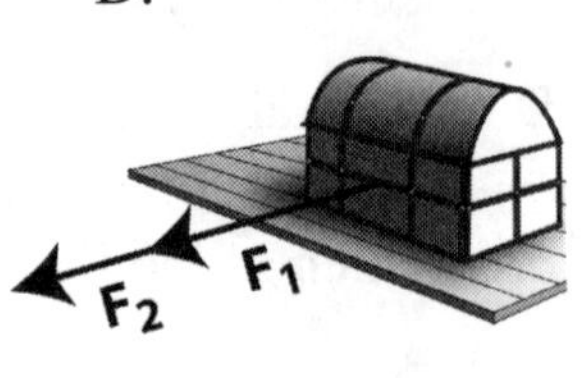

74. Which two molecules contain an equal number of atoms? SPS1 Ch 5

C_2H_6 $KMnO_4$ H_2SO_4 C_2H_3OH

A. H_2SO_4 and C_2H_3OH

B. C_2H_6 and $KMnO_4$

C. $KMnO_4$ and H_2SO_4

D. C_2H_6 and C_2H_3OH

75. Select the best description of a solvent. SPS6 Ch 9

A. substance in which other substances dissolve

B. a substance that dissolves in other substances

C. a liquid that has a high boiling point

D. a liquid that has a low boiling point

76. The diagram shows an electric circuit that uses a 120V source. It has a current of 10 amps. A toaster (R_1) and a blender (R_2) are plugged into the circuit. If the toaster creates a resistance of 5Ω, what is the resistance of the blender? SPS10 Ch 13

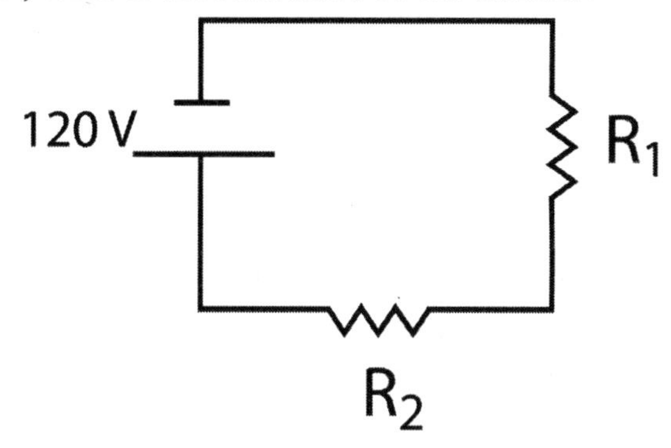

A. 12 Ω B. 17 Ω C. 7 Ω D. 60/59 Ω

77. Lead is a bluish-white metal that will burn in air to produce lead(II) oxide. It is highly malleable and was used by the Romans to make pipes to carry water ("plumbing" is taken from the Latin name for lead "plumbum"). It has a relatively low melting point of 327C and an atomic weight of 207.2 amu. Which property of lead is a chemical property? SPS4 Ch 5

A. high malleability

B. burns in air

C. low melting point

D. atomic weight is 207.2 amu

78. In order to determine the velocity of an object, what measurements must be made? SPS8 Ch 10

A. time and distance

B. time and distance and mass

C. time and distance and direction

D. time, distance and volume

79. Water boils at SPS6 Ch 9

A. 273.15 K B. 373.15 K C. 100F D. 212C

80. Rita wants to move a load of pumpkins from the garden to the barn. The pumpkins weigh 500 N. If Rita uses a wheelbarrow with a mechanical advantage of 5, how much effort force does she need to exert to move the pumpkins? SPS8 Ch 11

A. 1 N B. 10 N C. 100 N D. 1000 N

81. The diagram below shows a negatively charged body, X and a neutral body, Y. X and Y are far apart. SPS10 Ch 13

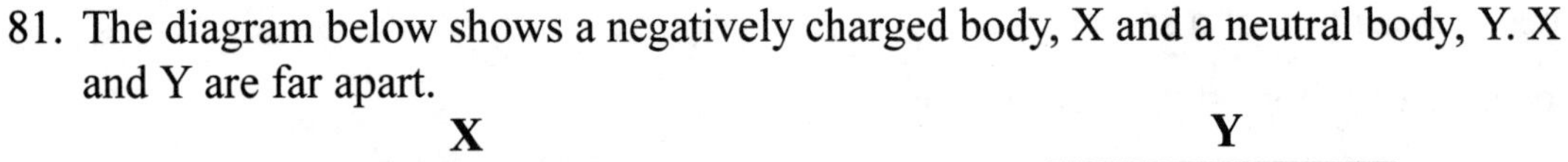

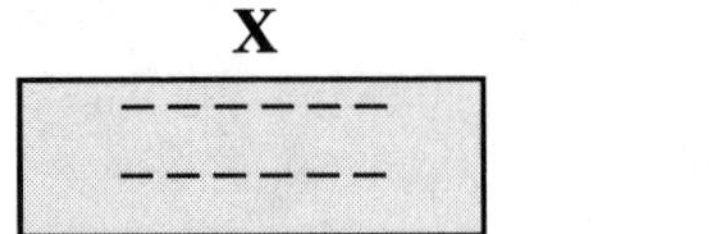

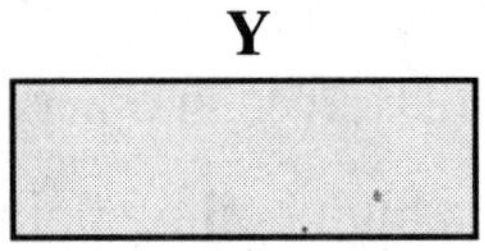

Select the drawing that shows the charges in bodies X and Y if they are brought very close together without being allowed to touch each other.

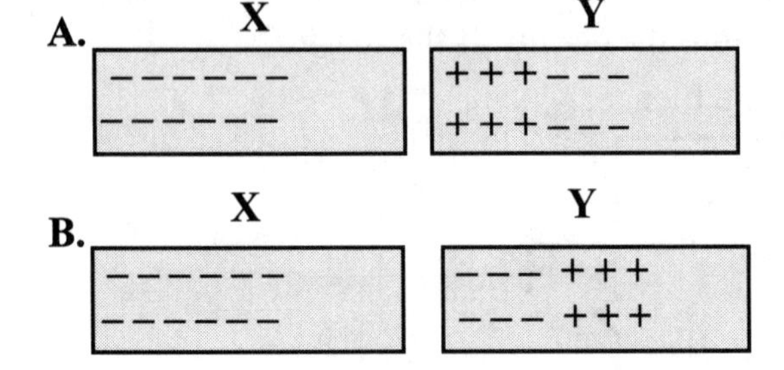

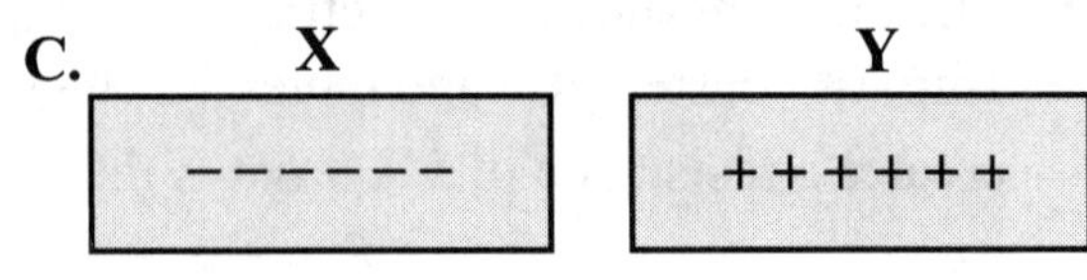

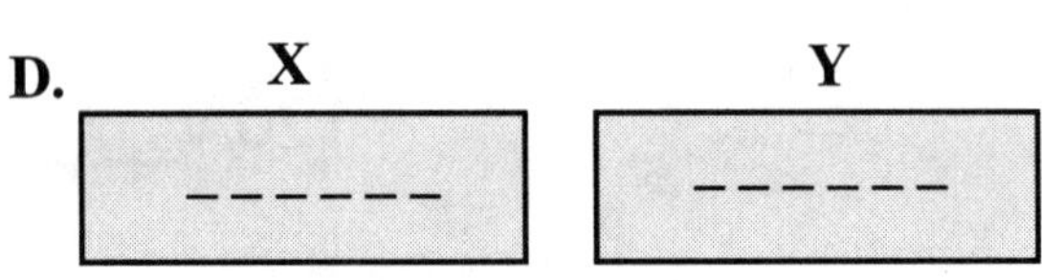

82. An analogy can be drawn between the work done by an electric current flowing through an electrical appliance and the work done by water flowing over a waterfall. In such an analogy, identify the property of the electric current that is analogous to the rate of flow of water over the fall. SPS10 Ch 13

A. power
B. voltage
C. amperage
D. resistance

83. Identify the types of fields that interact to produce motion in an electric motor. SPS10 Ch 13

A. an electrical field and a gravitational field
B. a magnetic field and an electrical field
C. two electrical fields
D. two magnetic fields

84. Identify the types of current that can be used to power appropriately designed electric motors. SPS10 Ch 13

A. direct current only
B. absolute current only
C. direct current or alternating current only
D. alternating current, direct current, or absolute current

85. Newton's Third Law of Motion is the law of action and reaction. It states that for every force there is an equal opposing force. For instance, a can of soda sitting on a table is pushed into the table by the force of gravity. The table pushes back with a force called the SPS8 Ch 10

A. oppositional force.
B. normal force.
C. inertial force.
D. frictional force.

86. On which surface will the force of static friction be easiest to overcome? SPS8 Ch 10

A. carpet B. hardwood floor C. grass D. gravel

87. Complete the following reaction. SPS2 Ch 7

$$NaOH\ (aq) + HCl\ (aq) \longrightarrow \_\_\_\_\_ + H_2O\ (l)$$

A. 2NaCl (s) B. NaCl (g) C. NaCl (aq) D. 2NaCl (aq)

88. The pipe organs found in churches have pipes of many different lengths, each of which creates a different note. What property is this method of sound production based on? SPS9 Ch 12

A. refraction of sound waves

B. destructive interference of sound waves

C. resonance of sound waves

D. cancellation of sound waves

89. Classify the following reaction: SPS2 Ch 7

$$CaCO_3\ (s) \longrightarrow CaO\ (s) + CO_2\ (g)$$

A. neutralization B. synthesis C. displacement D. decomposition

90. Which of the following will have the lowest pH? SPS6 Ch 9

A. rainwater B. distilled water C. soapy water D. bleach

Georgia Physical Science Post Test 2

SECTION 1

1. A double blind experiment is run to test the effects of a new chemical compound as a sunscreen. The scientists divide test subjects into two randomly chosen groups. One group will apply a base lotion containing the new compound. The other group will apply a base lotion without the compound. SCSh3 Ch 1

   A. The control group is the one receiving the base lotion only.

   B. The experimental group is the one receiving the base lotion only.

   C. The two groups should not be randomly chosen.

   D. Each group will be told which lotion they will receive.

Use the following information for questions 2 – 3.

The following information about the results of an experiment on goldfish is included in a research paper.

> Goldfish in the 12″ x 16″ tank were given the same amount of food as the goldfish in the 18″ x 20″ tank. Also, equal numbers of goldfish were present in each tank. Results indicate that goldfish raised in the 18″ x 20″ tank grew on average 1″ longer than the goldfish in the 12″ x 16″ tank.

2. Which of the following sentences is most likely to be in the conclusion of these results? SCSh3 Ch 1

   A. Goldfish growth was not affected by the size of their habitats.

   B. Goldfish grow larger as their habitat size increases.

   C. Diet is a deciding factor in goldfish growth.

   D. Dissolved oxygen is a deciding factor in goldfish growth.

3. Which of the following was the dependent variable in the previous experiment? SCSh3 Ch 1

A. goldfish diet
B. tank size
C. number of goldfish per tank
D. growth of goldfish

4. Water is often called the universal solvent. The pH of pure water is about 7. Adding bleach to pure water would make the pH SPS9 Ch 12

A. go down.
B. go up.
C. stay the same.
D. more temperature dependent.

5. Alexander Fleming conducted an experiment in 1928 in which he grew bacterial cultures in Petri dishes. Green mold contaminated the cultures. Fleming observed a ring around the mold where bacteria were absent. Identify the statement that is NOT a scientific hypothesis that could be tested, based on Fleming's observation. SCSh3 Ch 1

A. The mold releases a chemical that kills the bacteria.
B. The mold will kill both harmful and beneficial bacteria.
C. Other scientists were not curious about the effect of mold on bacteria.
D. Changing temperature will change the effect of the mold on the bacteria.

6. Ignacio is interested in purchasing a product that will help him reduce the amount of weeds he has in his yard. At the local home improvement store, he finds two brands of weed killer, Weeds Won't Stay and Wipe Out Weeds. The price of a 3 pound bag is the same for each brand. The graphs at right show the results of the two products based on the number of applications of each weed killer. According to the information in the graphs, which of the following conclusions is valid? SCSh3 Ch 4

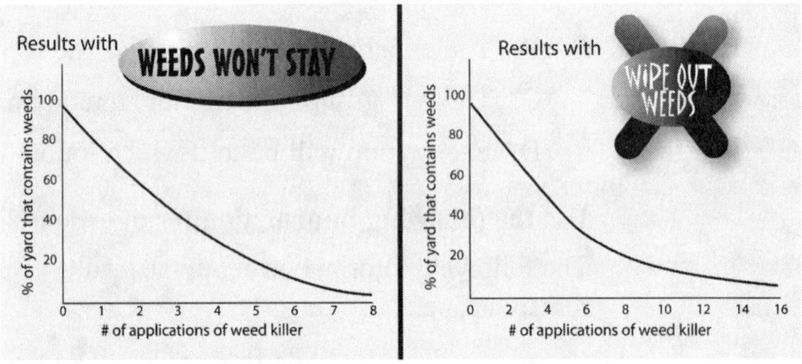

Results with WEEDS WON'T STAY
% of yard that contains weeds: 100, 80, 60, 40, 20
0 1 2 3 4 5 6 7 8
# of applications of weed killer

Results with WIPE OUT WEEDS
% of yard that contains weeds: 100, 80, 60, 40, 20
0 2 4 6 8 10 12 14 16
# of applications of weed killer

A. Weeds Won't Stay and Wipe Out Weeds are comparable and there is no advantage in one over the other.
B. Weeds Won't Stay is the better buy because it kills more weeds with fewer applications.
C. Wipe Out Weeds is more effective because it kills weeds more quickly.
D. Neither Wipe Out Weeds nor Weeds Won't Stay are effective because they both leave 20% of the yard infested with weeds.

7. A science class is conducting a laboratory activity that involves observing a chemical reaction. Each pair of students is given a flask closed with a rubber stopper and containing an unknown liquid. The students are told that the stopper needs to stay on the flask because the liquid inside will burn skin. Students are also told to keep their aprons and goggles on throughout the lab. Students will observe the chemical reaction that takes place when, after shaking the flask, the liquid is mixed with the air sitting above it. Which of the following actions displays safe laboratory procedures? SCSh2 Ch 2

   A. Instead of shaking the flask, Adam and Shalandra turn the flask upside down and balance it on the rubber stopper to ensure that the liquid mixes sufficiently with the air.

   B. After completing the experiment, Mandy and Raul carefully remove the stopper and waft the fumes produced by the liquid to try and determine its identity.

   C. Alka and Travis take turns shaking the flask throughout the experiment. They return the flask to a location designated by the teacher at the end of the lab.

   D. Since the rubber stopper is going to remain in the flask throughout the experiment, Matias and Tara take their goggles off so that they can see the reaction more clearly.

8. Select the best explanation of why scientists build on previous knowledge. SCSh1 Ch 3

   A. Scientists do not like new ideas.

   B. Scientists use previous knowledge to ask new questions.

   C. Scientists understand that most questions have been asked and answered.

   D. Scientists do not like to ask questions that build on previous knowledge.

9. In the 16th century, astronomer Tycho Brahe reported that the observed positions of the planets could not be explained by Copernicus' theory, which said the planets orbit the sun in circular paths. Mathematician and astronomer Johannes Kepler applied mathematics to Brahe's results and revised the theory to state the planets orbit the sun in elliptical, not circular, orbits. Identify the true statement that is illustrated by Kepler's work. SCSh1 Ch 3

   A. Mathematics is more credible than science.

   B. Scientific theories must be based on mathematics.

   C. Accurate observations always confirm scientific theories.

   D. Scientific theories are revised when they do not explain observations.

10. The following data shows a three-year period for plant populations in the savanna. What is the slope of the curve showing the population change of shrubs over the last two years? SCSh5 Ch 4

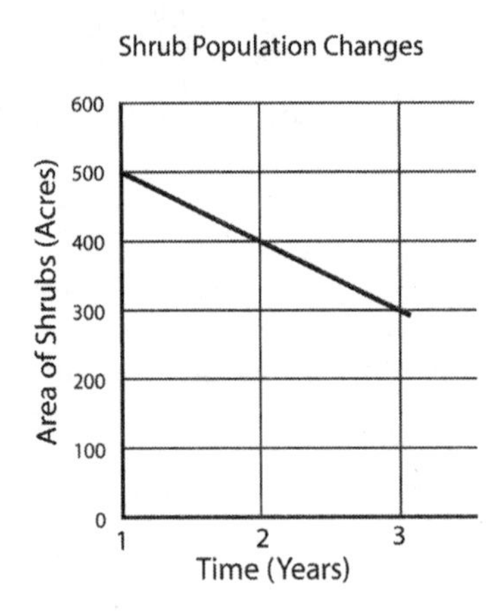

A. – 125 acres per year
B. –100 acres per year
C. 75 acres per year
D. 100 acres per year

11. Antoine Becquerel discovered natural radiation. Marie Curie and her husband Pierre built on Becquerel's work and won the Nobel Prize for their contribution to the understanding of radiation. Identify the way or ways the Curies could have learned about Becquerel's work SCSh8 Ch 1

A. collaboration with Becquerel
B. reading a paper in a scientific journal
C. hearing a presentation at a conference
D. Each of the above is a way the Curies could have learned about Becquerel's work.

12. Which of the following pieces of equipment would you use to heat a liquid on a hot plate? SCSh2 Ch 2

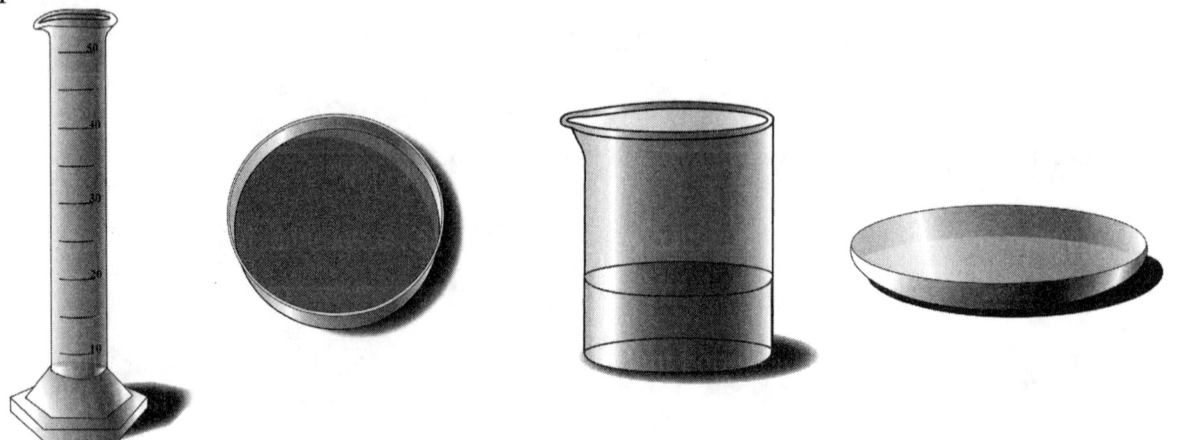

A. Graduated cylinder B. Petri dish C. Beaker D. Watch glass

13. Which of the equations below is the correct dimensional equation to convert 55 mph to feet/sec? SCSh5 Ch 4

A. $\frac{55\text{ mi}}{\text{hr}} \cdot \frac{1\text{ mi}}{5{,}280\text{ ft}} \cdot \frac{1\text{ hr}}{60\text{ min}} \cdot \frac{1\text{ min}}{60\text{ sec}}$

B. $\frac{55\text{ mi}}{\text{hr}} \cdot \frac{5{,}280\text{ ft}}{1\text{ mi}} \cdot \frac{1\text{ hr}}{60\text{ min}} \cdot \frac{1\text{ min}}{60\text{ sec}}$

C. $\frac{55\text{ mi}}{\text{hr}} \cdot \frac{1\text{ mi}}{5{,}280\text{ ft}} \cdot \frac{60\text{ min}}{1\text{ hr}} \cdot \frac{1\text{ min}}{60\text{ sec}}$

D. $\frac{55\text{ mi}}{\text{hr}} \cdot \frac{1\text{ mi}}{5{,}280\text{ ft}} \cdot \frac{60\text{ min}}{1\text{ hr}} \cdot \frac{60\text{ sec}}{1\text{ min}}$

14. Consider the graph shown right. A, B, and C each represent a different solute, and the curves represent solubility of the solute in water. Which of the following statements is most likely to be correct? SPS5 Ch 8

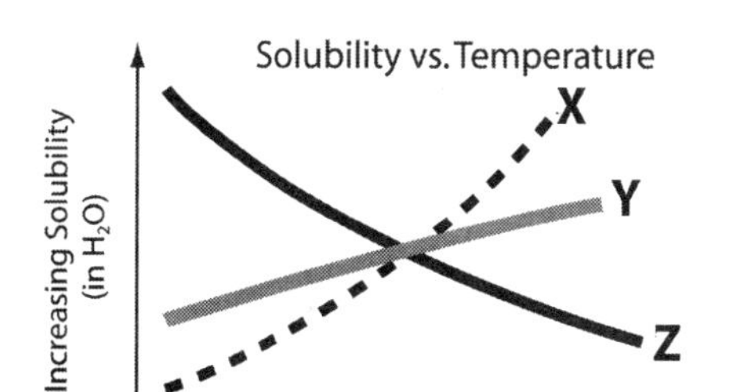

A. X, Y, and Z are gases

B. X, and Y are gases; Z is a solid

C. X, Y, and Z are solids

D. X and Y are solids; Z is a gas

15. Nuclear fission is a powerful energy source, used in both nuclear power plants and atomic bombs. Which statement is a possible description of the nuclear fission of an atom of U-235? SPS3 Ch 6

A. U-235 absorbs one neutron and splits into an atom of krypton-94 and an atom of barium-139.

B. U-235 absorbs one neutron and splits into an atom of krypton-94, an atom of barium-139 and three neutrons.

C. U-235 absorbs one neutron and splits into an atom of krypton-94, an atom of barium-139 and two neutrons.

D. U-235 absorbs one neutron and splits into an atom of krypton-94, an atom of barium-139 and one neutron.

16. Which of the following might be the effect of an unbalanced force acting on an object? SPS8 Ch 10

A. An unbalanced force can only cause an object to slow down.

B. An unbalanced force can cause an object to speed up only.

C. An unbalanced force can cause an object to speed up, slow down, or change direction.

D. An unbalanced force can only cause a change in direction of motion.

17. A collision between two billiard balls is described below. Ball 1 has a mass of 0.5 kg and a velocity of 6 m/s – north. It collides with ball 2, which has a mass of 0.5 kg and is stationary prior to the collision. After the collision ball 2 achieves a velocity of 5 m/s – north. If ball 1 continues to move north, what will be its speed? SPS8 Ch 10

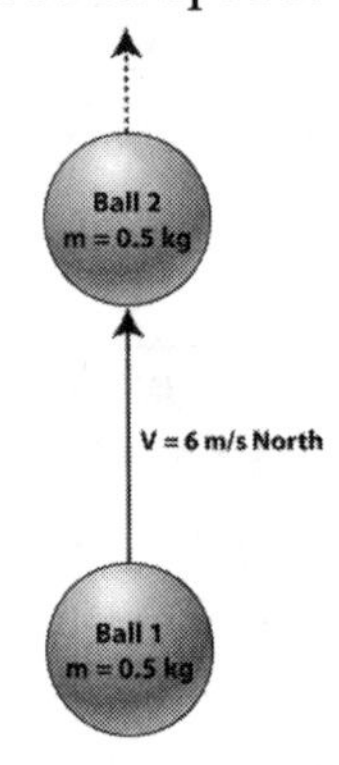

A. 0 m/s B. 1 m/s C. 0.5 m/s D. 1.5 m/s

18. The circuit below represents the wiring in a power strip. Identify the way the circuit is wired and, if one device breaks, the effect on the other devices. SPS10 Ch 13

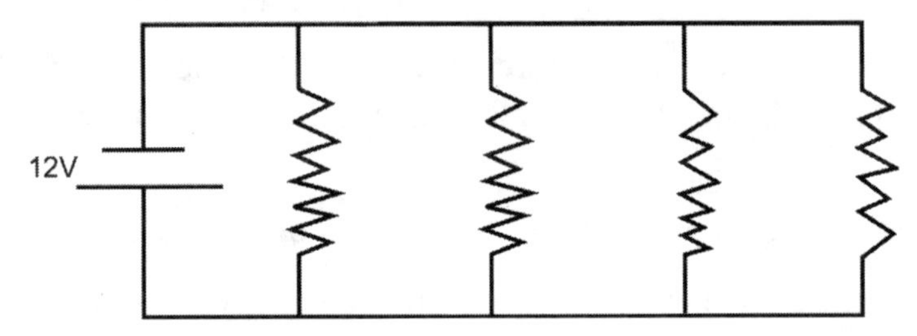

A. wired in parallel; other devices remain on

B. wired in parallel; other devices turn off

C. wired in series; other devices remain on

D. wired in series; other devices turn off

19. A bowling ball with a mass of 5.44 kg and a soccer ball with a mass of 0.43 kg are dropped from a 15 m platform. Identify the correct description of the acceleration of the bowling ball and the force with which it hits the ground, with respect to the soccer ball. SPS8 Ch 10

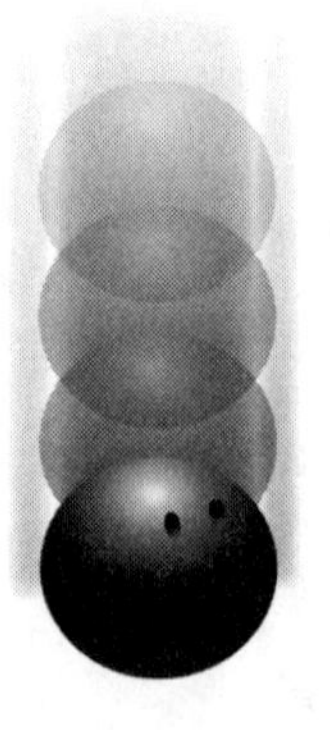

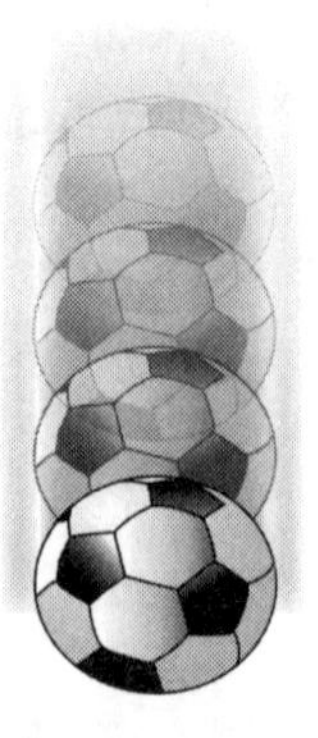

A. The force of the bowling ball is greater, and its acceleration is greater.

B. The force of the bowling ball is greater, and its acceleration is the same.

C. The force of the bowling ball is the same, and its acceleration is greater.

D. The force of the bowling ball is the same, and its acceleration is the same.

20. X-rays are representative of what type of energy? SPS9 Ch 12

    A. thermal B. nuclear C. electromagnetic D. chemical

21. There is a general tendency in nature toward a state of greater disorder. For example, a sugar cube placed in a glass of water will dissolve, and the sugar molecules will spread themselves throughout the glass. Which of the following is **NOT** an example of a system moving toward a state of greater disorder? SPS5 Ch 8

    A. evaporation
    B. photosynthesis
    C. weathering of rocks
    D. expansion of the universe

22. The diagram below shows a 400 N crate being pushed up an inclined plane by a force of 200 newtons. What is the mechanical advantage of this machine? SPS8 Ch 10

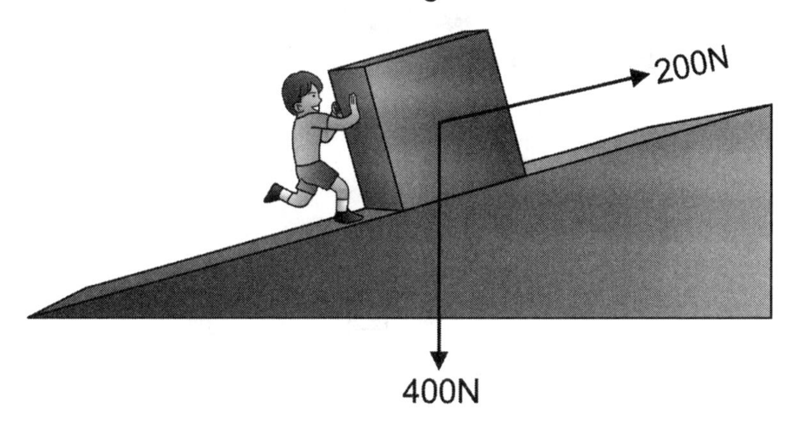

    A. 2 B. 1/2 C. 200 D. 600

23. Which of the following lists electromagnetic radiations from lowest to highest energies? SPS9 Ch 12

    A. Microwaves, radio waves, visible light, X-rays.
    B. Radio waves, infrared radiation, visible light, ultraviolet radiation.
    C. Radio waves, microwaves, ultraviolet radiation, visible light.
    D. Infrared radiation, visible light, gamma radiation, X-rays.

24. The angle of the stirring rod in the beaker appears to change at the surface of the water. This phenomenon is explained by which property of light? SPS9 Ch 12

A. scattering B. diffraction C. reflection D. refraction

25. Where are the halogens in the diagram below? SPS4 Ch 5

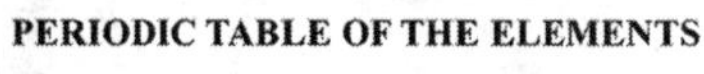

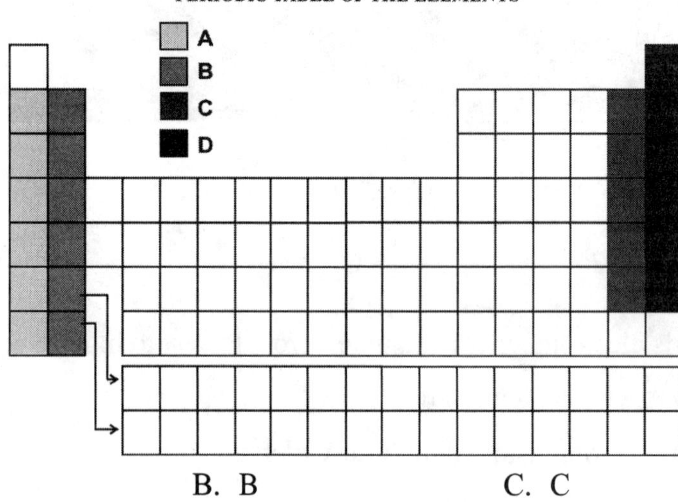

A. A B. B C. C D. D

26. Which of the following diagrams shows an ion with a positive charge? SPS1 Ch 5

A.

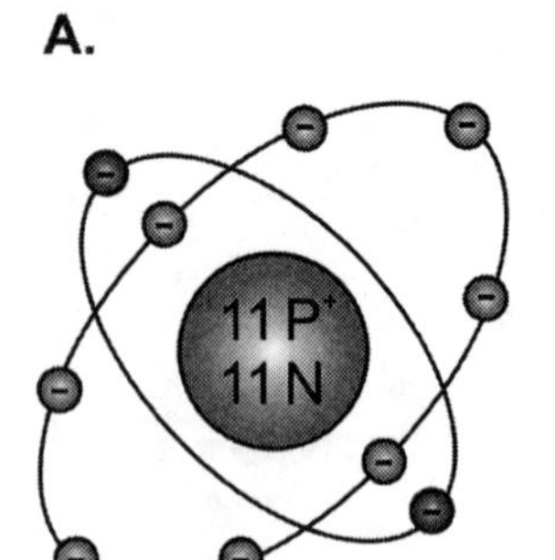

B.

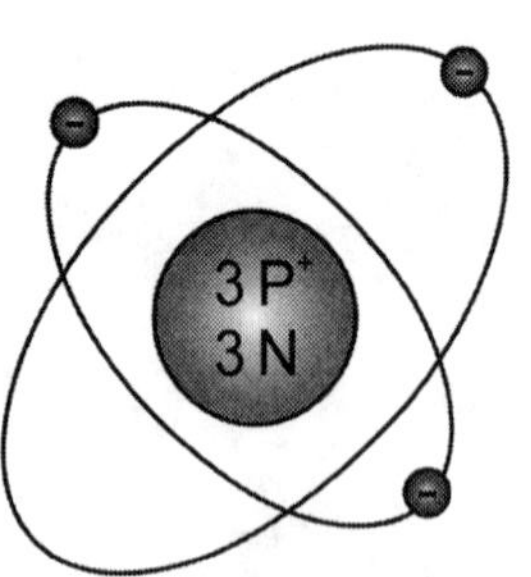

C.

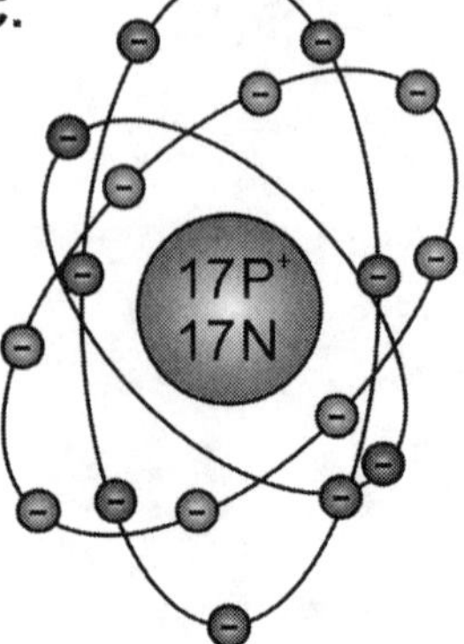

D.

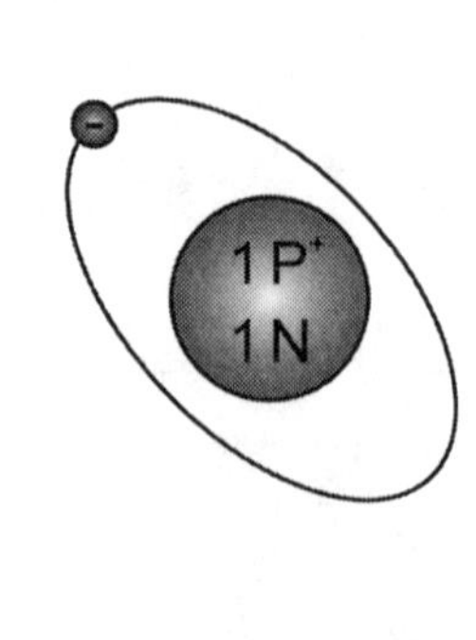

27. Noble gases are nonmetallic elements that do not readily react with other elements. What accounts for this non-reactivity? SPS4 Ch 5

    A. Noble gases have six valence electrons.

    B. Noble gases have eight valence electrons.

    C. Noble gases have ten valence electrons.

    D. Noble gases have two valence electrons.

28. Consider the pH scale shown right and the pH of the labeled items. Which answer choice lists items in order from the most basic to the most acidic? SPS6 Ch 9

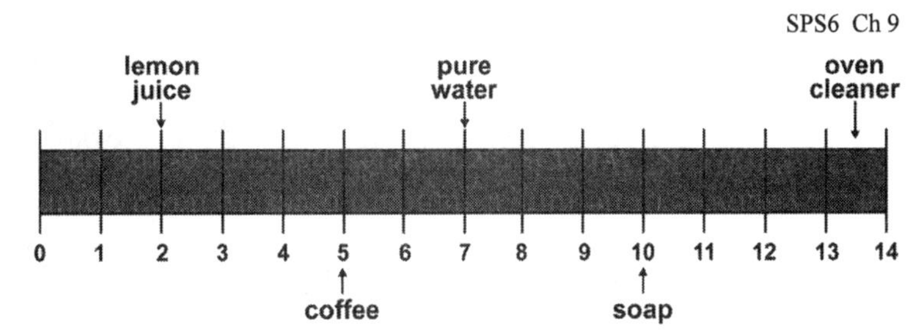

    A. oven cleaner, soap, coffee, lemon juice

    B. soap, oven cleaner, pure water, lemon juice

    C. lemon juice, coffee, soap, oven cleaner

    D. coffee, lemon juice, pure water, soap

29. If Henry walks out of school in the afternoon and the ground is wet, he will assume that it has rained. Henry is making a(n) SCSh3 Ch 1

    A. hypothesis. B. prediction. C. theory. D. inference.

30. Which of the following would inflict the most tissue damage if ingested? SPS3 Ch 6

    A. gamma B. beta particle C. alpha particle D. X-ray

31. Refer to the equation below. It is a/an ________ reaction. SPS2 Ch 7

$$P_4 + 5O_2 \longrightarrow P_4O_{10} \text{ + heat}$$

    A. exothermic B. neutralization C. endothermic D. decomposition

32. An ionic bond results from ______ electrons, while a covalent bond is a result of __________ electrons. SPS1 Ch 5

    A. shared, transferred

    B. transferred, shared

    C. outer, inner

    D. inner, outer

33. Two electrons are located 10^{-10} meters apart (the diameter of a typical atom). Identify the effect of the electrostatic force between the two electrons and its magnitude (size) compared to the gravitational force between the electrons. SPS8 Ch 10

A. The electrostatic force is repulsive and stronger.

B. The electrostatic force is repulsive and weaker.

C. The electrostatic force is attractive and stronger.

D. The electrostatic force is attractive and weaker.

34. You rubbed a balloon against your hair and touched it to a wall. The balloon stuck to the wall. Select the correct explanation of why the balloon stuck to the wall. SPS8 Ch 10

A. electrostatic forces between the particles of the balloon

B. magnetic forces between the particles of the wall

C. electrostatic forces between the particles of the balloon and the particles of the wall

D. magnetic forces between the particles of the balloon and the particles of the wall

35. A battery provides a total of 2.0 volts to a video game controller. If the controller has an operating resistance of 4.0 ohms, the current through the video game controller is SPS10 Ch 13

A. 2.0 A. B. 0.50 A. C. 8.0 A. D. 6.0 A.

36. Which of the following correctly places the phases of water in order from the most dense to the least dense? SPS5 Ch 8

A. solid, liquid, gas

B. solid, gas, liquid

C. liquid, solid, gas

D. liquid, gas, solid

37. The diagram below shows two identical spheres, X and Y, separated by distance *r*. Each sphere has the same mass, but they have opposite charges. SPS8 Ch 10

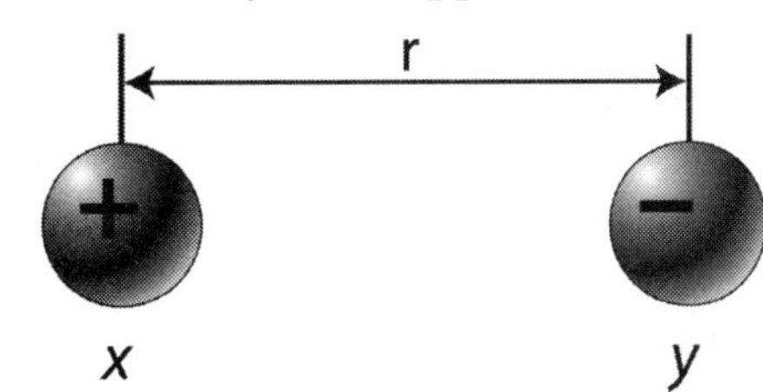

Which diagram below best represents the electrostatic force, F_e, and the gravitational force, F_g, acting on sphere X due to sphere Y?

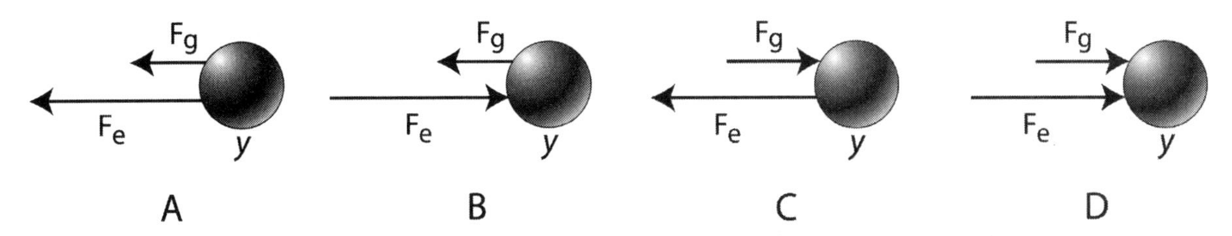

38. Which subatomic particle will be electrostatically attracted to a negatively charged object? SPS1 Ch 5

A. proton B. neutron C. electron D. a beta particle

39. Which two particles have approximately the same mass? SPS1 Ch 5

A. proton and electron
B. proton and neutron
C. electron and neutron
D. electron and alpha particle

40. When a potassium atom (K) forms a potassium ion (K^+), the atom SPS1 Ch 5

A. gains a proton.
B. gains an electron.
C. loses a proton.
D. loses an electron.

41. The figure shows the energy diagram for the decomposition of ozone. One curve indicates the energy pathway for the catalyzed reaction and the other curve represents the uncatalyzed reaction. Curve _____ represents the reaction that was aided by a catalyst, and has a _____ activation energy. SPS2 Ch 7

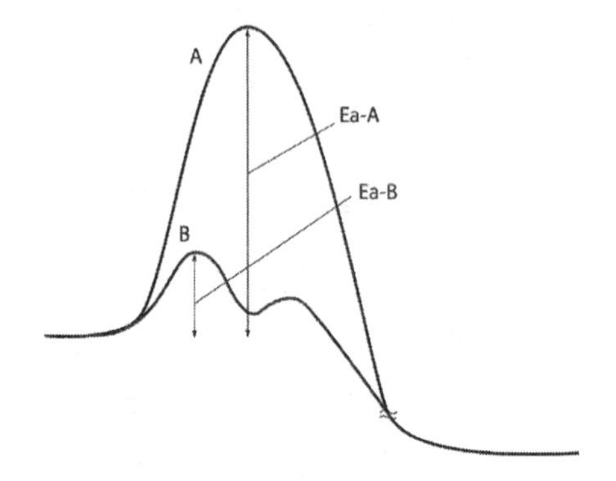

A. A, lower
B. B, lower
C. A, higher
D. B, higher

42. A sample of helium at standard pressure has a volume of 150.0 mL at a temperature of 200.0 K. What is the new temperature of the gas if the volume of the sample is decreased to 50 mL and the pressure is kept constant? SPS5 Ch 8

A. 66.67 K B. 100.0 K C. 600.0 K D. 666.7 K

43. The nucleus of an atom of gold-198 contains SPS1 Ch 5

A. 79 protons and 119 neutrons.
B. 79 protons and 118 neutrons.
C. 197 protons and 197 neutrons.
D. 198 protons and 198 neutrons.

44. Which substance represents a compound? SPS2 Ch 7

A. C B. O_2 C. CO_2 D. Co

45. Atoms in the _____ phase have the most distance between them. Atoms in the _____ phase have less distance between them, but are still mobile. SPS5 Ch 8

A. solid; gaseous
B. solid; liquid
C. gaseous; solid
D. gaseous; liquid

This is the halfway point in the test. With the approval of your instructor, you may take a 5-minute stretch break.

SECTION 2

46. Which process represents a chemical change? SPS2 Ch 7

A. melting of ice

B. corrosion of iron

C. evaporation of alcohol

D. crystallization of sugar

47. Which compound is a saturated hydrocarbon? SPS2 Ch 7

A.

```
H       H
 \     /
  C=C
 /     \
H       H
```

B.

H—C≡C—H

C.

```
    H   H
    |   |
H—C—C—H
    |   |
    H   H
```

D.

```
        H
        |
        C
      //  \
  H—C      C—H
     |      ||
  H—C      C—H
      \\  /
        C
        |
        H
```

48. Rubbing alcohol contains both 2-propanol and water. These liquids can be separated by the process of distillation because 2-propanol and water SPS6 Ch 9

A. mixed physically and retained their different boiling points.

B. mixed physically and have the same boiling point.

C. combined chemically and retained their different boiling points.

D. combined chemically and have the same boiling point.

49. The table below shows the values of the properties determined by analysis of substances 1, 2, 3, and 4. SPS4 Ch 5

Substance	Melting Point (°C)	Boiling Point (°C)	Conductivity
1	-90	-10	none
2	25	200	none
3	350	790	as solid
4	850	1325	in solution

Which substance is most likely an ionic compound?

A. 1 B. 2 C. 3 D. 4

50. Which graph best represents the electric force between a proton and an electron as a function of their distance of separation? SPS8 Ch 10

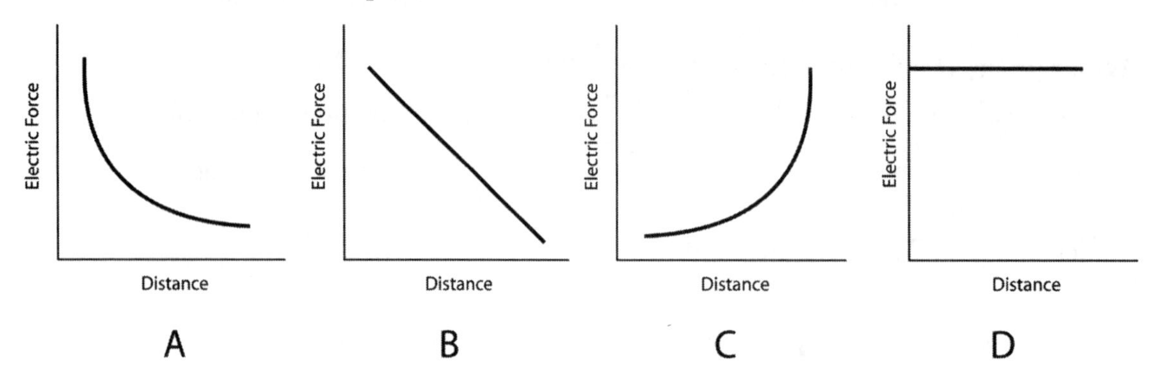

51. The diagram below shows a bar magnet. Which arrow best represents the direction of the needle of a compass placed at point A? SPS10 Ch 13

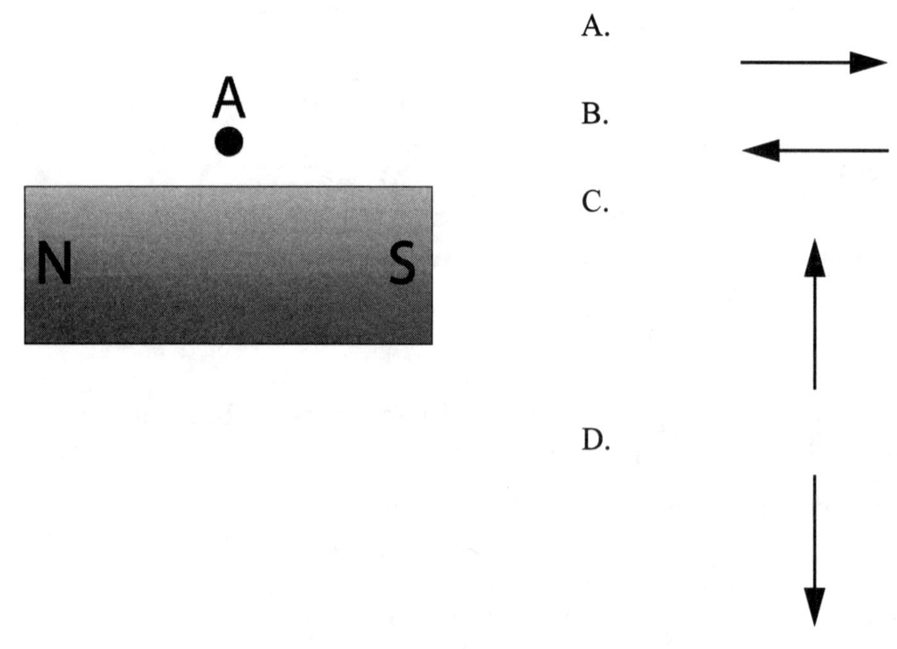

52. Four resistors are wired in series. R1=3 ohms, R2=4 ohms, R3=3 ohms, and R4=1 ohm. What is the total equivalent resistance of this 12V circuit? SPS10 Ch 13

A. 23/12 B. 23 C. 11 D. 132

53. Conductivity in a metal results from SPS1 Ch 5

A. high electronegativity.

B. having a large atomic radius.

C. highly mobile protons in the valence shell.

D. highly mobile electrons in the valence shell.

54. In which situation is the net force on the object equal to zero? SPS8 Ch 10

 A. a bicycle braking to a stop

 B. a pitched ball being hit by a bat

 C. a go-cart moving at constant speed on a straight, level road

 D. a satellite moving at constant speed around Earth in a circular orbit

55. The distances needed for a car to stop when traveling at different speeds are recorded in the table below. Identify the distance required for the car to stop from 45 miles per hour. SPS8 Ch 10

Speed (mph)	Stopping Distance (ft)
30	33
40	58
50	90
60	130
70	177
80	231

 A. 58 feet B. 73 feet C. 88 feet D. 90 feet

56. A scientific review describes all of the research performed in a given subject area for a defined period of time. What research goal could be best served by the use of a review? SCSh8 Ch 1

 A. To find updates on an event that happened last week.

 B. To find out what has been written on a broad subject area, like biology.

 C. to find out what has been written on a narrow subject area, like hummingbird migration.

 D. To find out where to find the work of a particular scientific author.

57. Lithium has an atomic number of 3. How many electrons are in its valence shell? SPS1 Ch 5

 A. 1 B. 2 C. 3 D. 8

58. Which of the following substances contain an acid? SPS6 Ch 9

 A. liquid soap B. vinegar C. baking soda D. flour

59. Each hydrogen atom has one electron and needs two to complete its first energy level. Since both hydrogen atoms are identical, neither atom dominates in the control of the electrons. The electrons are shared equally. This makes H_2 a(n) SPS1 Ch 5

 A. polar bonded molecule.

 B. non-polar covalent molecule.

 C. polar covalent compound.

 D. non-polar ionic compound.

60. Lead metal reacts with HCl by the following reaction. SPS2 Ch 7

$$Pb(s) + 2HCl(aq) \longrightarrow PbCl_2(s) + H_2(g)$$

The $PbCl_2$ product of the reaction is insoluble and adheres to the surface of the reacting lead as hydrogen gas escapes. The decrease in ________ leads the reaction rate to _______.

A. hydrogen, decrease

B. pressure, increase

C. surface area, decrease

D. pressure, decrease

61. The octane rating on gasoline is a measure of the knock resistance of the gasoline in the engine, during ignition. Octane rating is determined from percentages of isooctane (2,2,4-trimethylpentane) and heptane present in the gasoline. The higher the octane percentage, the less knocking. An 87-octane gasoline has the same knock resistance as a mixture of 87% isooctane and 13% heptane. Isooctane is drawn below. SPS1 Ch 5

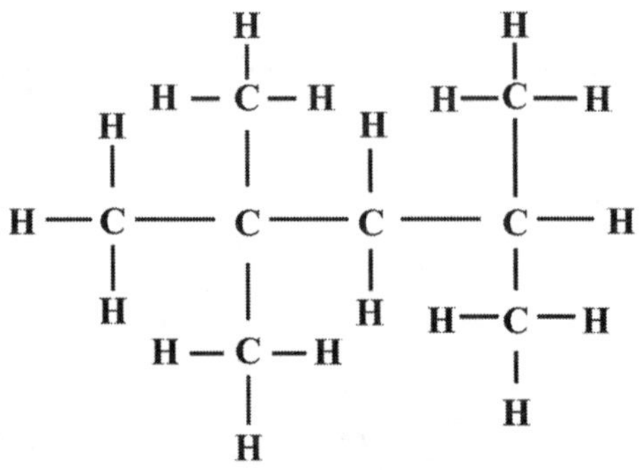

$(CH_3)_3\,C\,CH_2\,CH(CH_3)_2$

ISO OCTANE

Which of the following is correct statement?

A. Isooctane has 8 carbon atoms and is an aliphatic compound.

B. Isooctane has 8 carbon atoms and is an unsaturated hydrocarbon.

C. Isooctane has 5 carbon atoms and is an aliphatic compound.

D. Isooctane has 5 carbon atoms and is an unsaturated hydrocarbon.

62. Select the type of matter that includes solutions. SPS6 Ch 9

A. ionic compounds

B. molecular compounds

C. heterogeneous mixtures

D. homogeneous mixtures

63. A standard transistor radio battery is 9V. If the local store was out of 9V batteries, how could you construct an equivalent power source? SPS6 Ch 9

A. Securely connect three 1.5V batteries in series.

B. Securely connect six 1.5V batteries in series.

C. Securely connect six 1.5V batteries in parallel.

D. Securely connect nine 1.5V batteries in parallel.

64. Identify the correct description of a saturated solution. SPS6 Ch 9

A. a solution of an ionic solid in an ionic liquid

B. a solution that is composed of multiple solutes and solvents

C. a solid solution made by cooling a hot mixture of two liquid metals

D. a solution in which no more solute can dissolve at the given temperature

65. Latisha rubbed a balloon vigorously on her shirtsleeve. She then observed that small bits of paper stuck to the balloon. Select the best explanation of why the bits of paper stuck to the balloon. SPS10 Ch 13

A. The balloon was magnetized by friction.

B. The balloon was magnetized by conduction.

C. The balloon was electrically charged by friction.

D. The balloon was electrically charged by conduction.

66. Two objects attract each other electrically. Identify the statement that correctly describes the electrical charge of the two objects. SPS10 Ch 13

A. The objects must be like charged.

B. The objects must be oppositely charged.

C. The objects are either like charged or electrically neutral.

D. The objects are either oppositely charged or one is charged and one is neutral.

67. The diagram below shows a positively charged body, X, and a neutral body, Y. X and Y are not touching each other. SPS10 Ch 13

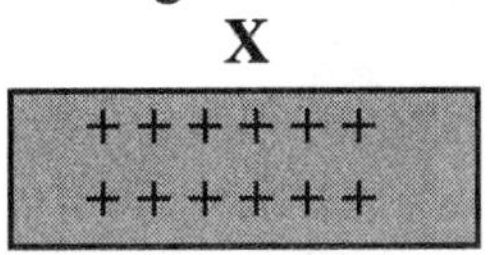

Select the drawing that shows the charge in bodies X and Y if they are brought into contact with each other.

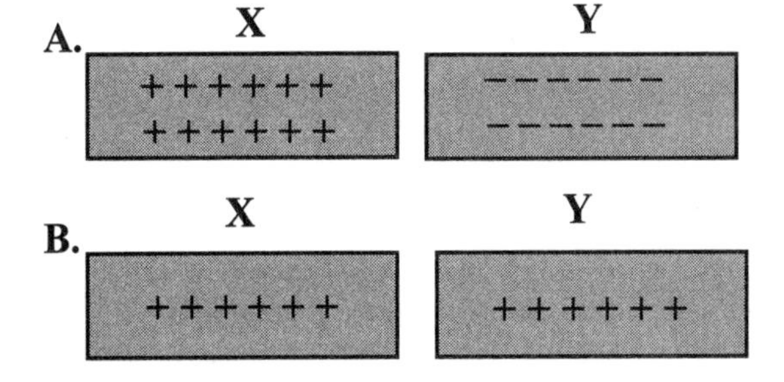

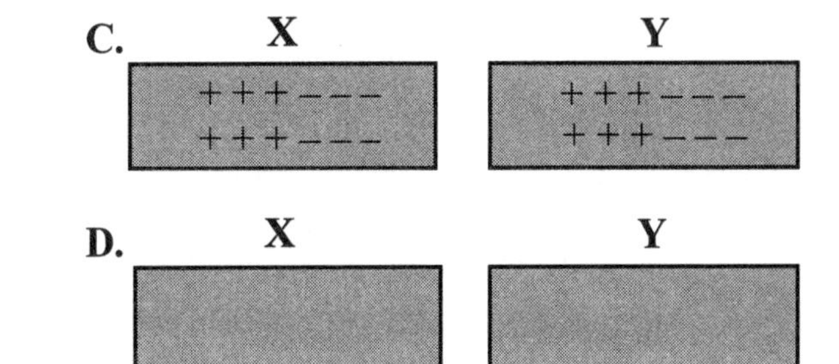

68. Identify the factor or factors that determine the amount of work that can be done by an electric current. SPS10 Ch 13

A. static only
B. voltage only
C. static and amperage
D. voltage and amperage

69. A voltage V is applied to a circuit with resistance R produces a current I. Identify the current when a voltage V is applied to a circuit of resistance 5R. SPS10 Ch 13

A. 0.2 I
B. I
C. 5 I
D. 10 I

70. Identify the types of current that can be produced by appropriately designed electric generators. SPS10 Ch 13

A. direct current only
B. absolute current only
C. direct current or alternating current only
D. alternating current, direct current, or absolute current

71. In this diagram, a weight rests in the middle of a supported board. Gravitational force pushes the weight toward the floor. What opposing force pushes upward on the weight? SPS8 Ch 10

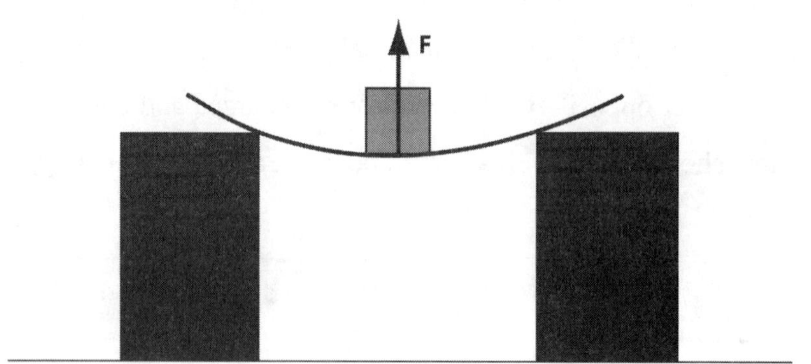

A. kinetic frictional force
B. rolling frictional force
C. inertial force
D. normal force

72. A wet cell battery is an example of stored potential energy. Once connected to a load, the stored energy is converted. Describe the conversion of energy in a battery powered clock. SPS7 Ch 8

A. chemical to electrical to mechanical
B. chemical to thermal to electrical
C. electrical to thermal to mechanical
D. chemical to electrical to thermal

73. A given radionuclide R_A has a half life of 100 years. Another radionuclide R_B has a half life of 25 years. One thousand kilograms of each material are placed in a hazardous waste receptacle. How much of each will be around after 100 years? SPS3 Ch 6

A. 500 kg R_A and 62.5 kg R_B
B. 500 kg R_A and 125 kg R_B
C. 250 kg R_A and 250 kg R_B
D. 250 kg R_A and 125 kg R_B

74. A gas law states: At a constant temperature, increasing the pressure on a gas will decrease its volume. Conversely, decreasing the pressure on a gas will allow it to expand and so increase its volume. Which gas law is this? SPS5 Ch 8

A. Newton's Third Law
B. Boyle's Law
C. Charles' Law
D. Avogadro's Law

75. A 200 mL gas sample is maintained at constant temperature, and pressurized from 10 atm to 50 atm. What is its new volume? SPS5 Ch 8

A. 40 mL
B. 2.5 mL
C. 0.4 mL
D. 1000 mL

76. Which of the following are alkali earth metals? SPS4 Ch 5

Mg, magnesium
Li, lithium
K, potassium
Ca, calcium
Sr, strontium

A. Be and Li
B. Li and K
C. Mg , Ca and Sr
D. Li, Ca and Sr

77. In what section of the periodic table can metalloids be found? SPS4 Ch 5

A. in the transistion metals
B. in the alkali metals
C. diagonally across the p-block
D. only in family VA of the p-block

78. Fission and fusion are two methods of producing nuclear energy. Which method is used in nuclear power reactors and why? SPS3 Ch 6

A. Fusion, because it creates less waste.
B. Fission, because it produces more energy per atom of fuel.
C. Fusion, because it produces more energy per atom of fuel.
D. Fission, because fusion requires a large input of energy in order to occur.

79. Which of the following reactions is most likely to occur spontaneously? SPS2 Ch 7

A. $F^{1-} + Cl^{1-} \longrightarrow FCl^{2-}$
C. $K^{1+} + Cl^{1-} \longrightarrow KCl$
B. $Ar_2 + Cl_2 \longrightarrow 2ArCl$
D. $Ne + Ar \longrightarrow NeAr$

80. Which of the following is an example of an endothermic process? SPS5 Ch 8

A. burning methane in oxygen
C. rusting of an iron nail
B. evaporation of water
D. decomposing food scraps

81. Heat will transfer from a high temperature area to a low temperature area by several different methods. The heat generated by a burning candle is an example of transfer by SPS7 Ch 8

A. convection.
B. combustion.
C. radiation.
D. conduction.

82. Energy is not destroyed, but rather converted from one form to another. Describe the conversion of energy in the following process. A power reactor utilizes the process of uranium fission to create electricity. SPS7 Ch 8

A. nuclear to electrical
B. nuclear to thermal to electrical
C. nuclear to mechanical to thermal to electrical
D. nuclear to thermal to mechanical to electrical

83. Which of the following common substances will have a pH that is closest to neutral? SPS6 Ch 9

A. lemonade
B. blood
C. baking soda
D. soapy water

84. A machine performs 60 J of work but uses 100 J of energy. What is the machine's efficiency? SPS8 Ch 10

A. 6000 J^2
B. 0.6
C. 60%
D. 167%

85. Sound waves are examples of what category of wave? SPS9 Ch 12

A. longitudinal
B. transverse
C. mechanical
D. both A and C

86. Salt is added to a beaker of boiling water until no more will dissolve. As the solution cools SCSh2 Ch 2

A. the concentration of salt in the water will increase.
B. the concentration of salt in the water will decrease.
C. the concentration of salt in the water will not change.
D. the concentration will go to zero as all of the salt crystallizes.

87. What is the best way to heat a beaker of water on a Bunsen burner? SCSh2 Ch 2

    A. Place the beaker on a covered tripod over the flame. Heat for 5 minutes, then add the water.

    B. Add water and boiling stones to the beaker, place on the covered tripod stand and then heat until boiling.

    C. Fill the beaker with water and boiling stones, then use metal tongs to hold it over the flame.

    D. Fill the beaker with water and boiling stones, then hold it over the flame using your hands.

88. What element must be present for combustion to occur? SPS2 Ch 7

    A. carbon B. nitrogen C. oxygen D. iron

89. A golf ball rolls across the green. It rolls from an area of grass that has been cropped short, to an area with longer tufts of grass (the rough) and finally into the leaves and branches at the edge of the golf course. In going over those three surfaces, what kinds of friction has the ball experienced? SPS8 Ch 10

    A. kinetic friction

    B. static friction

    C. kinetic and static friction

    D. kinetic and rolling friction

90. A rocket at the launching pad starts its engines. 1 minute later, it is traveling at 720 kilometers per hour. What is its acceleration over that time period? SPS8 Ch 10

    A. 12 km/min^2 B. 2 km/s^2 C. 720 km/hr^2 D. 0.2 km/min^2

N

O

P

Q

R

S

NOTES

NOTES

NOTES

NOTES

NOTES

NOTES

NOTES

NOTES